THE LAST TRIALS

OF

CLARENCE DARROW

THE LAST TRIALS

OF

CLARENCE DARROW

Donald McRae

wm

WILLIAM MORROW

An Imprint of HarperCollinsPublishers

FIRST EDITION

Designed by Lisa Stokes

Library of Congress Cataloging-in-Publication Data

McRae, Donald, 1961–
   The last trials of Clarence Darrow / Donald McRae. — 1st ed.
       p.   cm.
   Includes bibliographical references and index.
   ISBN 978-0-06-116149-0
   1. Darrow, Clarence, 1857–1938.   2. Lawyers—United States—Biography.   I. Title.
   KF373. D35M38   2009
   340. 092—dc22
   [B]   2008051237

09 10 11 12 13   OV/RRD   10 9 8 7 6 5 4 3 2 1

For Alison

Darrow was the most intriguing yet strangely contradictory man I ever knew. His compassion for the world was breathtaking. But he could break your heart, if you were not careful, because he was often more concerned with saving comparative strangers than thinking of those who loved him most. You see, Darrow didn't believe much in either—but he always said he wouldn't mind ending up in heaven or hell. He had just as many friends in both places.

—MARY FIELD PARTON

CONTENTS

THE LAST TRIALS
OF
CLARENCE DARROW

CHAPTER I

BACK IN THE LOOP

The Loop, Chicago, June 17, 1924

Darkness spread slowly across a city in tumult. It seeped through the burnt orange and faded red streaks of a sky that softened the stone buildings towering over her. Alone in the Loop on a summer evening, Mary Field Parton picked her way through the teeming streets, slipping quietly past the blurred faces and babbling voices. And the farther she walked the more she lowered her gaze, as if willing herself to become invisible. The dusk framed her own trepidation as she went to meet the man she had loved so long.

Clarence Darrow was America's greatest and most controversial criminal lawyer, a battered sixty-seven-year-old defender of the lost and the damned. It was Darrow's knack, and his fate, to be drawn to cases of such drama and dissent that they received saturation coverage across the country. His fame was enshrined, but Darrow was revered and hated in equal measure.

The lawyer's reputation skirted redemption and ruin again as he immersed himself in yet another complicated defense. Only one certainty remained. He would soon enter the courthouse and begin the most infamous murder trial of his long career.

Darrow's name, and those of Nathan Leopold and Richard Loeb, the two nineteen-year-old killers he would defend, echoed around Mary. Newspaper barkers, pressing hard to sell their final editions, shouted out rival headlines from adjoining street corners. "Darrow," they yelled, was "ready for the trial of the century." "Leopold and Loeb," who had confessed to the senseless murder of a young boy less than a month before, on May 21, faced their likely death sentence with "an eerie calm."

Mary did not share their serenity. She could not pull down a similar mask and ignore the threat her relationship with Darrow now posed to her own marriage. Her face, as plainly intelligent and practical as it appeared, felt on the brink of collapse. She could have broken down and cried on the sidewalk if she thought too closely about the risk she had taken in traveling from New York to be with Darrow, her former lover. But the tears did not fall. A longing to see Darrow kept her moving toward him. She wanted to hear why he had called her.

Their four-year affair had ended in 1912, but in the intervening twelve years, after the pain had ebbed, they had remained friends and exchanged regular letters. They had even met occasionally and buried their past feelings in talk of books and politics, as well as gossip about former members of their circle who had known of their illicit love. But this was different. Darrow had written to her in a way she could barely believe, intimating how much she still meant to him and that he needed to see her urgently.

His words were given a fierce charge by their jolting backdrop: a saga reeking of forbidden sex and murder. The curious case of Leopold and Loeb, the sons of two millionaires, fixated America and reached the world beyond in faraway cities like Paris and London. Two weeks earlier Darrow had urged Mary to travel to Chicago to see him, and to bear witness to a new kind of trial that would test the limits of their intelligence and compassion. As a writer, and the woman who had saved him once before, she felt compelled to be with him.

At the age of forty-six Mary was not some mindless fantasist or even a wretched wife. She had endured many difficulties with her husband, Lemuel Parton, but they had recently healed the raw patches of hurt in their marriage. Mary, having battled for so long to reconcile her inner self with her public roles as a wife and a mother, had found a newly settled life. If her ambitions had narrowed at the same time, she had also found an acceptance of the virtues of marriage and motherhood.

Those feelings had shifted again after Darrow's stark appeal to her. The familiar yearning for work and passion, for writing and recognition, seized her once more. Mary did not know if it was destiny or luck that Lem had already planned to be away the following month, on an expedition to Greenland and Labrador, and so it had been easy for her to convince him that she should seek a commission from a New York newspaper to write about Leopold and Loeb—and, of course, Darrow. On the inside she was tugged more by the beguiling fact that the man who had changed her life, and then hurt her, had called for her. Darrow had turned to her again, but Mary did not know what he might say when the moment came for them to be alone.

Even though the years seamed their faces, the same feelings lurked within her. If it did not move her in a way that once made her helpless to resist, Mary felt a pulse of the old desire. Her love for Darrow had changed, but it had not been entirely withered by marriage to Lem. There was an awkward irony, a jagged reminder she could not quite ignore, that her eleventh wedding anniversary was meant to be celebrated in two days' time—on June 19.

Mary could still not shut her mind to everyday responsibilities. As she expected to spend at least a week in Chicago, before returning later for an indeterminate period, she had brought her nine-year-old daughter, Margaret, with her from New York. Mary might have wished to become the writer of her deepest imagination, but she could never forget that she was a mother first. She felt relieved now that her daughter was safe with a friend in Chicago—another Margaret, the older and wiser Margaret Watson, who could guess the tangled feelings inside her.

The previous night Mary and her little girl had caught the express train to Chicago, the winningly named 20th Century, and hurtled through the blackness. Margaret had bounced up and down excitedly in the seat opposite hers. They had made faces at each other in the window and had laughed at their reflections in the gauzy yellow light cast by the gas lamps above their heads. As the steam train rocked and whistled they fell into its rhythm, and, eventually, as her frenetic talking lessened and her pale eyelids grew heavy, Margaret had allowed her mother to extinguish the lamp closest to her. Mary had settled her sweetheart down into the top bunk bed of their compartment and kissed her lightly on the cheek, before reaching for her pen.

In the gloom of the hushed carriage she had scribbled a few words on the page marked *Juin 16, Lundi*, in the diary she had bought earlier that year in Paris when time alone, and away from Lem, had revitalized her marriage. "Left for Chicago on 20th Century," she wrote. "Full of hope! Here is my start! Got a story from Darrow on this strange murder in Chicago—Loeb and Leopold, rich boys, precocious, everything to live for. Kill, brutally, a little boy of 14, 'for the thrill' they say. Whole country, foreign countries, avid for news . . . for explanation."

When Mary began her affair with Darrow sixteen years earlier, at the age of thirty, she had been among that first wave of women who, in a new century, fought against social convention and demanded sexual equality. Determined to open herself to different ideas, she now needed to understand the cold but entwined pair of postgraduates Darrow would try to save from death row. He believed that the roots of their murderous act lay in "a bizarre homosexual compact." But how could he explain away their terrible crime?

When she had looked through the diary, noting the blank pages where she had written nothing while ruminating over Darrow's letter, she saw the last entry before leaving for Chicago had been on May 22—Margaret's ninth birthday. The evening before, on May 21, 1924, around the same time that Leopold and Loeb had murdered fourteen-year-old Bobby Franks on a whim, she had written just two innocent lines: "Last night M is eight. Kissed my little eight-year-old for the last time!"

Everything had tilted on its axis since then. As one mother to another, she could only wonder how Flora Franks might feel whenever she thought of that same night, May 21, and the grisly image of two boys stripping her dead son, Bobby, and stuffing him into a cement culvert after they had taken his life as casually as she had snuffed out the light in a gas lamp.

Less than twenty-four hours later Mary hesitated as she neared Darrow's office. The *Chicago Daily Tribune* vendors, outnumbering their rivals in a furious circulation war, reminded her of Darrow's overwhelming presence. They hollered his name while waving copies of Chicago's most influential newspaper. A block away from Darrow's company, and ten minutes early for their appointment, she succumbed and bought the *Tribune*. Mary ducked into a doorway, opened up the paper, and shielded her face with it.

She skimmed through the psychiatric examinations Darrow's chosen doctors had carried out on Leopold and Loeb. "Somewhere in the lives of the two boys," the *Tribune* reported, "the doctors are sure they will come across some action which will give them the lead for which they are searching. It is said that the youths have a 'skeleton in the closet' which turned their minds toward the 'experiment' of taking the life of young Franks."

Despite her grand plans to write about the trial, Mary was unsettled by the way the modern world flashed past her, in a deluge of cruel sensation. Darrow was twenty-one years older than she, and she could only begin to wonder how he might find a way into the two frightening young minds that had shocked a city so accustomed to violence. The savagery of gangsters like Al Capone and Johnny Torrio had already turned Chicago into the slaughterhouse of America. There had been assassinations and bombings, abductions, and even castrations among the 177 killings that blighted the first six months of that year in the Loop. But one murder gripped the nation. It took two boys, with their sharp suits and gleaming hair, slicked back in a snappy Valentino-sheik style, to fix the public's minds on a new depravity—the thrill killing of a child.

Only Darrow, she thought, would be able to find the words of pity and understanding these murderers now needed if they were to escape death by hanging. Much of his genius in the courtroom, and greatness as a man, resided in his rare ability to find redemptive qualities in everyone. Darrow was a master at establishing a context of forgiveness in which to defend his clients. Hate the sin, he always said, but never the sinner.

He was not a handsome man—but a large and shambling figure with rounded shoulders and a face weathered by years of strife. His clothes were so creased that, even if they were newly bought from expensive stores in Chicago, it looked as if he had been sleeping in them for a week. Yet part of Darrow's attraction resided in his comfort with a look she described as "magnificently ugly" rather than tamely presentable. The romantic implication he enjoyed most was that he was far too busy saving lives to be bothered by the trifles of appearance.

Darrow was like no man she had ever met, and Mary compared him to Tolstoy and even Christ. He listened to people, whether they were dissidents or murderers, and tried to understand them. Darrow did not judge them, for he did not endorse vague concepts like free will or self-determination. He believed that personal destiny was shaped by heredity and the environment in which a person lived. External forces of poverty or wealth, love or loneliness, dictated a man or woman's actions. Individual responsibility was, therefore, less significant than social conditions.

On the page it looked a simplistic philosophy, but, in a courtroom, Darrow's rhetorical flourishes snared the listener in a way that made it possible to imagine what it must really be like to live the life of the accused. Darrow would often be so moved by his own words that, with tears rolling down his face, it looked as if he had lived through the same anguish himself. He knew what it meant to make an awful mistake. Darrow understood guilt and what it felt like to be accused by the world outside.

Leopold and Loeb, however, would test him like never before. They induced both a psychological and visceral horror while conjuring up a picture of F. Scott Fitzgerald's "flaming youth." Mary's parents, a fire-breathing Baptist of a father and a gentle Quaker for a mother, would have been outraged by the flappers and petting parties of this very different age group. Yet Leopold and Loeb exuded something blankly amoral. It appeared as if they would do anything to alleviate their boredom and reach some kind of stardom. In her last letter to Darrow, Mary had highlighted the words Fitzgerald had written four years before in *This Side of Paradise*: "Here was a new generation . . . grown up to find all Gods dead, all wars fought, all faiths in man shaken."

Mary quaked when she read that, before killing Bobby Franks with a chisel, Leopold claimed he and Loeb had been driven to act by Friedrich Nietzsche. The German philosopher's book *Beyond Good and Evil* had supposedly fueled his and Loeb's theories that they were "supermen" existing outside the boundaries of normal society. It was hard to remember how often Darrow had exhorted her to study Nietzsche. "You must read him," he insisted. "You will love him. With all of his egotism and hardness, he is honest and brave and how we do like that."

In the end, she did read Nietzsche. The title of one of his earlier books, *Human, All Too Human*, came to define Darrow's complex character for her. Beneath all the high-minded thinking and courtroom battling, the attorney showed a fleeting but almost unbearable tenderness. But she suspected they would not talk much about Nietzsche that steaming Tuesday night in the Loop.

Her life had once been wrapped around Darrow. She had felt that unvarnished appeal most clearly whenever he called her Molly. Darrow used that private name when he touched her or, more simply, brushed aside a strand of thick brown hair dangling over her face. Even then she called him "Darrow." She knew how much he hated the name "Clarence"; but she said "Darrow" in those secret whispers because it fitted both her lover and the courtroom icon he had been for so many years.

The newspaper fluttered in her hands, as if she could barely contain either herself or its more ghastly contents. Mary understood the urgency of the press for she had been a part-time journalist for years—first as a political reporter sympathetic to the struggle of labor unions against big business and then as a writer who harbored hopes that she might eventually produce books rather than articles. Her sense of a good story, combined with her close attachment to Darrow, meant that she had been able to write at length about him in the past. He had tried for years, after all, to persuade her to write his biography. This time, perhaps, they might even begin work together on a project that had long fascinated and tormented her. She could almost convince herself that her arrival in Chicago was propelled more by professional ambition than an emotional entanglement.

And then, giving into the truth, Mary trembled as she folded the paper away and glanced at her watch. It was almost time for her to meet Darrow.

D ARROW SAT SLUMPED in his chaotic office on North Dearborn Street. A silvery pall of smoke hung around his huge head as he took another drag from his last cigarette. He would soon push aside the piles of case reports and legal papers and saunter out into the sultry city where Mary waited. Darrow could see through the dirty blinds that the streets below were deeply shadowed as the sun sank over the evening rush of Chicago.

The end for him, as an attorney, was near. It was one more reason that he had plunged headlong into a perilous trial. Apart from the prospect of a last big payday, a gravy case that would free him from the courtroom forever and provide the liberty he craved to write full-time, there remained one overwhelming reason for him to risk everything. Darrow ached for redemption.

He was an old man now, and there were days, weeks, or months even, when he felt the full weight of his sixty-seven years bearing down on him. At such times the bleak task of this trial appeared overwhelming. Darrow complained often of being weary and decrepit, afflicted with rheumatism that made his joints throb and neuralgia that shredded the nerves in his face with random bursts of pain that could turn a wince into a yelp. In such a state he would never be able to force himself into the minds of two boys and explain what they had done.

But, since writing to Mary, Darrow had been galvanized. He now relished the ferocious odds against him. In a month's time he would take on the State of Illinois, as well as the mass stupidity of those across America who demanded vengeance against two mentally diseased boys.

As always with Darrow, personal ambition and pride also drove him. The days and nights he had shared with Mary in Los Angeles twelve years before, in a Californian summer thick with the disgrace of bribery and the threat of jail, continued to haunt him. Darrow longed to wipe the stain from his name with a bravura performance in defense of Leopold and Loeb. He would entrance those who heard him, making them think of something more forgiving than the death penalty.

Mary would understand. She knew better than anyone how far he had fallen during the last months of their affair. In January 1912 Darrow had stood on the doorstep of her rented apartment in downtown

L.A. with a gun in one pocket and a bottle of whiskey in the other. He had just learned that he was about to stand trial on the charge of bribing two members of a jury in a murder case that had gone disastrously wrong. Darrow looked utterly guilty; but Mary had rescued him that night.

It would take another fifteen months to break the spell of the legal scandal that almost finished him but, though their friendship held and a hung jury spared him prison, their affair had buckled. Ruby, Darrow's wife, who had been acutely aware of his relationship with Mary, felt vindicated. She had always insisted Darrow would never leave her—and she was right. He stayed locked inside their convenient marriage while Mary returned to Lemuel, who was about to propose. Mary had said yes almost instantly, and she opted for a different life with a husband she loved.

Her fleeting encounters with Darrow were no longer physical, and the new distance allowed Mary to see a flawed man more clearly. The trauma of Los Angeles scarred him, for he was still banned from practicing law in California and also remained outside the fold of the labor movement—a group that had once revered him as its champion. After the events in Los Angeles they had turned against him, and their rejection had forced him to switch from labor to criminal law. Mary, just like Ruby, stood publicly in his defense, but there were countless others who spurned or attacked him.

Edgar Lee Masters, the acclaimed poet who had been Darrow's senior partner in their Chicago law firm, was his most vocal enemy. Masters had reluctantly collected testimonials on Darrow's behalf during the bribery trials, yet his bitterness had intensified. Convinced that Darrow was "as crooked as a snake's tail," Masters vowed that "I'll make that son of a bitch the most detestable figure in American history."

Part of his campaign included the crude lampooning of Darrow in a poem, "On a Bust," which Masters published in 1915:

You can crawl
Hungry and subtle over Eden's wall,
And shame half grown up truth, or make a lie
Full grown as good. . .
A giant as we hoped, in truth a dwarf;
A barrel of slop that shines on Lethe's wharf. . .
One thing is sure, you will not long be dust
When this bronze will be broken as a bust
And given to the junkman to re-sell
You know this and the thought of it is hell!

The poetry was worn and clunky, but its vehemence hurt Darrow. He appeared modest in public, but, deep down, he considered himself a giant of the American courtroom. Masters's coarse depiction of a deceitful "dwarf" struck home. Darrow could dismiss it with a soft laugh, repeating his insistence that life, let alone a legacy beyond the grave, held scant meaning for him. But, as Mary knew, he secretly craved immortality.

He had often told Mary that he was at his very best when she was near him. Darrow was not sure if this sudden urge to see her came from his need to draw strength from her love—or if it was just his lusty superstition that Mary Field was his good luck charm. But now, in the gathering dusk, with only a few weeks left before the trial, he needed her.

After they had parted in 1913, Darrow had felt the intellectual and emotional loss most of all. He had written to her, in one of his many letters, to stress that he felt "lonelier all the time . . . how I wish I could see you . . . I am always the same as you knew me with my dreams and my loves and hates . . . and many, many of these are connected with the thought of you."

Until Leopold and Loeb, his old life had felt partly rehabilitated, partly smothered, partly tamed. It sometimes seemed that, just three years from seventy, he was waiting to die. The risk and danger, the fervor and adventure, had all gone. Even his latest book—*Crime, Its Cause and Treatment*—had been ignored on its publication the previous year. Yet Darrow considered it his best and most radical work, and he was certain this new trial would at last give his book, and its theories, a deserved platform.

There was something equally radical, Darrow thought, in insisting that even the rich have rights. But Leopold and Loeb did not seem to be in receipt of the same basic rights as ordinary criminals. For that reason their families had begged him to take the case, stressing that no lawyer in America could match his force of will or poetic eloquence. Only Darrow, they said, could withstand the hysterical public demand for retribution. Darrow, despite the might of his celebrity and the depth of his intellect, was easily flattered.

He knew their story by heart now, but it still perplexed and fascinated him because Leopold and Loeb were the exceedingly bright sons of two Jewish millionaires. Nathan Leopold Sr. was a box manufacturer and Albert Loeb the vice president of Sears & Roebuck. Their boys had more money and better prospects than almost anyone their age, and yet they squandered everything on a seeming fancy and an inexplicable motive. Loeb would consent to their continuing sexual relations on the condition that Leopold helped him pursue the perfect crime. They had moved from petty thieving and burglary to kidnapping and murder—and, now, they were threatened with execution.

Darrow's hatred of the death penalty was clear, but the *Chicago Sun-Times* told its readers he had sold out for a "cool million bucks." They branded him a hypocrite who had spent his whole life fighting for the rights of the poor and unfortunate—only to rush to defend the impossibly rich for a reprehensible murder. Darrow's protests that a more modest fee would be decided by the Bar Association after the trial were brushed aside. All his diligent work since the bribery scandal threatened to be shredded by this new controversy.

Compassion and hope remained. In his career-long battle against the death penalty Darrow had fought 102 such trials in which he had saved all but one of his clients from hanging. He had failed only in his very first case, when he represented a deranged man, Eugene Prendergast, who had assassinated Carter Harrison, the mayor of Chicago, in 1893. Prendergast hardly qualified as a client, for Darrow became involved only at the very end when, having attended the trial in a personal capacity, he took up the futile appeal after judgment had been passed. The loss of a man, even a certified lunatic, distressed him, for Darrow hated to see a mob quench their mindless thirst for vengeance.

He had once written to Mary that "Chicago is beautiful now and I am back in the old apartment with the lake and the park and green trees in front of me and the people so far below me and away from me that they don't hurt. And here I think of Nietzsche who says, 'How did I soar to the height where there are no more rabble' . . . [Nietzsche] is influencing me greatly against the rabble with its cruelty, its littleness, its prejudice, its hatred, its stupidity. That was not my reason for writing this—but I want to see you."

That desire illuminated him.

DARROW STUBBED out his cigarette and rose from his desk. It was almost dark outside, but he felt suddenly light and hopeful.

In the Loop the throng around Mary had parted and left a clear stretch of sidewalk. She glanced up as she turned the last corner and there he stood, having just left his office to meet her. Darrow looked a little older, but he now walked quickly toward her.

The closer he came, the more plainly she saw how tiredness and even vulnerability framed his familiar features. The cleft in his chin was as deep as the lines creasing his high forehead. She noticed that, even though his black suit was disheveled, he had combed his usually tousled hair.

They were less than a foot apart when, with his gray-green eyes shining in amusement, he stretched out his brawny hands. Mary was swallowed in his grasp as he studied her face. Her doubt slipped away, and she felt suddenly certain that she had been right to answer his call.

"Molly," he said softly, as if nothing had changed. He brushed his lips against her thick brown hair, piled around her head with an assortment of pins and clips. She felt him tug gently at one, hoping to bring it all tumbling down.

"Darrow . . . ," she said.

D INNER AT the LaSalle Hotel, late that night, was finally over. Darrow and Mary were among the last guests left in a restaurant the city's proudest socialites claimed could rival any of the most expensive establishments in America. The LaSalle provided glittering proof that Chicago could define style and sophistication rather than just gangland mayhem and murder. Midnight was closing in, and Darrow had been talking softly for hours. The case of Leopold and Loeb consumed him. Mary, too, had lost herself in his telling of a disturbing love story and murder. The boys sounded unhinged, but, in Darrow's version of their lives, they touched her. Yet Mary, as a mother, was still shocked by the depravity of a child's killing.

Darrow reminded her that he would defend two boys who were literally children themselves. He would enter the court in a matter of weeks to fight for their lives. He had to save them, he said, as Mary felt the force of his mercy. It was easy to remember why she had first fallen for him in this very city.

They spoke again of the night they had met in Chicago—and of the night that she had saved Darrow in Los Angeles. He had survived bribery and escaped suicide. And so here they were, back in the Loop where it all began, together again. It was late, and they were tired.

They stood up, Darrow allowing Mary to lead the way to the first-floor landing.

Darrow had booked and paid for her suite at the LaSalle. He had reserved it in her maiden name, which made her blush. But "Mary Field" was less a lie than an outdated truth, a wistful nod to the younger single woman she had been when they first shared a sexual, emotional, and intellectual intimacy.

He would see her up to the suite, he said. "Thank you, Darrow," Mary murmured, her fingers lacing through his as she held his hand tightly. Slowly, silently, they took two flights of stairs and then walked down an empty corridor toward the room.

CHAPTER 2

TANGLED TOGETHER

CLARENCE DARROW and Mary Field had made a private vow. Every intimate feeling that passed between them and the walls of the LaSalle would remain in their secret possession. The intrigue left Mary feeling briefly languid after his hasty departure from the hotel. She had hoped for a more leisurely time together, but Darrow was determined to slip away before he was seen by either a reporter or a curious member of the public who might ask him for news of the looming trial. He was also intensely preoccupied with the need to return to work, shrugging aside Mary's suggestion that he should rest before rushing back to his office.

In a pale dawn, with Chicago looking subdued and almost peaceful, she had risen from her hotel bed and thrown open the shutters. Mary gazed down at the city where she had met Darrow sixteen years before, not far from where she now leaned over the balcony. The man she loved most was the radical idealist from the past, the pioneering attorney who stood up to protect the innocent, the exploited, and the unjustly accused. Perhaps that was why he seemed so unsettled by the difficulties of representing Leopold and Loeb—for their cruel and random murder of an innocent boy was blurred by layers of extreme privilege and indolence. They did not look desperate. To most ordinary people they just looked evil.

Mary drifted back into the room and, sitting on the edge of a bedraggled bed, she scrawled a few lines in her French diary, on a page marked *Juin 17, Mardi*: "Darrow is brooding over this case that focuses millions of eyes upon him. Talks of pathology, philosophy, etc. Same old Darrow, giggling and chuckling at the human race whose elephant feet and ostrich head he so loathes—and pities."

When she held the diary up close to her face, Mary imagined that she could still smell the scent of Paris in its pages. Six months before she had bought that leather-bound volume marked 1924 in a bookshop tucked away on a hilly side street in Montmartre, from where you could clearly see the Eiffel Tower. She would never have dared believe, then, that she would be back with Darrow that very summer. But her journey from Paris to Chicago was tinged with pain.

Mary had remained in France after the uneasy European trip she and Lem had taken, along with little Margaret, in a bid to bolster their fragile marriage in late 1923. Their hopes of renewing their love in great old cities like Paris and Rome had soured amid petty squabbling and the bruising worry they shared about not having enough money to justify such an indulgent trip. Lem had eventually returned to America and by mid-December he'd begun working as an agency journalist in New York—after he and Mary had agreed that they would leave their former home in San Francisco and make a fresh start on the East Coast.

They had been apart over the Christmas of 1923, with Mary and Margaret staying in France, while Lem toiled away in his new job at the North American Newspaper Alliance. She could still remember her turbulent emotions when she sat down in a Parisian café to read the long, neatly typewritten letter he had sent to her in early January. "I want to pour my heart out to you," Lem wrote, "but I am determined not to load you down with my loneliness—I know you have enough of it as it is . . . but I feel I can do something, at last, to give you a hand over the rough places. Perhaps you are tired of hearing this but something seems to have happened to me—I've got a second wind, or something. I know we can pull something worthwhile out of this little game of life before we quit."

Lem made much of his rejuvenated spirit, claiming that a more interesting brand of journalism and a steady income had transformed him. He could now provide for her and Margaret in a way he had not done properly before, and he was determined that she should take advantage of his new stability. "Now, Mary darling, God knows I want you home but I am terribly anxious to have this European adventure yield you something better than it yielded us, before I left. You are now in Europe and you have an opportunity to see it comfortably, and without worry about money—so far as that is concerned our combined expenses now are less than they would be if you were keeping house here."

It was agreed that Margaret would spend a couple of months in Lausanne, Switzerland, where she'd attend a small school and stay with the family of Madame Chesaux, a kind and dignified woman. Mary visited her daughter every other weekend while spending the rest of her time, encouraged by Lem, traveling and reading and trying to discover exactly what caused the disconcerting emptiness she often felt back home in America.

In the early years with Lem, not long after she and Darrow had fallen apart, Mary had struggled most. After the initial thrill curdled, marriage had become an endless cycle of domesticity. Instead of writing, and stretching both her mind and her renown, she had been engulfed by the chores of running an ordinary family. She was ground down by the cleaning and cooking, the washing and the ironing, and resentful of the fact that Lem, a fellow journalist, could escape the house every morning for his newspaper's office while she toiled alone in a house deserted but for her and baby Margaret.

And yet Lem, as lovely and thoughtful as he could be with her, was unlike Darrow and Mary herself. He did not exult in his work. Rather, he was a dreamer and a drifter, a man who flitted between assignments as a necessary way of earning money while he plotted his next exotic trip abroad. She missed the fact that he was not driven like Darrow, or her. Lemuel's world was a quieter place than the fevered domain Darrow straddled with such purpose as he set about saving lives and railing against mass prejudice.

But there was a kindness in Lem that also consoled her—especially when they were separated by a space as wide and cold as the Atlantic Ocean. In January 1924 she had written to her husband to tell him, "Ah Lem, I must not ever be away from you. We love you so—Margaret and I. M talks of you constantly. At night M prays for her daddy and kisses me as proxy. I fold you in my arms, darling. I love you, love you always."

In that same letter, she also wrote that "Margaret is becoming a good sport and has learned to control herself to a great degree. She is polite and thoughtful."

When Margaret was in Switzerland, Mary missed both her little girl and her husband terribly, so much so that being back with them mattered more to her than the gorgeous sights of Florence or the darker mysteries of Paris. Her love for them was also far deeper than the fleeting melancholia that sometimes wafted over her during her more mundane days as a wife and a mother. And so she found acceptance and even contentment as Lem instructed her to make the most of Europe before they were reunited in New York for good.

"For God's sake," he'd also urged her, "write if you feel like it. Do whatever you want to do and don't give a hang about what anybody else did or didn't do. If I had your ability and my made-over attitude toward all this I'd be the greatest writer in America in a year."

He had reacted, with even more alacrity, to her passing suggestion that she might enter psychoanalysis on her return home in an effort to clear her mind. "I'm strong for your psycho-analysis scheme," Lem had gushed. "Why not go into Germany, brush up on your German and look over the big stuff that is being done there?"

She had laughed a little despairingly at his extravagant enthusiasm, expecting that he would eventually insist that she take a train from Berlin to Vienna so that she might see Sigmund Freud himself. Mary now knew, six months on, that she did not need even Freud to tell her what she understood so plainly about her intermittent unhappiness. It had taken her this long, almost a dozen years, to get over her furtive longing. She had felt lovesick, off and on, all that time as she pined for Darrow, and silently cursed him for not leaving Ruby, his wife, for her.

That dizzying realization had hit her as soon as she saw her old lover again on the shaded Chicago sidewalk the evening before, and he had reached out to touch her. She had felt a similar reawakening, this time in her desire to write, when he had first told her about Leopold and Loeb. It was obvious to Mary that this would be a trial like no other, with an unparalleled national interest in a distressing murder. Darrow stood at its very center, staring into the darkness in a poignant attempt to make sense of a seemingly pointless crime. She understood Darrow better than almost anyone and, with his help, Mary was convinced that she could write about the famous old attorney, and the mystifying killers he planned to save, in a way that would make her name.

And so those different urges, to be with Darrow and to finally forge ahead as a writer, had come together with almost shattering force. The thought of seeing him again, and writing like she had never written before, left her almost breathless amid the low but gathering rumble of the city outside. Chicago, like her, was about to come alive again on a new day.

I N THE SUMMER of 1908, years ago now, Mary had met Darrow for the first time. In the Loop she and her friend Helen Todd had attended a political rally on behalf of an exiled Russian revolutionary, Christian Rudovitz, whom Darrow represented in his fight against extradition. Darrow had taken to the stage and made it shockingly clear that Rudovitz faced certain execution if he was sent back to Moscow by the U.S. government. He spoke with such lucid empathy for a lost and hounded man that it felt to Mary as if Darrow's words pinned her to the floor.

She recovered quickly when Helen, who knew Darrow, guided her through the crammed hall toward the feted attorney. Darrow's strength and vitality were obvious even beneath the rumpled charm he presented to the world. Mary was struck by the piercing way in which he looked at her, the intensity of his scrutiny tempered only by the crinkled humor in his eyes when he heard that she, like Helen, described herself then as a social worker. He preferred meeting writers to social workers, but he was willing to bet she could string a few salty words together.

Mary fired back a salty quip as if to prove him right before, more pointedly, suggesting she was usually happier talking to radical poets than old-fashioned lawyers. Darrow roared in delight, telling her that inside every decent lawyer she would find "a wreck of a poet." She thought he looked more wrecked and poetic than any attorney she had ever seen—a joke that sharpened Darrow's already obvious interest. Mary made him laugh more than anyone had done in a long time, and he found himself helplessly attracted to her irreverent spirit. She, in turn, was intrigued by his beguiling mix of wit and compassion.

Soon after becoming lovers Mary and Darrow had visited Rudovitz in prison, as the U.S. authorities moved inexorably through the extradition process. There were no cameras with them, and no courtroom juries to seduce, but Darrow still wept in the presence of the broken Russian dissident. Mary had never witnessed such feeling in a man before, and it made her fall hard for Darrow. And there was something thrilling about the way in which he then gathered himself and went to work with redoubled ferocity in order to save Rudovitz's life. Darrow became a lion of the courtroom. As charismatic as he was fierce, he won that case which, like so many others both before and after they met, made national headlines.

Darrow had boosted Mary financially and emotionally at the outset of their affair when he made it possible for her to move from social work in Chicago to a more literary life in New York in 1909. He introduced her to his friend Theodore Dreiser, the esteemed novelist, who had then just begun a magazine in New York called *The Delineator*. Dreiser published her early short stories to some acclaim, and Darrow had reveled in her success. He flattered her in his regular letters from Chicago. "You have gone so far I can't see you anymore," he once pretended to complain. "You never were very large. We will have to stop praising you pretty soon, or you will lose your head— poor little Miss Field, when will you write your autobiography? Damn if I wouldn't like to see you Molly, dear."

He had seen Mary often, staying with her in New York or, during the wilder years of their affair, taking her around Chicago in flagrant defiance of Ruby. He even allowed both his wife and his mistress to sometimes share a dinner table with him and his bohemian friends. One evening he had shocked everyone by looking straight at his lover. "Mary, you're an awful clever girl," Darrow said boldly. "You're the cleverest woman I ever knew."

Ruby bristled as she turned to her rival and sneered: "You're the thousandeth!" When Mary looked at her in bemusement, Ruby snapped, "No, you're maybe the thousandeth and first woman he's told that to."

Darrow chuckled knowingly. Part of the pleasure for the old devil came from seeing two women glowering over him. But he knew how much more Mary had helped him than his past frivolous lovers. They shared an intuitive bond that translated into the deep trust that encouraged him to ask for her help on some clandestine legal matters.

In 1911, when he felt ill with worry about the fate of the McNamara brothers after they had firebombed the *L.A. Times* building and faced the death sentence, he asked Mary to visit the houses of potential jurors.

Accompanied by her six-year-old niece, Kay, she would knock on the front door of the juror under Darrow's furtive investigation and ask if the little girl could use their bathroom. Mary made it look like the sweetest of emergencies, and the juror or his wife could never resist. The door would swing open, her niece would be taken down the passage to the smallest room while Mary scanned any books, magazines, and personal items in the family home. Darrow placed particular emphasis on the juror's taste in books, for he believed that a man's character and outlook was defined by his choice of reading. He always insisted that Mary should study the shelves for a bible—which, for him, represented a surefire symbol of a man's moralistic bent.

In Chicago now, in the summer of Leopold and Loeb, Mary was sure that Darrow would win this even more difficult trial. He had already outlined to her a surprising, and still secret, switch of courtroom tactics. Such precious inside information would make her writing more compelling than anything produced by the hundreds of hard-bitten reporters who had descended on Chicago to cover the "trial of the century."

TWO DAYS after their dinner at the LaSalle, they met again in a discreet corner of the lounge at the Congress Hotel on Michigan Avenue. Mary looked exuberant while Darrow simply hunched over the small wooden table between them, grumbling and moaning about his exacting task. He felt pressured and disgruntled, and his mood was not improved by the contrasting vitality with which Mary pursued him about the case—and how he might help her.

At the outset of their reunion, the evening before last, Mary had appeared shy and tentative as he stretched out his giant hands to her.

She was now back to her old assertive self, that ambitious woman he had always admired as much as loved. But, suddenly, Darrow wanted some distance between his work and her plans. He had been taken aback by the arrival of a wire from Lemuel, urging Darrow to assist Mary as much as he could. It looked as if good old Lem was completely in on the game, hustling him, Mary's former lover, for an inside track on the biggest story in America.

She had been quite candid in telling Darrow that Lem knew all about her desire to write the definitive story on Leopold and Loeb. In fact, she had already persuaded an excited Lem to contact his boss, some anonymous old hack called Smith, about paying her expenses and selling her story through a national syndication if she "got the goods" on America's most infamous killers. It made Darrow wonder if they were using him—even more bluntly than he might have used her.

Leopold and Loeb's slyly grinning faces were splashed across the front pages of the country's newspapers as the public's fascination intensified. Mary had already told Darrow that she hoped he would provide her with exclusive access to his clients in their cells at the Cook County Jail. He had grunted evasively when she first mentioned that possibility, but now, as Mary pushed him harder, he brought the wall down. Darrow was too busy planning the salvation of Leopold and Loeb to involve himself in the resurrection of her journalistic career.

He told Mary firmly that he would not consider setting up a private meeting between her and the boys. The risk of alienating an already salivating pack of Chicago reporters meant that he could not be seen to give any special favors to a little-known writer from New York. And if their affair was revealed, stretching back to the indignity of the bribery trials in Los Angeles, he would be in an even deeper hole. Mary would have to make her own luck and go down to the prison and fight her way through the rest of the hacks. Most of them managed to talk to Leopold and Loeb in their cells—so what was stopping her?

Mary reeled in disappointment. She wanted to remind Darrow that she was a middle-aged mother, a woman who had been cast adrift from the reporting roughhouse for more than a decade, but he already knew everything about her. He seemed unwilling to offer anything beyond his own discreet insights into Leopold and Loeb while arranging a seat for her in the press box when the trial began. Darrow regarded these as sufficient sweeteners, even to a woman as important to him as Mary. He wanted to see her, to discuss the philosophical conundrums of the case and to share a little laughter as a way of alleviating the oppressive darkness that surrounded him. But Darrow had too much at stake to indulge any further distractions.

They would look for ways to continue their relationship, but, first, Darrow had to argue his case. The boys had to become his unwavering focus in the coming months and then, especially if victory secured his refurbished legacy, he would have as much time as he wanted to visit Mary in New York. Darrow was thrilled that she and Lem had moved from San Francisco. Ever since California had shredded his reputation, he'd avoided the entire state as if it were infested by the plague. But he visited New York often. There would be many opportunities to meet, and more appropriate times to consider their attachment within the constraints of marriage.

To Mary, however, the promise of occasional meetings in New York did not mark a new start. She was almost a generation older than the single and intensely romantic young writer she had been when first meeting Darrow. Their renewed intimacy, inevitably, had changed. She was married now. Darrow shrugged. He had been in that prison for many years. But he was still infused with passion—for saving lives and reading poetry, for writing and baseball, and for feisty and intelligent women like her.

Darrow had lost neither his sly old charm nor his maddening capacity to frustrate her. She could marvel again at his forceful intellect and wry insights into humanity. And yet, with Darrow's need for her briefly sated as he returned to his solitary preparation for court,

Mary herself brooded. Although she understood the need to allow him space while he searched through the intricacies of a mazelike case of psychological disturbance and futile violence, she wished he could be more attentive to her—and especially her desire to write about this epic trial.

Mary was a strong woman, and determined. She bit down on her rising anger and took another concerted crack at Darrow. Surely he could help her? She had come all the way from New York, in response to his letter, and so it was time he opened some doors for her in Chicago. He knew how much she wanted to write. And how could he forget her articles in support of him when the vultures of the press had circled over Darrow during his humiliating bribery trials in Los Angeles? Had he also not badgered her for years to write his biography? Could they not use this trial, with its seismic impact on modern America, as the starting point for serious work together?

Darrow was unmoved. He loved her, of course, but this was about more than just work. Darrow had lives to save rather than mere articles or biographies to plot. They could talk more of these ideas in the future, when the trial was over.

LATER THAT NIGHT, June 19, which just happened to be her wedding anniversary, Mary seethed to herself in the cheap hotel room she shared with her sleeping daughter. Margaret looked blissful in the dim light, but Mary felt tortured by the altered mood between her and Darrow. She knew he was at home with Ruby in their luxurious Midway apartment while her own husband would be out drinking with his new reporting buddies on Grove Street in Greenwich Village. In her agony and confusion she scribbled a single furious line in her diary: "Everything, everybody bores me!"

FOR THE NEXT few days that bitterness burned inside her. Mary was dejected and, as she wrote on Friday, June 20, "terribly disappointed Darrow will not give me anything. Nor will he 'see' my position. I too am employed to—oh hell!"

On Sunday morning she took Margaret back to the Congress Hotel to meet Darrow. But the old battler barely noticed her daughter. He could sit and listen for hours to a pair of killers, showing them nothing but generosity and understanding, excusing their madness and naïveté, but he was oblivious to her sweet young girl. "Margaret not a client, nor old enough to elicit any particular interest," Mary noted tartly. It seemed as if you needed to commit murder to hold Darrow's attention.

Two nights later they dined with the eminent sociologist Ellsworth Faris and an unknown female colleague, whom Mary dismissed as "some dame friend of Darrow's." The courtroom lion mutated into a lecherous old man. Mary had learned years ago, in Ruby's snarling company, that Darrow was at his worst when he jokingly played two women against each other. "With one his wisdom, his philosophy, his companionship is possible. Prof. Faris, too, disappointed. Darrow treats women as playmates rather than *work*mates. Great disappointment to me, especially when I know what richness of mind and experience he could share. Instead cheap, stale jokes! 'Soggy wit'! Dull puns. Invited to a banquet and taken to a one-arm lunch counter where there is poor food."

She ended the night by writing a dark and fretful note to her husband, telling Lem how Darrow had let her down badly and that all her hopes of a journalistic scoop had come to nothing. Her words were terse and brittle; but she and Margaret would leave soon for New York.

MARY HAD been both heartened and left rueful by Lem's reply. "My dearest," he wrote, "my disappointment at your failure to turn up anything exciting was offset by the news that you are coming home soon. . . . Don't be unhappy, Mary darling, about not turning up the big yarn. It was clearly understood that this was a short-end gamble and . . . that Darrow is in a position where it might be ruinous for him to play any favorite. . . . I think I would have realized more than you—from a newspaperman's viewpoint—the impossibility of his giving you a break."

Those supportive words then gave way to something more plaintive in Lem—which made Mary feel terrible for briefly abandoning him for Darrow. "I have been awfully disappointed not to have had a personal line or two from you," Lem chided her gently. "It seems a million years since I saw you. Believe me, Mary dear, I will be happy when you are back again. Love and a million kisses to dearest little Margaret. Wire me when you coming."

Darrow would not transform her life. She would have to find her own way, after all, within her writing and her family. Mary felt suddenly desperate to be home. Lem and Margaret were more important than any searing upsurge of journalistic ambition. She could still write about the case, and follow the trial, but from within the confines of her more ordinary world.

A T LEAST DARROW was gentle when she came to see him for the last time in Chicago. He might even have felt a trace of guilt, for his attention was unwavering and warm. She had helped steady him, he said, and he hoped she understood why he had not been able to do more for her in return. She nodded and smiled up at him. Of course, those doomed boys needed him.

They spoke more of the case, of his concerns and hopes, and she felt again the power of his humanity, his conciliatory search for understanding and forgiveness among people who seemed utterly damned to her. Leopold and Loeb were murderers but, as Darrow described them, they were still young boys who were tragically ill. If the State won and they were led down to the gallows, where the noose of a rope would tighten around their necks until the very life was squeezed from them, they would all be degraded. Darrow had to save them, to lend hope rather than hate to the future.

He made it sound so simple and profound, and so crushingly significant, that she began to forgive him herself. Mary knew why Darrow was loved and hated by so many. She certainly hated him less than almost anyone, and loved him more, but he was a devilishly contradictory man. Yet, as always with Darrow, in the end the good overshadowed the bad.

Darrow took her by the hand. He would save a seat for her in the press box, if she changed her mind. Mary shook her head. It was best this way. She planned to take Margaret to the country, and write, while he faced down the abrasive State prosecution. They would meet, and talk again, when the trial was over.

There was just time for a gesture from Darrow. He took her to the courthouse so that, for the benefit of her writing, she would at least have a strong image in her head of the room where the trial was due to unfold. On the corner of Dearborn and Hubbard streets, a couple of blocks from the green Chicago River that wound its way past the downtown skyscrapers, Darrow showed Mary how the giant stone slabs of the Criminal Courts Building had blackened. Years amid a dirty and roaring city, with soot belching from passing trains and steaming chimneys, had left their mark on its walls.

Darrow wore a baggy gray suit and a blue tie as, holding Mary's brown suitcase in his hand, they approached the waiting policemen who greeted him effusively. They had become accustomed to the celebrated attorney cutting through the courthouse to reach the jail where Leopold and Loeb were examined by a battery of America's leading psychiatrists. Instead of checking Mary's credentials, the officers waved them toward the rickety elevator. An equally rickety old man, who was probably Darrow's age, manned the elevator. He nodded solemnly at Darrow, warning the lawyer, as he did every time they met, that the elevator stopped one floor short of the courtroom. They would have to take the last flight of a creaking staircase by foot.

Darrow reacted with matching gravity. It was a risk they were prepared to take, he said, as he ushered Mary into the elevator. They stood at the back of the rattling old cage as the operator wearily pulled shut the rusted cantilevered gate. When the trial began, he would ferry hundreds of souls a day up and down a vault whose shadows were etched into his sunken features.

On a more ordinary day Mary might have giggled when Darrow winked at her as the old man muttered a ritual "Going up" on pressing the black button marked by a faded number 5. Once, in her first flush of love with Darrow, she would have been almost beside herself. But Mary felt as solemn as the elderly elevator rider whose rheumy eyes fixed on the small yellow light that shifted slowly from one number to the next as they crawled up the floors.

She and Darrow remained silent as the elevator struggled upward. But, somewhere between the third and fifth floors, their eyes met in a gaze that needed no words. In less than two hours Mary would board a ten o'clock train back to New York while Darrow returned to the defense of Leopold and Loeb.

In a shuddering halt, the elevator stopped and the attendant yanked back the gate. Darrow led the way up to the sixth floor, and then they were there, in the cramped room itself. Mary was surprised by its small size and shabby appearance. She expected a grand and stately setting to match a trial that would be followed around the world. Instead, she noticed the grimy windows and peeling paint where the walls met the cracked ceiling.

After a few minutes, when Mary saw him looking up at the courtroom clock, she tried to prize her suitcase from his grasp. Darrow refused but, when she persisted, he called over a young court attendant. Offering up the suitcase he asked the boy, who looked as young as the accused, whether he could arrange a cab for Mrs. Parton. When both the request and the case were accepted, Darrow drew Mary aside to thank her for coming to see him.

And then she pulled away and wished him good luck. Darrow promised quietly that he would be in touch. Mary nodded and then she headed for the exit. She took the stairs this time, partly to gather herself. But without Darrow to wink at her, or hold her in his gaze, Mary Field did not think she could bear to ride again in that strange little elevator with its strange little man. The boy holding her suitcase followed her down.

She held back the tears by smiling at the policemen lining the staircase. They would soon be needed to keep out the hordes, but as Mary descended in the opposite direction, they beamed at her. She thought she must have cut a curious figure—being the lone person to walk away from a courtroom that transfixed the rest of America. Her heart, however, lay elsewhere.

D ARROW TUSSLED with his own personal conundrum that night. The voracious newspaper interest in the trial made him wonder if any scrupulous reporter might have witnessed his early morning meeting with Mary, and their parting in court. Photographers were always milling around the building these days, and it suddenly seemed plausible that a passing snapper might have taken a random shot of them together—without realizing its true significance to their private lives.

He could not allow the risk of a shocked Ruby seeing a photograph of him and Mary in the morning paper. And so, not long after he had arrived home late that night, he told his wife that he had met Mary Field. Ruby was furious. How could he, she wailed, after all her past pain?

Darrow had resolved to avoid any protracted argument. He needed to conserve all his strength for the trial, and so he kept his version of their morning encounter as short as it was hazy. Mary Field, he reminded Ruby, was married to Lemuel Parton but she still sometimes worked as a journalist. She had arrived in Chicago with hopes of attending the trial, but there had been some kind of mix-up. Her press credential would be assigned to a different reporter.

Ruby could imagine the wretched Mary Field beseeching Darrow to intervene on her behalf and reclaim her place in the press box. She was outraged. How could he have even spoken to that woman? Darrow shook his head. He had simply arranged for a court official to help Mary and her luggage find a way back to the station.

Darrow said he had not been able to help himself when he had an affair with Mary in Los Angeles, but everything had since turned around. Twelve years had passed. He had seen Mary that morning, but now she was gone. Darrow looked down. He could not stand it any longer. He had to now face the most important trial of his life. He could not see whether his wife had begun to cry, but he turned away from Ruby. There was nothing more to be said. They both knew that Mary Field had once saved his life.

CHAPTER 3

DARROW ON TRIAL

1110 Ingraham Street, Los Angeles, January 1912

TWELVE YEARS EARLIER Clarence Darrow had shivered when he reached the small rented apartment where he wanted to kill himself. The drizzle had turned to steady rain as soon as he turned left on Wilshire Boulevard and trudged south to an obscure back-street. Darrow pulled his sodden coat around him so that he might conceal the bottle and the gun he carried in separate pockets. He did not want to frighten Mary Field even before he made it through the front door.

The contrasting weight, with a snub-nosed revolver in one pocket and the bulkier whiskey bottle in the other, made him feel unbalanced. Darrow, a man who could make people weep with his poetic celebrations of love and forgiveness over violence and hate, was a stranger to guns. He had also never been much of a drinker—even if he eventually became an eloquent opponent of Prohibition. Darrow usually restricted himself to the odd glass after a meal when he would chink the ice in his whiskey as he read aloud from Tolstoy or Swinburne, Burns or Ibsen.

He hammered at the door. Mary, knowing it would be Darrow, had not bothered to cover the white nightgown floating around her. Her hair was already unpinned and, had it been any other man, she would have rushed away to make herself decent. Darrow was different. Mary had loved him for three and a half years. She smiled as she saw the familiarly rumpled coat.

But Darrow's craggy face was pale and stricken. He looked more haunted than at any other time during the many months she had watched the spirit being sucked from him. She and Darrow had begun to fall apart as a consequence, but at least she had still been able to get him to laugh or, sometimes, even make love to her.

Mary took his hand and led him into the $35-a-month apartment she and her younger sister Sara had used the last twelve weeks. Sara described it in a letter to her own married lover, the corporate lawyer and aspiring poet Charles Erskine Scott Wood, as a temporary home where "everything disappears—the bed walks into the wall and becomes more paneling; the coffee pot goes up the chimney; the couch slides out the window." To Darrow it looked like the place where his life would soon disappear with a bullet in the head.

He slumped on a wooden chair in the kitchen. A bare bulb, dangling from a thin black cord, shone down on the round table where they had shared so many emotional meals in recent months. Mary stared helplessly as Darrow unbuckled the belt around his midriff. A few drops of rain slid down his crevassed face like the tears she felt would soon fall from her.

Darrow hesitated before he reached into his left pocket and brought out the whiskey. He looked at Mary as he placed the bottle in the middle of the table. In the uneasy silence she turned to the cupboard and found two glasses. Darrow half filled each with a large shot. His hand quivered as, with one gulping swallow, he drained his glass.

"Mary," he said suddenly, "I'm going to kill myself."

Darrow pulled out the gun and placed it on the table.

Mary's hand rose to her throat, as if to lessen the constriction in her breathing. She managed to utter one word: *"Why?"*

The truth spilled from Darrow's jabbering mouth. Words crowded together, but he spoke enough sense for Mary to understand the gist of it. Darrow was about to be indicted on a charge of bribing two members of the jury sitting in judgment of his clients, the politically radical McNamara brothers who had been charged with firebombing the *Los Angeles Times* building and, in the process, murdering twenty people. The district attorney's office was now determined to send Darrow to prison; and he could not bear the shame.

Mary knelt at his side. At her touch, first on his arm, and then his face, he began to cry. The tears descended into the most painful sobbing she had ever heard. Darrow was fifty-four. Mary was only thirty-three. The sight of an aging man breaking apart was almost too much to bear.

Eventually, after he had wiped his eyes on her nightgown and then blown his nose more appropriately into a handkerchief, Mary poured him another drink. She swallowed a mouthful from her own glass. The whiskey burned in her breast and made her eyes water in a way that matched Darrow's glistening stare.

She had to get him away from the gun. But she feared an abrupt movement might propel Darrow further down into the darkness. He looked ready to do his worst.

As she tried to think more clearly Mary stretched up to the top bookshelf where she kept her cigarettes. Darrow might have made ceaseless quips about the futility of life, but the man she loved was consumed by a desire to save his clients from the gallows. It was easier to believe that he might resort to anything, even bribery, before he gave up on anyone—especially himself.

Why would such a passionate man, with all the pleasure he found in the jousting and sharing of ideas and feelings, kill himself? She lit two cigarettes, one for Darrow and one for herself. As the smoke curled above their heads, the trauma of the last months lay between them.

Darrow had always been reluctant to act on behalf of the Mc-Namaras—John Joseph, or J. J., an admired union leader, and his younger brother James, a surly print worker, who were accused of detonating their dynamite at 1:00 A.M. on October 1, 1910. The blast intensified the bitter war between labor and big business, for the *Los Angeles Times* was published by the virulently antiunion millionaire and former Civil War commander General Harrison Gray Otis.

The day after the explosion Otis had written an overblown editorial: "O you anarchic scum, you cowardly murderers, you leeches upon honest labor, you midnight assassins, you whose hands are dripping with the innocent blood of your victims, you against whom the wails of poor widows and the cries of fatherless children are ascending to the Great White Throne, go, mingle with the crowd on the street corners, look upon the crumbled and blackened walls, look at the ruins wherein are buried the calcined remains of those whom you murdered."

Darrow had spent almost eighteen years in the service of labor—winning a string of widely publicized cases for militant strikers and their unions. He had saved proud and charismatic men from trumped-up charges of murder and insurrection. And even when his clients were guilty, he had explained their desperate actions by illuminating the oppression of many company bosses and state politicians. Darrow had become a hero to ordinary workers, who knew that he had been enticed away from the corporate world by Eugene Debs, the revered socialist and labor leader.

In 1894 Darrow had worked for the mighty Chicago and North-Western Railway conglomerate, and he initially represented the company against Debs and the national rail union in the Pullman strike. But the lawyer had been so struck by the morality of the strikers that he crossed the tracks—resigning from his lucrative position to defend

Debs against his former employers in court. He now knew, however, that many of the union men he represented were deeply flawed or even corrupt.

Debs, in turn, had become disillusioned with Darrow's own weaknesses—particularly his greed for fame. Darrow was so wedded to the labor movement, however, that whenever another epic battle occurred, he was called upon to fight the case. The most dramatic of his recent trials had seen him successfully defend Big Bill Haywood, the general secretary of the Western Federation of Miners, against a charge that he had murdered Frank Steunenberg, the former governor of Idaho. An exhausted Darrow became dangerously ill in its aftermath. He swore to Ruby that he would never again become embroiled in such a dispute.

And so Darrow initially resisted the plea of labor leaders to save the McNamaras. But Samuel Gompers, the president of the American Federation of Labor, threatened him that "you will go down in history as a traitor to the great cause of labor if now, in our greatest hour of need, you refuse."

He finally succumbed. Although the brothers had pleaded not guilty at a first hearing on June 6, Darrow soon realized that their claim to have been framed by conniving capitalists did not square with the evidence against them. The prosecution had obtained a copy of their detailed plan to bomb the *L.A. Times*. Darrow turned to Jim McNamara in his prison cell and said grimly: "My God, you left a trail a mile wide."

Only a change of plea, to guilty, might save them from hanging; yet such a U-turn would enrage labor activists and ruin his credibility. Darrow finally decided that two lives mattered more than his own name. He would do whatever was needed to save the McNamaras from the scaffold.

Darrow veered between recklessness and despair—and that gathering abandonment spilled into his private life. A few weeks earlier he'd been apprehensive when Mary had written to tell him that she had accepted invitations from New York's *American Magazine* and *Organized Labor* to travel to California to cover the trial. She saw it as an opportunity for them to spend precious time together. Darrow, however, shuddered at the danger. He loved Mary, but his stoical wife, Ruby, offered stability. Darrow had been divorced once before, from Jessie, the mother of his only child, Paul. The thought of another breakup unsettled him.

He knew he was being tailed by prosecution detectives, searching for any dirt they might uncover, and so he urged Mary not to join him. But she took the train to L.A. with renewed determination to report on the trial. Her spirited personality had always attracted him, and so, soon after Mary's arrival, he ignored his previous concerns and began to use her rented apartment as his hideaway.

Darrow also developed designs on her sister—which he obviously kept secret from Mary. On October 25, 1911, Sara wrote to Wood, her lover, to tell him that she had rebuffed two attempts by Darrow to seduce her. Wood was brutal in his response, deriding his former friend as vain and selfish and warning that "when Darrow is sated he will treat [Mary] like a sucked orange. When his personal fears are aroused he will run from her—and if necessary deny her thrice before the cock crows."

Yet, amid his darkest "personal fears," Darrow had run to Mary. With a gun and a bottle in hand, he looked up at the woman he loved. She spoke about the need for him to go on and asked Darrow to remove the revolver. He wanted her to keep talking and so he pocketed it with a sigh. Mary spoke of God and forgiveness, drawing on her family's religious past. It almost made him laugh, for she knew the depth of his agnostic views, having sat through more of his lectures on the subject than either of them cared to remember.

She might have been better quoting Nietzsche at Darrow, ramming home the point that what did not kill him would simply make him stronger. But the thought of standing in the dock before they shut him away in prison for years made him shake rather than glimmer with the zeal of a Nietzschean superman. He was comforted, instead, by her soothing voice.

That same calm had prompted Darrow to write to Mary when she was still in New York and say, "I miss you all the time. No-one else is so bright and clear and sympathetic, to say nothing of sweet and dear. Am tired and hungry and wish you were here to eat and drink and be with me and talk to me with your low, sweet, kind, sympathetic voice."

Sara, an avowed suffragette, was less easily smitten. She often said Mary had a first-rate mind until she met Darrow—but now she peddled a third-rate imitation of his own first-class mind. Yet Mary loved him utterly, and that love protected Darrow. As she talked of the need for him to fight for the legacy he had built up over decades, he noticed that she did not ask if the charge was true. Mary knew him too well. She knew how much he wanted to win. And she also knew how far he would go to save two young men from hanging.

The bottle of whiskey was half drunk by the time the rainy blackness gave way to a wintry dawn. Mary, looking unusually beautiful for a supposedly plain woman, covered Darrow's hand with her own.

"You must go on," she urged.

Darrow looked up at her, softening a shrug with the faintest of smiles. "Well, Molly," he said, "maybe you're right . . ."

O N JANUARY 30, 1912, when he was formally indicted on two counts of bribery, Darrow was a husk of his usually imposing self. The police had confirmed that, in late November, he had been present when they snared two men who were about to pass the first installment of a bribe to a potential juror in the McNamara trial. Bert Franklin, Darrow's jury investigator, was caught handing over $500 to George Lockwood in the street, on the understanding that a further

$3,500 would be paid to him should there be an acquittal. Just before he was intercepted by a policeman an agitated Darrow ran across the street—shouting at Franklin.

Now that Darrow had been charged, Franklin, hoping to earn immunity for himself by testifying in court, made a smiling prediction: "I shall not go to the penitentiary—of that I am positive. I shall tell the truth, no matter where it hits."

Darrow appointed Earl Rogers, a celebrated if often controversial attorney, as the head of his legal team. He was in need of expert help as he faced two separate trials—first for bribing Lockwood and then for attempting the same crime in relation to Robert Bain. Another key member of his defense team would testify against Darrow. John Harrington, his chief investigator, alleged that Darrow had shown him a roll of bills amounting to $10,000, suggesting that if they could turn a couple of jurors, then they might save the McNamaras.

Gloom obliterated everything in Darrow's world. In mid-February he received Debs's reply to his request that the union leader might "do something to bring the old-time people to my support. It is awfully hard to be deserted in this crisis by those who should stand by me." Debs did not spare Darrow. He revealed that a recent investigation into the attorney's character "was anything but flattering . . . others have concluded that you loved money too well to be trusted by the people."

As the weeks passed Darrow was so consumed by fear and self-pity that it felt to Mary as if the romantic phase of their life had ended. He still needed her, and she continued to be his closest confidante, but their relationship had been swallowed whole by the bribery scandal. Mary moved out of Ingraham Street and rented a cottage in Corte Madera—twelve miles north of San Francisco. She would return to Los Angeles for Darrow's trial, but until then she would await her old lover's occasional visits. It seemed to Sara as if Mary had "been 'widowed,' the widowhood of the living death of all she held dear. Slowly her idol has crumbled to dust before her eyes."

O N THE FIRST day of his trial, May 15, 1912, Darrow arrived at the Los Angeles Federal Court at exactly 9:30 A.M. It was a gorgeous Wednesday morning, but his face twitched in agitation. Twenty minutes later Ruby Darrow and an unexpected companion took the same walk up the sunlit hill leading to the building where her husband was about to be tried for bribery. Ruby paused to face a pack of reporters. In answer to their shouted questions she attempted a smile and insisted that "I shall attend court constantly with my husband. I believe it may help him to know that I am here. I have but one thought—to help my husband in any way I can."

The *Los Angeles Herald* confirmed innocently that Ruby was accompanied by "Miss Mary Field of San Francisco, who is making a stay in this city and spends most of her time with Mrs. Darrow."

In truth, the two women hated each other. Ruby was dressed in an expensive suit while Mary wore a baggy dress bought from a secondhand shop in San Francisco. Her bonnet, which she had owned for years, was large and unflattering. Ruby, meanwhile, wore a new and elegant hat topped by a ribbon. She was taller than Mary—even if, on closer inspection, she looked more tremulous than her rival. That contrast was evident in the way that, as they parted silently at the courtroom door, Mary headed purposefully for the press gallery while Ruby walked hesitantly toward a seat just behind Darrow. She was visibly affected by his defeated demeanor.

Mary glanced in Darrow's direction before she turned back to her open notepad. She could hardly bear to see the way Ruby stroked the back of Darrow's neck. He sat there mutely, making no move to acknowledge Ruby's tenderness. But the sight of husband and wife together underlined Mary's sense of her own invisibility. Here, she was merely a reporter.

The affair caused Ruby piercing hurt, for Darrow had insisted on seeing Mary, arguing with conviction that her journalism was a weapon they could call upon. Yet Ruby understood that, with Darrow under such strain, it would not last. Mary Field might win him a few more supporters, but in the end, once this grim business was done, she would be gone from their lives.

Mary's polemical report on that first day in court was published in *Organized Labor*. She compared the case of "The People against Clarence Darrow" to the persecution of both Jesus and Socrates—and suggested that the indictment of Darrow explained "why today there is rejoicing in stock markets, in business offices, in financial circles, in wealthy clubs." That emotive defense pointed to the very tactics Darrow would use later when trying to divert the focus from his likely guilt to his record as a passionate defender of the poor and the exploited.

Yet in a blistering opening address, the district attorney, John Fredericks, derided Darrow as a liar and a hypocrite. Fredericks confirmed that the State would present concrete evidence to prove unequivocally that, on November 28, 1911, Darrow had given money to Bert Franklin with the clear instruction that he should bribe jury member George Lockwood. Darrow flushed and fumbled with a cigarette paper as he listened.

The first week of the trial followed that same desolate pattern, and the arrival of Franklin on the stand merely escalated the battle. He testified that Darrow had first discussed bribery with him on October 5. They had decided to offer $4,000 as a bribe—with a further $1,000 to be paid to Franklin for each successful sting. Stressing that Darrow had underlined his determination to win the McNamara trial, "for it would probably be the last one he would ever try of that importance," Franklin portrayed a man who would do anything to secure victory.

The tension between Darrow and his own attorney was at its worst on the morning that cross-examination of Franklin was due to begin. Earl Rogers had disappeared on an all-night bender, and a frantic hunt spread across his drinking haunts. His staff eventually brought Rogers back to court—where, incredibly, he then produced one of his more robust performances without quite shredding Franklin's muddled if damaging evidence.

John Harrington, Darrow's fellow lawyer who had assisted him during the McNamara defense, also took advantage of the immunity on offer to testify against the accused. He confirmed that Darrow had approached him with corruption in mind.

"Do you mean to say," Rogers sneered, "that Mr. Darrow showed you a roll of bills and told you that he was going to bribe witnesses with it, or jurors with it?"

"He didn't use the word 'bribe,'" Harrington said coolly. "He used the word 'reach.'"

One witness after another presented his testimony before cross-examination—usually conducted by the flamboyant Rogers but, occasionally, by Darrow himself.

Eventually, six weeks after the trial began, Darrow walked to the witness stand. He began answering questions on July 29, 1912, and Mary Field suggested in the *San Francisco Bulletin* that "the calmest person in the room was the defendant himself. . . . The old Darrow, the great Darrow, the philosopher, the poet, the dreamer of days when courts and jails shall be no more, the life-long exponent of non-resistance answered his enemies."

Darrow, in fact, was a tentative witness. There was none of his natural connection with the jury as he parried questions with difficulty. Unable to prowl around the courtroom or lean casually over the railing in front of the jurors, as if he might be one of them, Darrow merely survived his examination.

Yet his mood lightened. On Sunday, August 4, the day after Mary's glowing newspaper account, he started to believe there might be a chance of victory. He took his lover out for the day. As they drove through the countryside he sang loudly and practiced segments of the speech he would make in his closing summary. Darrow was simply trying to regain his speaking rhythm, but his words gripped Mary.

Those feelings of hope spread among his friends and there was talk of a celebratory party should Darrow save himself. Mary stepped back from the surge of wild optimism. More painfully, the thought of

Ruby embracing Darrow was too distressing to ignore. "Ruby's too much for me," she wrote. "I shall all my life avoid her. I hate to lose Darrow's presence but I hate more to be out of harmony with my environment."

O N AUGUST 14, 1912, the clamor to hear Darrow speak was so intense that thousands of people had to be blocked from storming the courtroom. Mary, at the front of the press row, and Ruby, just behind her husband, looked on as court officials tried to restore order.

Pushing and heaving continued in the corridor outside but, with the courtroom door sealed, the accused stood up. Darrow was home at last, in charge of his own destiny, with the full force of his delivery ready to burst from his chest. He felt like a swaggering lion again, but control was everything. Darrow needed to draw in each member of the jury so seductively that it felt as if he was talking to them alone. He thrust his hands deep into his pockets and wandered over to where they sat in two rows of six.

"Gentlemen of the jury," he murmured as a wisp of hair dangled over his furrowed forehead, "it is not easy to argue a case of importance, even when you are talking about someone else."

As he scanned the faces of each juror, silence enveloped the courtroom. "An experience like this has never come to me before. Of course I cannot say I will get along with it. But I have felt, gentlemen, by the patience you have given this case for all these weeks, that you would be willing to listen to me. I might not argue it as well as I would some other case, but I felt I ought to say something to you twelve men besides what I said on the witness stand. In the first place I am a defendant, charged with a serious crime. I have been looking into the penitentiary for six or seven months, waiting for you twelve men to say whether I shall go or not."

Mary unclenched her fists, for she had inadvertently bunched them before his speech began. But now, as the real Darrow reappeared, she relaxed a little. Although it soon became obvious that he would skirt his own possible guilt, in favor of a broader theme, his words pulled the jury toward him.

"What am I on trial for?" he wondered. Darrow allowed his listeners to ponder the question as they looked into his sad old face.

"You have been listening here for three months. If you don't know, then you are not as intelligent as I believe. I am not on trial for having sought to bribe a man named Lockwood."

Darrow let his claim hang in the air, and Mary marveled again at his ability to flatter an audience just as he cajoled them into absorbing his perspective as if it were their own thinking. The suicidal man was now fighting for his life. His voice, which until then had been hushed and conversational, became louder as he swept aside the facts and made his listeners remember his wider legacy. He allowed his words to be framed by anger and hurt.

"I am on trial because I have been a lover of the poor, a friend of the oppressed, because I have stood by labor all these years, and have brought down upon my head the wrath of the criminal interests in this country. Whether guilty or innocent of the crime charged in the indictment, that is the reason I am here."

Darrow proceeded to claim, with startling audacity, that he was the victim of "as vicious and cruel a plot as was ever used against any American citizen."

Mary noticed how people around her shook their heads in sympathy. "Suppose I am guilty of bribery?" Darrow asked. His pause, this time, was barely noticeable for he did not want attention to drift back to the main charge. "Is that why I am prosecuted in this court? Is that why, by the most infamous methods known to the law, these men, the real enemies of society, are trying to get me inside the penitentiary?"

Darrow walked along the line of jurymen, looking each in the eye, reminding them of the extent of their responsibility in judging him. "No, that isn't it, and you twelve know it," he said as he reached the end of the railing. He stood quite still, less than two feet from the jurors. "These men are interested in getting me. They have concocted all sorts of schemes for the sake of getting me out of the way. Do you suppose they care what laws I might have broken? I have committed one crime, one crime which cannot be forgiven. I have stood for the weak and the poor. I have stood for the men who toil. And therefore I have stood against them, and now is their chance."

Darrow dropped his voice wearily and then, gazing at each juror, he seemed to take sustenance from their company. "All right, gentlemen, I am in your hands, not theirs, just yet."

He then shocked them. "They would stop my voice with the penitentiary!" Darrow cried out, his words raw and hoarse. "If you send me to prison, within the gray, dim walls of San Quentin there will brood a silence more ominous and eloquent than any words my poor lips could ever frame."

Darrow suggested that other defenders of labor would rise up in his place—but then underlined his own significance by suggesting, "I have been, perhaps, interested in more cases for the weak and poor than any other lawyer in America. But I am pretty nearly done, anyhow."

The word *done* sounded like the announcement of his actual death. Some jurors began to weep openly.

"Gentlemen, I could tell you that I did this bribery, and you would still turn me loose. I have been thirty-five or thirty-six years in this profession and I tell you I never saw or heard a case where any American jury convicted anybody, even the humblest, upon such testimony."

Darrow shook his head again. "There are other things in the world besides bribery. There are other crimes that are worse. It is a fouler crime to bear false witness against your fellow man, whether you do it from a witness chair, or in a cowardly way in an address to a jury. Suppose you thought I was guilty? Would you dare, as honest men protecting society, say by your verdict that scoundrels like Franklin and

Harrington should be saved from their own sins by charging those sins to someone else? If so, gentlemen, when you go back to your homes you had better kiss your wives a fond good-bye and take your little children more tenderly in your arms than ever before, because, though today it is my turn, tomorrow it may be yours."

Darrow only touched upon the more awkward facts. "I am about as fitted for jury bribing as a Methodist preacher. If you twelve men think that I would pick out a place a block from my office—and send a man with money in his hand in broad daylight to go down on the street-corner to pass $4,000—why, find me guilty. I certainly belong in some state institution.

"Gentlemen, I have been human. I have done both good and evil. But I hope that when the last reckoning is made the good will over-balance the evil, and if it does then I have done well. I hope it will so overbalance it that you jurors will believe it not to the interest of the state to have me spend the rest of my life in prison."

Darrow spoke seamlessly, without notes, for over two hours that afternoon. He then talked all through the next morning. While arguing that he had little motive for bribery, especially when a plea bargain was in the midst of being struck with the prosecuting attorneys, Darrow produced an emotive pitch for his own salvation. "If you should convict me, there will be people to applaud the act. But if in your judgment and your wisdom and your humanity you believe me innocent, and return a verdict of 'not guilty' in this case, I know that from thousands and tens of thousands and, yea, perhaps millions of the weak and the poor and the helpless throughout the world will come give thanks to this jury for saving my liberty and my name."

THE CHIEF PROSECUTOR, John Fredericks, acknowledged to the jury that "you have listened to one of the most marvellous addresses ever delivered in the courtroom"—but he urged them to concentrate on the facts. "You cannot make any mistake when you find

Clarence Darrow guilty of this crime. And if you do not, the result of that verdict will not end in your lifetime or mine."

On August 17, 1912, as they awaited judgment, Ruby sobbed into her handkerchief while Darrow paced the courtroom—sometimes looking up at Mary's encouraging gaze. But it took the jurors less than twenty-five minutes to return to court, just before ten that Saturday morning.

Judge Hutton turned to the twelve men. "Gentlemen of the jury, have you agreed on a verdict?"

The jury foreman, M. R. Williams, stood up. "We have, Your Honor," he said in a clear voice.

"You may read it," Hutton said.

The break between the voices of the two men lasted for a few seconds. And then Williams gave his answer in a ragged cry: "Not guilty!"

An unknown carpenter jumped over the press barrier and into Mary's lap, embracing her without any knowledge of her intimate closeness to Darrow. The attorney, meanwhile, was locked in a clinch with his wife.

Judge Hutton was one of the last to reach him—after Darrow had been engulfed by friends and members of the jury. Mary watched as the judge clasped her lover's hands. "There will be hallelujahs from millions of voices through the length and breadth of this land," Hutton promised.

The party lasted until midnight, and Mary Field spent most of those fourteen hours in close proximity to Darrow—squeezing his hand and looking at him lovingly whenever Ruby was prized away from him long enough. One resonant photograph survives from that day. As a seated Ruby and Darrow read congratulatory telegrams from across America, Mary Field stands between them, looking joyously down at the man she loved more than any other.

Darrow could not sustain his good cheer for long. His second trial, for the attempted bribery of Robert Bain, was next. He knew that his first acquittal would mean little in another sapping ordeal. Darrow was also about to lose Mary. She had been instructed by *Organized Labor* to head for Indianapolis to cover a different case featuring union officials who had been charged with a conspiracy to transport dynamite over federal territory.

Mary and Darrow wrote to each other almost every other day, but much of their correspondence was swamped by the fact that she had received scant payment for her journalism—and Darrow had to send her money from his own depleted savings. "You can always count on me," he promised on October 22, 1912, after he had wired her another $100. Darrow expressed sorrow for her plight but insisted she stand up to her employers. "Don't be discouraged about yourself. You don't need them. None of us need anyone—except love. You don't need them and you must work out your own life. I do miss you."

A month later, on November 28, after Mary suffered yet more trouble, Darrow wrote: "I am sorry for you dear girl—but what the hell. We just have to stand things." He was sharply aware of his own travails and, by December 26, 1912, in desperate need of her. "The case goes on from Jan 6. Come as soon as you can. . . ." He wrote again two days later, urging her to travel via San Francisco so that she could try to uncover more information to help his defense.

The same pattern of insecurity and neediness ate away at their relationship. And although Mary returned for his second trial, she was worn down and disenchanted. She was also, disconcertingly for Darrow, diverted by her increasing closeness to Lemuel Parton, a liberal journalist who had worked for the *Chicago Tribune*, the *Los Angeles Herald*, and the *San Francisco Bulletin*. After they had met in 1911, Lemuel had pursued her quietly, writing thoughtfully to Mary at her most fragile. They were now seen often together in public.

Darrow's salvation was tarnished. In the second trial, a deadlocked jury could not reach a verdict. On March 8, 1913, six jurors believed Darrow was guilty, and six argued that he should be freed once more. They pored over the facts again and again until two more men were convinced of Darrow's guilt. The count stood at 8–4 in favor of jailing him for the bribery of Bain. But the four who stood in defense of Darrow would not be moved. A mistrial was declared—and, eventually, the prospect of a third trial was quashed.

It seemed to an exhausted and dejected Darrow as if he had lost his reputation, his money, and his love. He returned to Chicago a bankrupt and broken man.

CHAPTER 4

THE BRIDGE OF SIGHS

THE "TRIAL of the century" closed in on Darrow. At the age of sixty-seven he would soon take to the courtroom floor to defend Leopold and Loeb, the nineteen-year-old killers who captivated and enraged America. He had spent the past eleven years clawing back respectability. Darrow had won many widely publicized cases and, in the process, reestablished himself as a powerhouse of American law. But nothing he had achieved could blot out the memory of his own bribery trials.

Darrow had been at his very lowest then—and he still carried the cowed image in his head. He had avoided a third bribery trial only by agreeing to a life ban from practicing law in California. Although he claimed that stipulation to be a pleasure rather than a punishment, his professional standing was undermined by the statewide injunction. He had also returned to Chicago in the spring of 1913, after the worst two years of his life, reeling from a $20,000 debt.

During the first miserable week of his homecoming, Darrow was reduced to selling many of his cherished first-edition books in order to buy food. He kept his spirit intact by swearing to Ruby that his "lawyering" days were over. Yet while he generated income through public speaking—starting with a lecture on Nietzsche before a sold-out crowd at Chicago's Garrick Theater in the summer of 1913—Darrow eventually slunk back to the law.

He was persuaded, both by economic necessity and a fervent admirer in Peter Sissman, who ran a modest firm in the Loop, to return to work. But he could not practice again as a labor attorney. He had also lost his lucrative contract with William Randolph Hearst's newspaper empire, and so, at the age of fifty-six, Darrow had to build himself back up from the bottom of a very different field—criminal law.

His private obsession with the shame of L.A. had now been replaced in the public mind with something of his former renown. But he still carried a secret longing that, in his more troubled moments, made him wonder if he was echoing King Lear as, on a blasted heath, the mad old man lamented, "I fear I am not in my perfect mind." Darrow yearned for a case in which he could peel away the scars and show the world a parting view of the courtroom king.

And, now, the trial for which he had waited so long had come— even if there was only horror and madness in the killing of Bobby Franks. Darrow risked widespread contempt by defending the unrepentant sons of millionaires, but he still turned to face the zealots who demanded the execution of Nathan Leopold and Richard Loeb.

Darrow had first met them on Monday June 2, 1924, on the morning after their confession and had been struck by the boys' unsettling coolness. "Mr. Darrow," Loeb grinned, "your slightest wish will be law to me." Yet the strutting Loeb also predicted, "This thing will be the making of me. I'll spend a few years in jail and I'll be released. I'll come out to a new life."

Loeb's apparent certainty in his early release, and likely ascent to stardom, displayed a bravado any jury would love to cut down. Leopold was harder to read. At that first meeting he had claimed to be "overpowered" by the attorney's intellect. Darrow joked in response, promising to quote Leopold the next time some oily prosecutor accused him of being a stupid old buzzard.

Leopold insisted that intelligence was the virtue he revered most and explained how, as a law student, he had followed the attorney's career closely. Darrow remembered later that Leopold used words like *hero* and *superman* to convey his awe. But the longer he talked, the more Darrow noticed a peculiar detachment in Leopold's voice. The actual words conveyed pleasure and empathy, but something remote held Leopold in check—as if he was unable to show genuine emotion. It revealed a chilling self-containment.

The attorney did not know it then but the young killer had been struck by a loaded thought as he stared at the famous old man. So this was Darrow? The magisterial Darrow of whom he had read so much, the greatest criminal lawyer in America? He cut an unexpectedly shabby figure. How long had he worn the egg-stained tie hanging loosely around his neck? And then there was the face, a worn canvas of seams and crevices that told of all the suffering Darrow had seen over so many years. He looked desperately old to someone as young as Leopold.

When Darrow asked him why they had chosen to end another human life, Leopold had answered calmly: "The killing was an experiment. It is just as easy to justify such a death as it is to justify an entomologist killing a beetle on a pin."

Leopold had also made a sweeping reference to Fyodor Dostoevsky's *Crime and Punishment*, in which the young murderer, Raskolnikov, regards himself as a superhuman above conventional morality. Yet any literary cachet was shredded by the savage image of little Bobby as a helpless bug skewered by a pin. Darrow had tried hard not to judge either boy. It was, after all, easy to like Loeb. But Darrow had warmed more to the gauche Leopold. Although his hair was slicked down and he wore sleekly expensive suits, like Loeb, Leopold was shorter and more awkward. His bulging eyes were matched by a jutting Adam's apple that evoked his recent puberty.

"You newspaper folk," Leopold laughed with reporters, "you've made me out a conceited smarty, haven't you?" Leopold had fed them a set of allusions—from Aristotle to Oscar Wilde—which distinguished him from every other criminal they had written about in their heated prose. Darrow was more concerned by the reaction of ordinary people. He was sure a jury would want to hang anyone who referred to himself as a "Nietzschean colossus" and spoke of his refined taste for squirrel meat. A smart-ass killer, and the son of a millionaire, would not work on the stand.

In their separate confessions Loeb insisted his partner had landed the deadly blows—a claim that outraged Leopold as he protested that he had merely witnessed the death. He also derided Loeb as a weakling for confessing first. Darrow, at that first meeting, had quickly put an end to their bickering. As neither boy claimed to be innocent of planning the killing, or of the abduction and disposal of the body, their spat was redundant. The evidence against them suggested that they could both expect to hang from a rope until their necks broke—unless they did as he instructed.

They bowed their heads in shining supplication when Darrow gave them his orders. "Be polite," he said, "be courteous. But say nothing and refuse to answer any more questions."

In the ensuing two weeks Leopold and Loeb had struggled to heed those instructions. They had jabbered and yakked to the stream of journalists who visited them in prison. At least they mostly restricted their conversation to animated musing on subjects stretching from philosophy and literature to fashion and fame. They avoided direct discussion of the murder, so Darrow allowed the informal interviews to continue. The benefits of enthralled reporters writing about the boys were obvious.

Leopold was routinely described as a genius, with an IQ of 210, while Loeb was said to be the University of Michigan's youngest-ever graduate. Loeb was obsessed by cheap American crime paperbacks—in contrast to Leopold, who read heavyweight philosophical tracts from Europe, often in the original German or Italian. Leopold was also an ornithologist who, six months before the murder, had presented a landmark paper and film capturing the rare Kirtland's warbler rearing its young. The press, meanwhile, depicted Loeb as a good-looking wisecracker with a string of girlfriends.

The young men became startlingly famous. Public absorption in the case was such that Darrow could see a bizarre new phenomenon developing in their wake. As the world's first celebrity-killers they would come to define a dismal modern archetype. Despite his hard-won cynicism after so much exposure to the stupidity that existed in real life, Darrow was staggered that the boys had received various marriage proposals from supposedly smitten young women. He was also disturbed by the way in which "Dickie" Loeb and "Babe" Leopold often seemed more concerned with the way they were depicted in print than ending up on death row.

Their case was unlike any other Darrow had encountered in forty years at the bar. Something strange and elliptical inside Leopold and Loeb awakened an interest in him that transcended the professional processes of his role as their attorney. Perhaps his own need for redemption shaped his longing to find some good amid the violent evil of May 21, 1924—that cold spring day when they had taken a life with such malice.

Criminal Courts Building, Dearborn Street, Chicago, July 21, 1924

Less than a month had passed since Darrow's reunion with Mary Field Parton had ended discreetly on the fifth floor of Chicago's Criminal Courts Building. On that same landing he now turned right toward the corridor that linked the rear of the courthouse to the adjacent Cook County Jail.

Stretching from one building to the other, and high above a dank alley, the thin passageway was called "The Bridge of Sighs" by Chicago's crime reporters in an ironic reference to its far more beautiful equivalent in Venice. Leopold had been delighted by the name, because it gave him the chance to show off his knowledge of Italian architecture and a Byronic passion for poetry. Before their first crossing of the prison bridge, on a hot day the previous month when the boys had been arraigned for kidnap and murder, Leopold had reminded his august attorney of Byron's epic poem *Childe Harold's Pilgrimage*. In a typically romantic burst, the poet suggested that the *Ponte dei Sospiri* echoed with the sighs of condemned prisoners who were taken from Venice's jail and across the bridge to the adjoining palace where they would be executed.

Leopold recited two lines out loud:

I stood in Venice on the Bridge of Sighs
A palace and a prison on each hand

The attorney was unsurprised by Leopold's erudition. Darrow already knew the young murderer had spent months planning his translation of Pietro Aretino's *I Ragionamenti* for a limited-edition publication of the Italian satirist's treatise on perversion. "Aretino has great literary value, if one can get over the first feeling of revulsion and disgust that it is absolute filth," Leopold had told a bemused Robert Crowe, the state attorney, during his confession.

Darrow was a sufficiently cultured realist to make a telling point to Leopold. By the time Byron evoked the Bridge of Sighs in 1818, the Italians had long since abandoned execution. In contrast, over a hundred years later, the thirst for Leopold and Loeb's hanging was so fierce that individual citizens put themselves forward for the task.

A man from Oakland, Nebraska, had written to the Chicago press to say, "I am prepared to give them a picturesque send-off in real Western style—high hat, boots and quick service. I like thrills myself, but nothing would give me a greater thrill than a hanging party with me to spring the trap." Among many others, a pensioner from Grand Rapids, Michigan, had also offered to pay $100 for the privilege of acting as hangman to Leopold and Loeb. "I am sixty-six years old, but I am game. I have no use for such fiends."

Leopold and Loeb did not look much like fiends to Darrow. He only needed to talk to the boys, to see the way that Babe looked at Dickie, to understand that they too pulsed with human feeling. Leopold lived in a lost world, skirted by dark delusions, but he also displayed a gifted young mind. Darrow talked to him with pleasure about Nietzsche, about *Beyond Good and Evil* and *Human, All Too Human*; yet he felt sadness when Leopold said he had not really meant to harm Bobby Franks. The young ornithologist had become embroiled in murder only out of a twisted sense of love.

Darrow was uncertain whether his own favorite line from Nietzsche—"What is done out of love always takes place beyond good and evil"—could resonate between murderers. Even if they shared some of his intellectual interests, the boys had seemingly obliterated all claims to love and compassion when they brought a chisel down on the helpless head of a child.

Once he had reached the far end of the bridge, Darrow was met by a guard who showed him into a cramped prison office. He had a little more time to rehearse his unexpected switch of tactics while he waited for his clients to be led from their cells.

Darrow had decided that, soon after they entered court together that morning, he would change their plea. Instead of claiming to be "not guilty," on the basis of a diminished sense of responsibility, Leopold and Loeb would admit their guilt. He would also stress to the judge that, if they were spared execution, his clients did not expect to be released from prison again. His plan was fraught with danger. Darrow needed to make judicial history by persuading an American court that mental illness should be regarded as grounds to commute the death penalty. Although that ruling was applied under British and European law, no attorney had achieved such a precedent in the United States. Darrow was about to cross unknown territory, in the company of two strange and broken boys.

NATHAN LEOPOLD and Richard Loeb had been born into Chicago families whose great wealth contrasted with the limited affection they offered their youngest sons. The boys, in one of the coincidences binding them together, had both been raised by a succession of governesses. Babe, as Nathan had long been called, spoke most to Darrow about Miss Wantz, a German governess who had looked after him for eight years in the plush family home.

Miss Wantz, known as "Sweetie," often encouraged Nathan and his elder brother, Sam, to bathe with her when their parents were out. Sweetie, in her late twenties when Babe was ten, showed her breasts to the boys, describing them as *"ballen,"* or balls, while her nipples were "strawberries." Babe was disturbed, but he could not discuss his feelings at school for he was one of only two boys in his year at Miss Spaides's School for Girls—where his parents had inexplicably enrolled him when he was only five.

Sweetie also encouraged Babe and Sam to wrestle naked with her, and, if they had been "good," she would allow each boy to place his small penis between her comparatively hefty legs. The brothers blushed and looked at her in confusion when she told them that she was frightened of men and preferred the company of children. Babe stressed solemnly that he had never experienced an erection while he grappled furtively with his governess.

She was eventually dismissed when Babe was twelve. He had, by then, moved to Harvard School and, seven years on, he told the psychiatrists Darrow had hired to examine him that "I realized I was not like all the other children, that I had wealthy parents, that I lived on Michigan Avenue and had a nurse who accompanied me to and from school." But, at fifteen, after taking additional exams in

German and Greek, he graduated from high school at the head of his class.

In his yearbook, his older classmates wrote a message beneath his photograph: "Nathan Leopold, the crazy 'bird' of the school. The avicular member of the Fifth class is forever harping on birds, their advantages, and their twitterings. . . . 'Flea' is the proud owner of a large museum of birds, bugs, antiquities and souvenirs. A favorite remark of this crazed genius every other Monday morning: *'Oh! Only sixteen 'A's.'* "

A less ambivalent insight into their feelings toward Leopold could be seen in the additional set of words they attributed to him: "Of course, I am the great Nathan. When I open my lips, let no dog bark."

Babe told the psychiatrists that, aged thirteen, he'd enjoyed his first sexual contact with another boy, Henry, who, in a private show, had ejaculated in front of the startled bird-watcher. Leopold confessed to being "extremely interested" in the act. He and Henry had used a spoon to scrape the semen into a jar that Leopold resolved, in a scientific experiment, to monitor over the coming days. But he was soon too busy experimenting on himself.

By the time Babe was fifteen, Dickie Loeb, who was seven months younger, had begun to obsess him. Despite an initial wariness toward each other, they hung around in the same small gang of boys—and joined enthusiastically in group sessions of mutual masturbation. Darrow noticed that the psychiatrists were unsettled by Leopold's natural sexual preference for boys. They concluded that "since he has a marked sex drive, and has not been able to satisfy it in normal heterosexual relations, this has undoubtedly been a profound, upsetting condition on his whole life."

For Darrow, however, Leopold seemed less distressed by his sexual orientation than the painful double life he and Loeb were forced to lead. It did not help that the boy he revered as superhuman felt compelled to conform to the high jinks of a straight freshman. Yet Loeb admitted to the psychiatrists that he felt "less potent" than most boys his age and had a low libido: "I could get along easily without it. The sexual act is rather unimportant to me."

Dickie had also lived under the sway of a claustrophobic relationship with his governess—a Canadian woman, Miss Struthers, who was twenty-eight when she first entered the Loeb household. He had been only four and a half then, but she immediately focused her energies on stretching him mentally and choosing his friends. And though there was a less overt sexual theme to their intense contact, her control was such that the young boy learned to lie as his only way of subverting Miss Struthers's domination. Yet her rigorous tutoring enabled Dickie to graduate from high school and begin college at the age of fourteen.

As the youngest in his class he lied regularly about his sexual exploits. But, to provide some proof to his boasting, Loeb occasionally accompanied his older friends on boozy visits to brothels. He and Leopold struck out together on a similar adventure. They drove around downtown Chicago in one of the Loeb family cars and picked up two working girls who might have raised their prices had they known that their trembling young clients were million-dollar heirs. But the boys settled on a $3 trick.

Leopold was shocked when his girl, having already hitched up her skirt and removed her panties in the backseat, drawled "Let's go."

He tried to mimic Loeb's diligent efforts with the second girl in the front seat. It was hopeless. "I was unable to perform," he told his doctors sadly. After they had dropped off the unimpressed women, who held on to their $6, Leopold made Loeb swear he would never breathe a word of his humiliation. Loeb grinned and nodded, for they already shared far more personal secrets.

Their sexual relationship had begun on a night train from Chicago to Charlevoix, in Michigan, where the Loeb family stayed at their mansion every summer. In a private Pullman berth, Loeb had climbed into Leopold's bunk rather than his own. It was then, with the steam train rolling through the dark, that Leopold was first able to lie on top of Loeb. Although he did nothing more than place his erect penis between Loeb's thighs, he was ecstatic. Leopold said that it gave him greater pleasure than anything else he had ever done.

Sex permeated their vacation, with Leopold's delight only curbed when they were discovered in bed by another boy from Chicago, Hamlin Buchman, who worked in Charlevoix during the holidays. Buchman was a few years older than Leopold and Loeb, and on returning to Chicago he gossiped openly about their relationship.

Having entered the University of Chicago at sixteen, Leopold took three majors, in English, Latin, and psychology—and extra courses in Italian, Russian, Spanish, both ancient and modern Greek, and Sanskrit. After a year he transferred to the University of Michigan so that he could continue his intimate friendship with Loeb, who was already a registered student in Ann Arbor. But in the fall of 1922, when Loeb expressed interest in joining the Zeta Beta Tau fraternity house, there was concern about his sexuality. Senior members of the fraternity contacted his elder brother, Allan, to discuss the stories connecting him and Leopold. Allan traveled to Ann Arbor and, having been reassured by Dick, convinced the frat boys there was no reason for alarm.

Loeb was offered membership on the strict condition he renounce his friendship with Leopold. Although he promised Leopold that he would continue to see him furtively, he had little compunction in declaring to the fraternity that he had severed their relationship. The public humiliation scarred Leopold, and he became an even more introverted figure—finding refuge in bird-watching and clandestine drinking bouts with Loeb.

His lover, meanwhile, was primarily interested in the imaginary life he had conjured for himself as a criminal mastermind. Loeb's murky reveries, which he called "picturization," allowed him to indulge his fantasies—the most pleasant of which saw him outwit the world's most famous detectives to the awe of his criminal partners.

In real life, and during college vacations, he and Leopold would slip out of their family homes late at night and set about a chosen crime. Loeb enjoyed planning outrageous escapades and then returning later to observe the impact of their delinquency. After they had set fire to a shack, Loeb was thrilled to join a crowd of onlookers at the burning site. He contributed animatedly to discussions about the possible identity and motives of the arsonists. His rapture was complete when firefighting experts trawled through the charred remains in a futile search for clues.

He also exulted in the lies he let slip to a casual girlfriend—boasting that he was a bootlegger who had evaded an armed group of policemen. Loeb used his revolver to shoot a hole in one of his shirts, which he produced later as evidence of his dangerous life to the shocked girl. As she begged him not to place himself at such risk again, Loeb shivered with the bliss of his ruse.

As each new kick intensified, so the seriousness of their plotting escalated. They decided to rob the fraternity house in Ann Arbor that had so humiliated Leopold. Just before 3:00 A.M. on Sunday, November 11, 1923, they pulled black woollen masks over their heads. Leopold and Loeb checked their coat pockets—containing flashlights, two loaded revolvers, a coil of rope, and a chisel whose blade was bound with tape so that its thick wooden handle could be used as a weapon to bludgeon anyone who disturbed them. They had also sworn to each other that they would not hesitate to fire their guns should the situation become desperate.

Yet, in their fear, they merely stole $74, a camera, a portable Underwood typewriter, a watch, and a fountain pen. It was not a crime fit for Nietzschean supermen. On the long drive back to Chicago, as they drank from a bottle of bourbon, Leopold stressed his disgust for the weakness Loeb had shown throughout the nervy theft. He said he was tempted to pull over and shoot Loeb in the head. They had threatened to kill each other before and so Loeb shouted back angrily. He was sick of Leopold's dumb idea that they were acting on behalf of Nietzsche or some other dead philosopher Loeb would never read. Leopold had hardly been cool and strong in the dark.

The conflict ceased only after they had stopped for breakfast at a roadside diner. Slowly, as they resumed their journey, they worked out what they might do next. Leopold would continue their life of crime, but he expected some tenderness in return. He opened himself to further ridicule, but Loeb did not mock him as he spoke of his desire for companionship. The boys forged a verbal contract that their relationship would move to a new level of intimacy and risk. They thrashed out the crude details as the sun lit up a crisp fall morning in Illinois.

Leopold could expect physical relief with Loeb on regular occasions. In exchange, he would work as Loeb's criminal accomplice and do anything demanded of him. As they drifted through the sleek suburbs of Chicago's South Side, feeling exhausted but elated in their revived unity, Loeb outlined his desire to plot the perfect crime. If they truly were supermen, then they needed to prove it by plunging into far more significant criminal acts than arson or burglary. He was sure they could carry out the perfect kidnapping and extract a sizable ransom from their victim's family.

Loeb sank back into his passenger seat in satisfaction as they slid to a halt outside his vast home on Ellis Avenue in Kenwood. "Kidnapping," he said with a light laugh, "and then . . ."

He had one more word to say, a word that defined the most terrible yet intimate crime of all. It was the final act that would bind them together forever. Loeb said the word so quietly and seductively that Leopold trembled in the gentle afternoon sunshine: *"Murder."*

DARROW REALIZED that Leopold and Loeb had to avoid a judgment passed down by a jury. Public outrage meant it would be almost impossible to convince twelve men to buck the nationwide demand for hanging. By reversing their plea to guilty, the trial would effectively become a hearing before a single man, Judge John Caverly, whom Darrow hoped would possess the courage to administer a sentence fueled by personal compassion rather than mass vengeance.

He had scoured the records and seen that Caverly had presided over the sentencing of five men to death—but on each occasion it had been a jury-led verdict. He could only hope that the fair-minded attitude he detected in Caverly would carry over into an emotive trial. Caverly himself had signed the writ of habeas corpus forcing State Attorney Crowe to allow Darrow his first access to Leopold and Loeb.

Robert Crowe, an arch Republican with his eye on the next Cook County election, was so determined to secure a populist hanging that Darrow knew it was crucial their unexpected new plea should be prepared as covertly and late as possible. Darrow's main concern, however, was that his unpredictable clients might not cooperate. He was unsure how Leopold, in particular, might react to his plea for mercy—and a humble acceptance of life behind bars. There were times when the curiously cool ornithologist looked as if he might consider being hanged as a pleasingly experimental way to end his life. Amid the delirious press coverage, and their surreal fame, could they accept a change of plea and a starkly anonymous life behind bars?

His long wait in the prison office ended when the door opened and four wardens steered Leopold and Loeb to the chairs opposite him. The guards withdrew to the corridor outside while Darrow looked at the two boys. A change had come over both of them. With the trial just over ninety minutes away, they looked much younger and more withdrawn. There was nothing of Leopold's arrogant wordplay or Loeb's languid élan. They sat quietly in front of Darrow.

He began cautiously, suggesting that he would ask them to agree to a strange course of action. Darrow outlined the reasons for them to alter their plea in precise detail, but he urged them to understand the significance of his strategy. "Hell boys," he said, "it's the *only* way." He explained that the alternative would give Crowe "two bites at the apple" of execution—for he would try them first for murder, and then, even if they were acquitted, he could pursue the same verdict on the kidnapping charge. By pleading guilty they would take Crowe by surprise and hopefully force a more benevolent sentence of life imprisonment.

After Leopold and Loeb nodded their mute agreement, Darrow stared at them in amazement. They had surrendered themselves to his care. After all their conceit and defiance, the boys had given into a frightening truth. Only Darrow could save them now.

SOON AFTER nine thirty that morning, Darrow went back into court with the anguished fathers of the accused—Nathan Leopold Sr. and Albert Loeb. They were accompanied by his cocounsels, the brothers Benjamin and Walter Bachrach, and Jacob Loeb—Dick's uncle and the man who, in early June, had fallen on his knees and begged the attorney to save the boys.

As the others settled in their seats Darrow shared a couple of wisecracks with his adversary. Robert Crowe, or "Fighting Bob" as he was known in Chicago, looked dapper in a dark blue suit as he chewed on his fat cigar. He slapped his illustrious opponent on the back and reminisced about their past battles in court. But they both knew that this would be a struggle like no other.

Darrow bore no malice toward his shorter, bullnecked opponent. They actually shared many convictions as Crowe was not far behind Darrow in his support for black Americans—citing "the plight of the Negro" as one of the worst injustices afflicting the country. They were also united in opposition to the draconian laws of Prohibition, arguing that the ban on alcohol was an assault on individual liberty that increased rather than undercut criminal activity.

But Crowe and Darrow parted furiously over the death penalty. Whereas Darrow believed that a state execution was a calculated crime even more inhumane than murder, the attorney for Cook County was a vehement supporter of hanging as a deterrent. It was obvious that judgment in the Leopold and Loeb trial would have a lasting impact on death row legislation—and Crowe had no intention of losing.

At precisely 10:00 A.M., the court clerk, Fred Scherer, noting the arrival of Judge Caverly, cut through the excitable rumble of noise with a high-pitched cry. *"Hear ye! Hear ye!"*

A low hush settled over proceedings. "This honorable branch of the Criminal Court is now in session pursuant to adjournment," Scherer intoned in the pompous legalese that Darrow hated. The lawyer raised his thick eyebrows toward the ceiling and settled back in his chair.

Judge Caverly rapped his dark gavel hard against its wooden sound block. "The case of the People against Nathan Leopold Jr. and Richard Loeb," Caverly said with due ceremony.

Scherer answered him with a theatrical echo: "Nathan Leopold and Richard Loeb!"

The accused, in typically elegant suits, were led in by a police escort. They commanded attention in a way that not even Darrow and Crowe could rival. Scores of press cameras clicked and whirred, flashlights popping softly, as Leopold and Loeb took their places next to Darrow. Buoyed by the attention of the photographers they smiled confidently at their old attorney.

But their expressions tightened when Caverly turned to the defense counsel. "You may proceed, gentlemen," he said impassively.

Darrow remained in his chair a moment longer and then, slowly and thoughtfully, he stood up and approached the bench. He tugged at both suspenders holding up his suit trousers and scanned the court, shaking his large head in apparent weariness. "Of course," he said in a voice quiet enough to make even the judge lean forward in anticipation, "it is unnecessary to say that this case has given us many perplexities and sleepless nights." He paused again, as if he might need to swallow a yawn of exhaustion.

"Nobody is more aware than we are of what this means and the responsibility that is upon us. This case has attracted very unusual interest on account of the weird, uncanny and terrible nature of the homicide. We have meant to consider it from the standpoint of the defendants, but we must also consider it, first of all, from the standpoint of their families—and by the families I include all three . . ."

Darrow ducked his head gently toward Jacob Franks, looking gaunt and ghostly. His wife, Flora, had been too ill with grief to attend court. Darrow then gazed at both the Leopold and Loeb families, reminding everyone that they, too, were locked in misery.

He moved ahead more swiftly now and, having seized the concentration of the court, debunked any suggestion that the defense would attempt to free his clients. "We want to state frankly that no-one in this case believes these defendants should be released. We believe that they should be permanently isolated from society and if we lawyers thought differently their families would not permit us to do otherwise. . . . After long reflection and thorough discussion we have determined to make a motion in this court for each of the defendants in each of the cases to withdraw our plea of not guilty and enter a plea of guilty."

Crowe flew from the prosecution benches, his angry objection merging with the noise that swept across the court. Darrow stood as motionless as he was silent amid the hubbub. He had been expecting the incredulous voices and wordless exclamations. "Quiet in the courtroom!" roared a frantic clerk. "Quiet! Order! The courtroom will come to order! Quiet please!"

The judge motioned Darrow to continue and, reluctantly, Crowe sank back into his seat. As he moved closer toward Caverly, conveying the dramatic intimacy of his decision, Darrow spoke as though they were the only two men in the room. "Your Honor, we dislike having to throw this burden upon the court. We know its seriousness and its gravity, but a court can no more shirk responsibility than attorneys.

"The statute provides that evidence may be offered in mitigation of the punishment. We shall ask that we may be permitted to offer evidence as to the mental condition of these young men, to show the degree of responsibility they had and also to offer evidence as to the youth of these defendants and the fact of a plea of guilty as further mitigation of the penalties in this case."

Darrow looked up meaningfully at the man who would decide whether or not Leopold and Loeb would be hanged. "With that," he said, his voice softening in tone, "we throw ourselves upon the mercy of this Court, and this Court alone."

"Does that go for both the cases?" Caverly asked, referring to the abduction and the murder of Bobby Franks.

"Both cases," Darrow replied.

Caverly turned to the accused. "Nathan Leopold Jr., will you please rise?"

The shorter of two dark-haired boys, Leopold stood up as if he might be a nervous student about to answer a difficult question.

"If your plea is guilty," the judge warned, "the court may sentence you to death." The words hung in the oppressive air. "The court may sentence you to the penitentiary for the term of your natural life."

There was a longer pause before Caverly asked Leopold another question: "Now, realizing the consequences of your plea, do you still desire to plead guilty?"

"I do," Leopold answered firmly.

"Let the plea of guilty be entered, Mr. Clerk, in indictment number 33623, charging Nathan Leopold Jr. with murder."

As if this were a stage routine they had rehearsed countless times, Leopold moved back one pace so that Loeb, with a single step of his own, could stand in his place. He went through the same ritual and when asked, "Do you still desire to plead guilty?" Loeb's reply could be clearly heard.

"Yes, sir."

"Mr. Clerk, let the records show that Richard Loeb has pleaded guilty in indictment number 33623 and, also, that both defendants have been warned by the court."

The boys swapped positions twice more so that they could both affirm their guilt in relation to the charge of kidnapping.

With Darrow's work seemingly done for the day, his colleague Benjamin Bachrach emerged with a practical suggestion to simplify the judge's task. As Leopold and Loeb had been interviewed by eighteen psychiatrists, or "alienists" as they were mostly called, Bachrach argued that the medical men should meet first, in private, to formulate an overview of the defendants' mental state. Fourteen of the psychiatrists, led by Harold Hulbert and Karl Bowman, had been retained by the defense. As a means of avoiding excessive repetition of their findings, and the "vaudevillian spectacle" that often accompanied a standoff in court between rival specialists, Bachrach suggested that a single psychiatric report would suffice.

Crowe was indignant. "Is there a plea of guilty entered here by two sane men," he snorted, "or is the defense entering a plea of guilty by two insane men? If there is any contention they are insane, I ask Your Honor to call a jury of twelve men so that the state may demonstrate beyond all reasonable doubt that these boys are sane."

Bachrach repeated Darrow's line that mental illness would be used only as grounds to avoid the death sentence—a tactic that infuriated Crowe. He rejected any meeting between the experts and insisted that, as originally planned, he would call up to a hundred witnesses to testify to the crime's savage brutality. Caverly upheld the prosecution's right to proceed and ruled that court would reconvene at 10:00 A.M. that Wednesday morning. The jury box would remain empty.

Crowe had been expecting the trial to commence after a recess lasting two weeks—rather than just two days. But there was nothing he could do as he watched the suddenly grinning boys shake hands with Darrow before they were shackled again with steel cuffs. "Led back to their cells in the county jail," the *New York Times* observed, "Leopold and Loeb's attitude was one of released tension. They laughed and talked eagerly with reporters, but refrained from discussing their trial. Leopold scorned the suggestion that either he or Loeb had been nervous, but both admitted they were glad the first day was over. They were curious about the ticking telegraph instruments in the court. They wanted to know how big a crowd waited outside and whether the headlines that told of them would be as big as the ones that announced the World War armistice."

Fighting Bob said their arrogance turned his stomach. After the jolt of being temporarily outwitted by Darrow, the state's attorney had regained much of his certainty when he spoke to reporters late that afternoon. "The fact that the two murderers have thrown themselves upon the mercy of the Court does not in any way alleviate the enormity of the crime they committed. The state is going to prove not only that these boys are guilty but that they are absolutely sane and should be hanged."

THAT MONDAY evening, in a rented summer house in Sneden's Landing, in the deceptively wooded countryside just outside New York, on the western bank of the Hudson River, Mary Field Parton sat alone at a desk. She missed Lemuel, her husband, on his third night away from home. He had sailed across the high seas somewhere, crossing waters that became colder the farther his ship, *The Sylvia*, moved toward Greenland. Lem had left at noon on Saturday, July 19, and she and Margaret had headed back to Sneden's Landing where they would spend the next two months of summer while he was on his adventure in pursuit of the Arctic explorer Donald Macmillan.

They had taken the little launch from Dobbs Ferry, piloted across that stretch of the Hudson by fiery old Captain Hill, who cursed all the "summer folk" coming from the city to while away the hot and steamy months in and around Sneden's and Palisades, the beautiful hamlet that was less than a dozen miles from New York. The old house that would be their temporary home was large and rambling, especially for a suddenly single mother and her only child; and the country seemed eerily quiet in contrast to the strident city.

Mary had worked all day on her writing, agonizing over every word in the first of a series she called her "Franks Stories." She homed in on the three different boys, little Bobby and Leopold and Loeb, and tried hard to make sense of everything Darrow had told her about them in Chicago. She aimed to submit the best of her stories to the *Nation*, New York's weekly periodical of culture and politics, and resume her career. The work was slow and difficult, for she felt out of practice and lacking in confidence, and it did not help that all day Margaret had kept gamboling in from the sunny yard outside to tell of her latest discovery or mishap.

"M takes much patience and I fail her often," Mary admitted in her diary. "She is at the incoherent, silly age—oh little darling—and full of lies too." In the past, when she had resented the relentless demands placed on her time and energy, she had sometimes railed in letters against "little Maggot"—the cruel but secret name she foisted on her helpless daughter. It was not a name she could bring herself to write anymore, so shocked was she at her own former frustration, and the kind of rage to which she had sometimes succumbed.

Mary had once snapped and slapped Margaret in the face—and spent the rest of the day in such wretched remorse that she eventually went to her daughter's school and found her so that she could make a tearful apology. She had then treated them both to children's theater tickets that afternoon and tried to feel a little better about herself as both a mother and a person.

The early years had been hard, with Margaret wetting her bed until she was six and suffering from regular asthma attacks—which had eased only when they moved away from foggy San Francisco to New York. But none of this seemed to matter when Mary attempted to write about poor Bobby Franks, who had been just five years older than Margaret. When her nine-year-old girl had rushed in to disturb her that afternoon, with her face dripping after she had run, screaming happily, through the fountain of water arcing from a faulty faucet, Mary had swallowed her irritation.

Margaret was alive and beautiful to her mother, who could marvel at her good fortune in contrast to Flora Franks, a woman almost made mad with grief by the murder of her son. It did not help Mary's writing much, but she could respond sweetly in that moment to Margaret.

Now, with her little girl blissfully asleep in bed upstairs, and her work put away for the day, Mary felt the ache that sometimes came to her at night. In Lem's absence, and with the loss of Darrow gnawing at her, it was more acute than usual. A single window in the ghostly white house opposite was ablaze with light, and she could hear the tinkling of a piano drifting across the way, accentuating her melancholia. "Terrible loneliness comes," she wrote. "The fireflies seem like little creatures carrying my loneliness, haunting someone."

She found some solace in reading Darrow's latest book, *Crime: Its Cause and Treatment*, which he had given to her in Chicago while lamenting its neglect by the critics. Mary was sure that they would now be poring over its contents, trying to find some additional clues for his imminent brief in defense of the seemingly indefensible. It comforted her to read his words, hearing a little of his old voice in her head, as she became caught up again in his world.

It was, typically of Darrow, full of contradictions. There were pages when it seemed as if man was bereft of hope, as if the pain and injustice of life was fixed and immovable, and she wanted to cry out and ask him why he even bothered to write the damn book when everything was swamped in futility. But then Darrow would produce a chapter in which he wrote tenderly of how humanity might help the criminal, and she felt moved by his benevolence and understanding: "Dear Darrow, bigger than his pessimism."

He might have disappointed her painfully in Chicago, and part of her wished that she were there now, in that wild and heaving city, with Darrow himself, drinking and carousing and analyzing the day's events in court, but she also believed something far more momentous was unfolding. It was better, even amid her solitude, for him to be surging forward so that he might do something extraordinary during the coming days and weeks in court.

Even Margaret, missing her father, and liking little about Darrow, had sensed the momentous pull of the trial. Mary had turned her back on the religious hysteria with which her Baptist preacher of a father had shrouded her own unhappy childhood, but she allowed Margaret to find comfort in nightly prayers at bedtime. Margaret had recently begun to include Darrow's name, and those of Leopold and Loeb, in her soft beseeching. It seemed as strange as it was touching for Mary to hear Margaret hope that Darrow would look "less lonely" in front of the judge, that he would look "beautiful sad" as he pleaded for the lives of the lost boys.

Margaret would be on her knees, at her bedside, the glow of the candle illuminating her face in a gauzy flicker, as she prayed: "Dear Jesus and God. Watch over dear daddy. Please put Leopold and Loeb in the right place. Take care of my cat and doll and bring daddy home safely. Amen."

Outside, the piano trilled and rippled softly while the fireflies shimmered in the sultry darkness. Mary felt redeemed by Margaret's goodness, and by her own recognition of it. But a little thrill also rippled through her at the thought of Darrow, the magical old devil, as he hunkered down in Chicago, alongside two killers in the dock.

BEYOND GOOD AND EVIL

MURDER ALWAYS concentrated his mind. It left room for little else as Darrow confronted its brutal consequences. And so, moving with fierce purpose into the enclosed world of a murder trial, he embraced a courtroom strategy freighted with risk. In his attempt to force a legal precedent, Darrow had decided there was no point in detailed cross-examination of the long line of prosecution witnesses. Leopold and Loeb had already confessed to abduction and murder; Darrow's only objective was convincing the judge that their mental illness provided sufficient mitigation to overturn the death penalty. Allowing huge chunks of the prosecution case to remain unchallenged, he would rely on expert testimony supplied by the psychiatrists who had examined his clients.

Robert Crowe, meanwhile, was determined to quash any psychiatric input into the trial. He knew U.S. legislation operated in black-and-white parameters that did not recognize any shifting gradients of mental illness. A person standing trial was either sane or insane—and the latter were always tried before a jury. That simplistic reduction denied the existence of any complex shading where the accused might blur the boundaries between "normality" and "madness."

Psychiatry was still not considered a medical science in legal circles. It was regarded, instead, with intense suspicion. If the judge ruled against Darrow, and declared such evidence to be inadmissible, the case would be lost in an instant. He had no other defense to offer up against Crowe's demand for execution, beyond pointing to the youth of the boys and begging for mercy.

Darrow's approach was derided by some as a desperate gamble, as if he were just some clammy casino dreamer at the spinning roulette wheel. His fluid intelligence, however, was as formidable as Crowe's aggressive pursuit of vengeance. He knew that the world's leading psychiatrists had begun to penetrate the bleaker mysteries of the mind. Darrow also sensed that the American public was increasingly prepared to engage with a subject they had found to be previously alarming or inexplicable.

In the aftermath of World War I, which had ended five years earlier, many returning soldiers had suffered from acute psychological distress after the grotesque scenes they had witnessed in the trenches and on the battlefields of Europe. There was a new understanding that it was possible, after all, for "ordinary" people to suffer from mental disorders. The great pioneering psychiatrists had also achieved a level of fame that meant their opinions were now pursued in a case as strange and sinister as the murder of Bobby Franks.

Even the coarsely hustling press of Chicago had realized that this would be a trial fought largely over the mental state of Leopold and Loeb. William Randolph Hearst, the newspaper baron who published both the *Herald and Examiner* and *Chicago's American*, had sparked a bidding war with the *Chicago Daily Tribune* in an effort to entice Sigmund Freud from Vienna so that he might comment on the trial. The *Tribune* had already sent a telegram to Freud, offering him $25,000 or "anything you name." Hearst had promised to surpass that first figure while chartering an ocean liner to transport the sickly psychoanalyst in luxurious privacy across the Atlantic.

Freud was not seduced. "I cannot be supposed to provide expert opinion about persons and a deed when I only have newspaper reports to go on and have no opportunity to make a personal examination," he retorted. And just in case either side piped up to announce that they had arranged for Freud to meet Leopold and Loeb, the sixty-eight-year-old Austrian psychiatrist stressed that he was too ill to make the journey under any circumstances.

Darrow was certain Freud would have been fascinated by the undertow of juvenile crime and sexual psychology that lay at the heart of a contract his clients had conjured up more than seven months before. The fact that Leopold and Loeb, in the end, resorted to abduction and murder told Darrow how far their minds had slipped into the shadows.

The dark swirl of sex and violence explained why, on Wednesday July 23, 1924, the first real day of the trial, Darrow was met by a vast throng of spectators who had begun to gather outside the Criminal Court just after dawn. Thousands of men and women clogged the entrance as they fought for their place in a snaking mass. The tumult was more like the buildup to a prizefight than a judicial hearing. Boxing was the most popular sport in America, and Jack Dempsey, the Manassa Mauler, remained heavyweight champion of the world. Darrow thought that only Dempsey, or maybe Babe Ruth in the midst of a World Series for the Yankees, would understand his emotions as he entered a contest that gripped a nation.

Betting on the trial had spread across America, as if this were a sport rather than a judgment that would decide life or death for two nineteen-year-old boys. Darrow did not care for the overwhelming odds in favor of a hanging verdict. He had a different outcome in mind.

At the scheduled start of 10:00 A.M., the heat of a midsummer morning could be felt in a packed courtroom. It was predicted that lunchtime temperatures would exceed 90 degrees as the first scent of human sweat mingled with the salty tang from the fish market below. Judge Caverly had readied himself by placing a large metal fan just to the right of his bench. There was no such respite for the attorneys. Darrow, in deference to the magnitude of the trial, broke with personal tradition by wearing his jacket during initial proceedings. Crowe, perspiring heavily, was too uncomfortable to do the same. He peeled off his jacket.

During Crowe's protracted opening address, damp rings formed down the back and under the armpits of his white cotton shirt. His spectacles steamed up regularly. He soon turned a disadvantage into a telling routine as he repeatedly framed his withering accusations by using a handkerchief to wipe clean his misty glasses.

The fury of his attack on the accused was even more intense than the stifling atmosphere. Pointing angrily at Leopold and Loeb, he told Caverly it was crucial not to forget "that they are both sons of highly respected and prominent citizens of this community; that their parents gave them every advantage wealth and indulgence could give to boys. . . . [But] the evidence will show that in October or November of last year these two defendants entered into a conspiracy, the purpose of which was to gain money, and in order to gain it they were ready and willing to commit a cold-blooded murder. The state will show to Your Honor by facts and circumstances, by witnesses, by exhibits, by documents, that these men are guilty of the most cruel, cowardly, dastardly murder ever committed in the annals of American jurisprudence."

Crowe cleaned his glasses again as that overwhelming claim enveloped the court. Replacing them carefully on the bridge of his nose, he gathered himself for the concluding sentence of an emotive address. "In the name of the people, the womanhood, the fatherhood, and the children of the State of Illinois, we are going to demand the death penalty for both of these cold-blooded, cruel and vicious murderers."

Darrow waited for Crowe to return gravely to his seat. He then pushed back his chair, hauled himself up, and strolled over to the front of the court. As he made the short walk he sank his hands deep in his pockets. It looked as if he were about to take a leisurely detour from the serious business of murder and opt for a gentle amble past the rows of gleaming fish in the market before settling on a lunchtime recess. He then looked up at Caverly and, in contrast to Crowe's hectoring, began to talk as if he had just entered a private conversation with the judge.

"A death in any situation is horrible," he said with deceptive simplicity, "but when it comes to the question of murder it is doubly horrible. But there are degrees, perhaps, of atrocity."

Darrow dug his thumbs beneath his suspenders at chest level. He held on to the two strands of elastic as if clinging to a parachute that would allow him to float gently down from a dizzying murder and secure the safety of both boys. Darrow knew that he could not evade the grisly fall of Bobby Franks, but, as deftly and quietly as possible, he tried to soften the impact.

"Instead of this being one of the worst," he said of Bobby's sudden death, "it is perhaps one of the least painful and of the smallest inducement. Bad enough, of course, but anybody with any experience of criminal trials knows what it is to be branded as the greatest, most important and atrocious killing that ever happened in the State of Illinois or in all the United States."

Crowe almost yelped with indignation. "Objection!" he shouted. "Objection, Your Honor!" His learned friend, he said scornfully, was being unnecessarily "argumentative" and "improper." An opening address was meant to merely outline the evidence that would be put forward—rather than provide a platform for any grand speech making.

"Your Honor," Darrow sighed majestically, "it comes with poor grace from counsel after, for more than an hour, he sought to stir up feelings in this community."

Caverly advised Darrow to restrict his comments to the proof he intended to produce, but he allowed him to continue after waving away Crowe's excessive objection. "Very well, Your Honor," Darrow muttered. "I am aware at this time it isn't a proper statement but I felt outraged."

Darrow had noted how Crowe always referred to Leopold and Loeb as "men," often lingering over the word. Crowe had also argued that the $10,000 ransom had been one of their prime motivations in killing Bobby Franks. Darrow thought it was an erroneous charge he could easily demolish; but, first, he needed to underline that Leopold and Loeb were still boys.

"We shall insist in this case, Your Honor, that, terrible as this is, terrible as any killing is, it would be without precedent if two boys of this age should be hanged by the neck until dead, and it would in no way bring back Robert Franks or add to the peace and security of this community. I insist it would be without precedent if, on a plea of guilty, this should be done.

"When this case is presented, I know the court will take it calmly and honestly, in consideration of the community and in consideration of the lives of these two boys . . . we know Your Honor will do what is just, fair, and merciful—and a court must always interpret justice and mercy together."

Darrow, an attorney of such expansive eloquence that he could spin out an address in court so that he spoke for days rather than just hours, had restricted himself to a five-minute opening statement.

His concise tactics also applied to the cross-examination. Darrow chose only to ask a few plain questions of a single witness out of the fifteen men and women who were called by the prosecution that first day. These ranged from the painfully moving—Jacob and Flora Franks—to anodyne or obscure witnesses who imparted fleeting confirmation of undisputed details.

The same pattern continued throughout the opening week as Darrow mostly declined to cross-examine. He was determined to avoid unnecessary repetition of the actual murder and to fly through the prosecution evidence as quickly as possible. Leopold and Loeb, meanwhile, derived immense enjoyment from their joint role at the heart of proceedings. Before one ordinary morning in court, Loeb gathered the newsmen around him as, with mock gravity, he revealed that he would make a significant statement. "We are united in one great and profound hope for today," he said sagely. "Mr. Leopold and I have gone over the matter and have come to a mutual decision. We have just one hope: that it will be a damned sight cooler than it was yesterday."

Leopold smirked at his lover's wit. When asked if it was the kind of summer day that made him want to play tennis, he answered dryly. "Thank you, no. It so happens that I have another more pressing engagement." Encouraged by the laughter, he offered another polite crack. "Perhaps another time?"

On July 30, a week into the trial, the prosecution presented its final witness. The one-sided questioning was over. Darrow and the defense could now offer evidence in mitigation of abduction and murder—as long as Judge Caverly upheld their right to proceed. There was real uncertainty as to how he might react to psychiatrists as Darrow's star witnesses. Fighting Bob Crowe was ready to shoot them down and bring a sharply focused defense to a catastrophic end.

Even Darrow, attempting to appear laconic, felt the poisonous tension in court. It was impossible to know which way Caverly, his face half shielded by the rotating fan on his desk, would lean. But how else, beyond calling on medical expertise, could Darrow illuminate the murky depths of two murderous minds?

TEN WEEKS earlier, on the day they committed the murder now described as the crime of the century, gray clouds had hung threateningly over Chicago. Nathan Leopold felt calm as he slid behind the wheel of his bright red Willys-Knight touring car at 7:15 on the morning of May 21, 1924. He drove steadily through Kenwood toward the University of Chicago, where his first lecture, at 8:00 A.M. that Wednesday, was in criminal law.

Although there had been times during the past months when he wished that their planning of an otherwise random murder had not unfolded so swiftly, he had resolved to embrace the adventure with as much relish as the boy he loved. Loeb had been quiet during the last few days, but Leopold attributed the change to the absolute focus he expected from a Nietzschean superman.

Their original plan had been driven by Loeb's suggestion that they should kill Hamlin Buchman, who spread such venom after he'd caught them in bed together in Charlevoix. Loeb said that they should kidnap him first and deliver a ransom note to his family. After killing Buchman they would then collect the money. It would be sweet vengeance. But Leopold argued that Buchman was too physically imposing to overpower. He had also left Chicago for a more distant college, so the entire scheme seemed impractical. Loeb, reluctantly, gave into Leopold's logic and accepted a more useful suggestion: that they choose a younger boy as their victim.

Loeb pointed out that his own ten-year-old brother, Tommy, was small and weak and, obviously, the son of a multimillionaire. Leopold quaked at the coldness of this suggestion and was reassured when Loeb eventually dismissed the killing of little Tommy as a sick joke. They drew up a more feasible list of boys, aged between ten and fifteen, who came from wealthy families and whom they knew. It was essential that they should be able to persuade the victim to get into their car without any suspicious delay.

Although Leopold had no compunction about killing, he was acutely aware of the risk. But Loeb simply lost himself in the intricate details of the plot. Leopold's own vehicle was far too dashing not to be noticed, and, even though car rental was still a rare activity in 1924, they were determined to do their work as anonymously as possible. The first steps entailed constructing a false identity, in the name of Morton D. Ballard, opening a bank account, and taking a room at a luxury Chicago hotel to use as a temporary address. Then, in a dummy run and using Ballard's name, Leopold hired a car, which they subsequently returned on schedule.

Leopold would visit the same rental showroom later that morning, but first, and unusually for a murderer, he had three lectures to attend. After criminal law, he took French at 9:00 A.M. with Susan Lurie, a sweet girl he dated occasionally. He liked to read French poetry to Susan when they went picnicking in Jackson Park—for she was one of the few women he knew who appreciated his love of language. Leopold often referred to Susan as his girlfriend as she inadvertently covered his passion for Richard Loeb. Once he and Susan completed their French class he had another law lecture at ten.

The morning drifted past pleasantly, with Leopold thinking little of the challenge that lay ahead. He met Loeb soon after eleven and, after renting the same pale blue Willys-Knight, again in the name of Morton D. Ballard, they ate lunch without mentioning their plans. They knew what they needed to do—even if, as yet, they were unsure whom they might kill.

During the preceding months they had kept a few boys in mind. Their close friend, Dick Rubel, had come close to topping their list. They were convinced that, unlike the beefy Buchman, they would easily knock him unconscious with the wooden handle of their chisel. Loeb fantasized that he would be asked to be a pallbearer at the funeral—a detail that appealed enormously to him. But he was unconvinced Rubel's father would pay any ransom.

Leopold still believed that they should target a smaller boy. Johnny Levinson, a nine-year-old in the same fourth-grade class as Tommy Loeb, stood out. He cut a puny figure and his father, Salmon Levinson, a venerable attorney, would surely stump up a $10,000 ransom. But there were other candidates who could be even more appropriate. It all depended on fate, and just who might wander past on a breezy Wednesday afternoon.

They returned to Leopold's house around one to drop off his car and switch the chisel and the rope, as well as a gag and hydrochloric acid, from one trunk to the other. In the blue rental car they then drove to Jackson Park and killed another hour. Minute by suffocating minute, they counted down the clock. And then, at 2:15, they slid to a stop a block away from the exclusive Harvard Preparatory School. Leopold stayed in the car while Loeb wandered over to "reconnoitre" the playground. It was a word, full of mystery and danger, that he had always loved in his crime paperbacks.

Loeb had hardly begun his reconnaissance when, as school ended at 2:30, Johnny Levinson came into view. Loeb moved in fast.

Johnny shrank back. He had not forgotten how, at Tommy's recent birthday party, he had been frightened when Loeb spanked him. When asked what he was doing that afternoon, Johnny muttered that he was off to play baseball. Loeb watched the kid scuttle away.

Another of their potential victims, eleven-year-old Armand Deutsch, or "Ardie," usually cut across the playground on his way home. He was one of Leopold's bird-watching students, and his grandfather, Julius Rosenwald, the chairman of Sears & Roebuck, had long been one of Chicago's richest men. The only downside to killing his grandson would be that Albert Loeb, Richard's father, was sure to be affected—for he was Rosenwald's immediate deputy. But if the boy crossed their path, they might still take him.

Ardie Deutsch got lucky. Instead of walking past Loeb, Ardie turned the opposite way and headed for the school's main entrance. The family chauffeur would drive him to the dentist that afternoon for his six-month checkup. He was safe.

Leopold joined Loeb on foot, and they again spotted Johnny Levinson, hanging around with a few other boys on a vacant lot. They dared not approach him yet. After they had returned to Leopold's house on Greenwood Avenue, so that they could pick up the powerful field glasses he used on his birding jaunts, Loeb drove to a drugstore, where he found the Levinsons' home address in the local directory.

They went in search of him, even stopping outside his house on Forty-first Street, before heading down Ellis Avenue again. By 5:15 the light had begun to fade, and there was a distinct chill in the air as they continued their measured cruise. As they rolled past the Harvard playground, their blue car slowed. Boys were still waving their mitts or swinging their bats as the game meandered to a close.

Leopold and Loeb were completely silent. Just one of them needed to back down before the other would collapse in relief. But their secret contract bound them together. Neither felt ready to risk the other's contempt if they failed to turn fantasy into reality. And so they drove on, coasting toward the crime that would consume them.

Meanwhile, nine-year-old Irving Hartman, the son of a wealthy furniture dealer, ambled past a suburban garden. He slowed alongside a bed of swaying tulips. Irving caught a glimpse of Bobby Franks across the street before he bent down to sniff the colorful flowers.

Bobby, wearing a beige cap and jacket, short trousers, and woollen golf socks with a checkerboard edge, carried a light coat and his schoolbag. He had spent the last two hours umpiring the Harvard ball game and, with a long day drawing to a close, he'd ducked away before the end. He wanted to get home, just two and a half blocks away, before it got too cold.

Seven or eight houses farther along, in a seemingly lazy arc, the Willys-Knight swung around the wide and empty street in a slow U-turn. The car now aimed south, in the direction of the two boys, who were diagonally opposite each other and lost in their own small worlds.

Leopold and Loeb gazed at Bobby Franks. They had discussed him before and, only a couple of weeks earlier, Loeb had played tennis on the Frankses' court. Loeb pointed, and Leopold nodded. This was their moment. This was the climax toward which they been building for years together.

There was still time to pull back, to take a foot off the brake and slip into reverse, and so they hesitated again. And then, in a muffled voice, Loeb gave the word: *"Go!"*

The blue car rolled forward, bearing down on Bobby Franks.

Loeb wound down the window. "Hey, Bob!" he yelled with an inviting wave. As Bobby turned toward them, Loeb slipped into the backseat and opened the front door on the passenger side. He offered him a ride back home. Bobby shook his head. He would rather walk. But Loeb was insistent. There was something else he wanted to discuss with Bobby. The fourteen-year-old faltered, unsure why two boys five years older than him would want to talk. Loeb had his answer ready. He wanted to tell him about a new tennis racket.

Bobby nodded his agreement. He and Dick Loeb had tennis, if nothing else, in common. Bobby crossed the street and ducked down to take the passenger seat next to Leopold at the wheel. He carefully placed his bag and coat on his lap and then closed the door behind him.

By the time Irving Hartman looked up from the tulips he noticed that Bobby Franks had disappeared. All he could see was a blue car driving south down Ellis before, at the next block, it turned east onto Fiftieth Street.

With the street deserted on both sides, the car slowed. A hand curled around the taped blade of the chisel. The heavy wooden handle was raised high in the air and then brought down hard against the side of Bobby's head. He moaned when he was hit again, and then again.

Bobby Franks did not scream. He whimpered and twitched as blood seeped from his scalp.

Leopold was the first to speak, his voice rasping. "Oh God, this is awful," he wailed. "I didn't know it would be like this."

Loeb pulled the stricken boy over the front seat and dumped him on the floor near his feet. He forced the gag roughly into Bobby Franks's sagging mouth.

The muffled sounds soon became as still as his now motionless body. Leopold did not need to ask if Bobby Franks was dead. The way that Loeb covered him with a heavy tarp, bought for that specific purpose, told him there would be no need for the rope later that evening.

After driving for thirty minutes, they picked up a couple of sandwiches and root beers at a roadside shack. As they did not want to leave the body unattended they ate in the front seat of the car. The color had returned to Leopold's face. Devoid of remorse he began to revel in Loeb's wisecracking celebration of their new status as real killers.

Loeb suggested they drive farther along that desolate stretch of road toward Gary until they found a convenient spot where they could partially strip the body. Twenty minutes later they had removed Bobby Franks's shoes, socks, and short trousers. They left on his underwear and shirt and hid the shoes and his belt in the long grass.

Amid the strangeness of their work Leopold stopped at a drugstore along the highway and called home. He told his father that he would be back around ten and not to worry about dinner, as he had just eaten. There was something incredible, he thought, about their ability to stay rooted in the everyday while a warm corpse lay on the floor of their rented car. Perhaps, in true Nietzschean style, they really did operate outside the boundaries of conventional morality.

The headlights of the car were ablaze by the time they reached Wolf Lake. A mile inland from Lake Michigan, and straddling the Illinois and Indiana border, the eight-hundred-acre lake attracted a myriad selection of birds. Leopold, who knew the area well from his birding classes, drove across the narrow stretch of marshland that led to the cement culvert they had chosen weeks earlier. The funnel-shaped culvert, tucked beneath a railroad track, diverted water from Wolf Lake and down a channel that ran west into the smaller Hyde Lake.

From the car they carried the corpse almost two hundred yards to the culvert. Once the body was completely naked they poured acid over the face and genitals. In their attempt to obscure the identity of Bobby Franks, they also sprinkled hydrochloric acid over a surgical scar on his abdomen. He still looked pretty much like the same boy in the glare of their flashlights, but they did not dare loiter. They pushed him headfirst into the culvert. It was difficult and awkward, and, even using their feet to try and ram the body deeper, they did not stash it far enough in the funnel.

Early the following morning a Polish workman would spot a foot sticking out of the culvert, after flowing water had dislodged the body. If they had been less anxious, they might have concealed the corpse for weeks or even months. But under a starless sky Leopold had bent down to gather Bobby's remaining clothes, which he stashed inside the tarp cover. It was then that his distinctive glasses slipped from his coat to the ground—to be discovered by another workman. Those spectacles would lead the police from Wolf Lake to Leopold's home.

Oblivious to his mistake, Leopold and Loeb drove back to Chicago where, at another drugstore, they found the number for the Franks home. Jacob and Flora, Bobby's parents, were distraught. Their son had been missing for almost five hours. Calling himself "Johnson," Leopold told Flora that Bobby had been kidnapped but that he was still alive. She and her husband would receive further instructions in the morning.

Flora screamed: "What do you want?" But Leopold hung up, without answering.

They then mailed their ransom letter by special delivery. Leopold had typed it up a few days before on plain white paper. He'd used the portable Underwood typewriter they had stolen in Ann Arbor. They scrawled Jacob Franks's name and address on the front of the envelope, but the letter inside was as anonymous as it was specific—demanding that $10,000 should be obtained in used notes to guarantee the safe release of their son. Further instructions would follow.

Leopold and Loeb burned Bobby's clothes and tried to wipe clean the red stains on the rental car floor. Returning to Leopold's house on Greenwood Avenue, the boys had a few drinks with his father. Once Nathan Sr. went to bed around 11:30, they kept drinking and playing Casino. At 1:30 A.M., Leopold finally drove Loeb back home in his own car to Ellis Avenue—the same long road where they had abducted Bobby Franks. But before they parted they had one last task. As Leopold's red car sped south on Greenwood Avenue, Loeb opened the passenger window and hurled an object into the black sky.

It landed with a clatter. A night watchman, Bernard Hunt, looked up in surprise. He wandered across the street toward the dark shape, turned it over, and saw, beneath a yellow streetlight, a wooden-handled chisel streaked with blood.

The following morning, having slept solidly for five hours, Leopold attended another 8:00 A.M. lecture in criminal law. He then called his friend George Porter Lewis to ask if he would be able to run his Thursday afternoon bird class—which Leopold gave every week to thirteen young girls from U-High. Lewis, a keen ornithologist, agreed that he would take the girls on a brief bird-watching tour while Leopold attended to business as important as it was personal.

That business focused on their complicated ransom plot. They phoned Jacob Franks and informed him that a Yellow Cab would soon arrive at his home. He needed to take a cigar box, containing $10,000, in the cab to a drugstore on East Sixty-third Street where he would be called again with additional instructions. At the drugstore he would be told to catch a Michigan Central train at 3:18. In the rear compartment, he'd find another note telling him that, after passing a specific station, he needed to throw the box out of the window once he saw a redbrick factory on the left.

Rather than being a masterpiece of detailed planning, Loeb's unwieldy plot exposed them to human error and bad timing. The plan went wrong when, soon after 2:30 that afternoon, Loeb, identifying himself as George Johnson, phoned through the first set of orders. As soon as their short conversation ended, Jacob Franks, in his agitation, forgot the address of the drugstore. Although he'd collected the money, he suddenly had no idea where he was meant to go next. Leopold and Loeb's luck was already running out.

When the phone rang a few minutes later, Franks immediately picked it up—in the hope it might be "Johnson" with a reminder of the address. It was, however, his brother-in-law, Edwin Gresham, calling from a funeral home with some devastating news. The body of a young boy had been discovered. The corpse was described as belonging to a boy of about eleven, weighing a hundred pounds, who wore glasses and whose face was badly discolored. Jacob and Flora Franks had already been phoned by a reporter, Alvin Goldstein, but they were convinced it could not be their son. Bobby was fourteen. He weighed ninety pounds and had never worn glasses in his life. Gresham had still gone to double-check and, as he told Jacob, the evidence was unmistakable. The dead boy was indeed Bobby. Gresham did not know it then, but the glasses that had been placed on the bridge of his nephew's nose belonged to Nathan Leopold.

A few minutes later a Yellow Cab turned up at the Franks home. But Jacob was going nowhere. His son was dead. His wife was hysterical. And he was on the point of collapse.

Leopold and Loeb phoned the drugstore at 1465 East Sixty-third Street on two separate occasions. Each time their request to speak to Jacob Franks was met with incomprehension. When they stepped back out onto the street, their disappointment turned to shock. The afternoon editions carried headline news that a young boy had been found dead at Wolf Lake.

They were not downcast for long. Early the next morning Leopold and Loeb were thrilled by the wild speculation surrounding the death of Bobby Franks. Yet Leopold became increasingly uneasy that his lover might engage in open and passionate discussions about the murder. Indeed, less than twenty-four hours after the collapse of their ransom plan, Loeb approached his friend Howard Mayer, a correspondent for the *Evening American*.

After they had discussed animated theories about the case, Loeb argued that the killer sounded far too smart to have planned an actual meeting with Franks at the still unknown drugstore. Loeb was certain he would have phoned through another set of instructions. Mayer was impressed by the sharp thinking, but he resisted Loeb's suggestion that they visit all the drugstores along East Sixty-third to see if any had received a call from the mysterious George Johnson. It sounded like a long shot to Mayer.

Loeb was undeterred. When they were joined by James Mulroy and Goldstein, the young journalist who had phoned the Franks family from the funeral parlor, he repeated his apparent hunch. Mulroy and Goldstein, who both worked for the *Chicago Daily News*, took the bait and persuaded Mayer to drive them over to Sixty-third in his father's car. Naturally, it did not take Loeb long to find the correct pharmacy.

A porter confirmed that both he and the pharmacist had each taken a call the previous afternoon from a caller asking for Mr. Franks. Loeb was ecstatic. "This is the place!" he yelped.

When he was complimented, he smiled and said he had always suspected his love of detective stories would pay off one day. Once they had phoned through the details of their scoop, the reporters allowed Loeb to accompany them to the funeral home where, in fascination, he listened to their interviews with the coroner and a weeping Jacob Franks—a broken father who could not understand why anyone would murder a lovely boy like Bobby.

His wrenching words affected the young hacks who had briefly forgotten Loeb's own reaction when, in the car, they asked him if he'd known Bobby Franks.

Loeb had answered with a grin and a nod. They had pressed him further. What had little Bobby been like? "If I was going to murder anybody," Loeb had said oddly, "I would murder a cocky little son of a bitch like Bobby Franks."

That Friday night Leopold and Loeb went dancing with their two favorite girls—Susan Lurie and Germaine Reinhardt—at the Edgewater Beach Hotel. Germaine, or "Patches" as she was known, was a slender and beautiful flapper with a dashing bob. Patches, Loeb liked to brag, was "pretty much a shocker." He claimed that he often had sex with her—although Leopold knew that Loeb lied constantly about jumping in and out of bed with girls.

Leopold had a different kind of hold over Loeb, a hold that even a racy girl like Patches would never dare imagine. And so he felt no jealousy as, instead of fixating on Dickie Loeb, he danced with Susan, a dark-haired serious girl who regarded Leopold as an intriguingly delicate soul. He was only briefly startled when he heard Loeb greet a passing acquaintance with a smirking remark: "Congratulations, you've just enjoyed the treat of shaking hands with a murderer." Leopold remained more cryptic. After he and Susan left the hotel, he claimed he would be happy to be hit by a bolt of lightning because he had already experienced all that life could offer.

Six days later Leopold was given an unexpectedly new experience. On Thursday, May 29, the police made their first breakthrough. Almer Coe & Company, the only opticians in Chicago to supply the Bobrow horn-rimmed spectacles found near the cement culvert at Wolf Lake, confirmed that they had sold just three pairs among their fifty-four thousand previous orders. The police established that one set belonged to a middle-aged woman and another to a man who had been on business in Europe for weeks. The third pair had been made for Nathan F. Leopold of 4754 Greenwood Avenue.

A confident Leopold had already been interviewed by the police the previous Sunday when, in a routine procedure, they spoke to all those ornithologists known to frequent Wolf Lake. Once they received the optician's records, Leopold was pulled in for more abrasive questioning. He used the alibi he and Loeb had agreed they would rely on for at least the first week after the murder—claiming that, on May 21, rather than murdering Bobby Franks, they had picked up a couple of unknown girls and got drunk with them. Loeb was called in, and as eight days had passed since the murder, he reverted to their second plan and claimed to have forgotten what they had done on the afternoon that Bobby Franks died.

That inconsistency gave State Attorney Crowe the opportunity to drive a further wedge between them. Cranking up the ferocity of his questioning through a long night, he put the squeeze on the boys. Loeb cracked first, and then, more furiously, Leopold followed. They each confessed, reliving the criminal trail they had taken from arson and burglary to abduction and murder.

The fortunate families who had just avoided the wretched fate of Jacob and Flora Franks expressed their disbelief. "My wife and I dined at the Loebs' a month ago," Salmon Levinson said. As the father of Johnny Levinson, whom Leopold and Loeb had stalked, the liberal attorney was bewildered. "Richard was there and delighted us with his charming personality. I regarded him as one of the finest youths I've ever known. His confession of this awful crime is simply unbelievable to all who knew him."

Susan Lurie, meanwhile, praised Nathan Leopold's keen mind and gentle nature. She told the *Chicago Daily Tribune* that in the sensational days following the killing they had discussed the case both thoughtfully and, occasionally, humorously. "I said to him, 'You confess the Franks murder and I'll claim the reward.' We had been reading a newspaper about it. He laughed and said it would be a good joke. What a grim joke it turned out to be."

She also provided a more tender insight into the character of the boy she also called Babe. "He is a superior intellect [and] an unusual man. His brain works fast. Babe was always fond of studying emotions. He would go back over conversations and attempt to trace our line of thought, our reasoning. It was all very interesting. He was good company but, of course, I had no intimation of the terrible things that were whirling through his wonderful brain."

Susan also knew Babe's special friend. "Dick Loeb read detective stories," she revealed to the *New York Times*. "He had many such magazines and was fascinated by mysteries. He was with Babe frequently and of course I met him. He was the one interested in police matters. I think Dick was a fine boy, but not of the same brilliance as Babe. Dick was more down-to-earth. But the Babe I knew was never one to discuss mysteries of crime, although he was long on talking of mysteries of the mind."

CHAPTER 6

COURTROOM WHISPERS

C LARENCE DARROW, more than anyone, understood the severity of their struggle. If the judge could be persuaded to allow expert psychiatric testimony, they would have taken a significant step toward the salvation of Leopold and Loeb—and Darrow's own redemption. But a different response from Judge John Caverly might spell the end for the boys. Darrow's own image as a courtroom giant would also turn to dust.

His first witness, Dr. William White, the director of St. Elizabeth's Hospital for the Insane in the District of Columbia, took the stand. He had barely managed to state his name and position when, for the prosecution, Crowe shouted out another strident objection—any detailed assessment of their mental condition was inadmissible for, in order to avoid a trial by jury, the defendants had not claimed insanity. Darrow and his team, Benjamin and Walter Bachrach, two attorney brothers related to the Loeb family, shook their heads in disdain. They were weary of Fighting Bob.

The prosecution's objection was groundless, Benjamin Bachrach explained, because Judge Caverly presided over a hearing rather than a trial. The guilt of the defendants had already been accepted. Their sentencing was the only outstanding issue and the use of psychological evidence, to determine the extent of their culpability, would be shown to be fundamental.

Caverly looked unconvinced. He was not about to make any early ruling in favor of the defense.

Crowe saw the judge's hesitancy as a sign to ratchet up his attack. "What is the defense trying to do here?" he roared. "They denied that their clients are insane and are now spending tens of thousands of dollars to get them off on the grounds of insanity. Are they attempting to avoid a trial before twelve men that would hang them? Are they trying to produce a situation where they can get a trial before one man that they think won't hang them?"

Caverly, sensitive to any charge that he might be dazzled by Darrow, responded sharply. If the defendants appeared insane at any point during his deliberations, then he would immediately terminate the hearing and set a new date for a trial by jury. He had still to decide whether or not they were mentally fit to have changed their plea to guilty.

When the defense emphasized that only trained psychiatrists who had examined Leopold and Loeb could help the judge reach that decision, Crowe was indignant. "You do not take a microscope and look into a murderer's head. You look at the facts surrounding the case. Did he kill the man because the man had debauched his wife? If that is so, then there is mitigation. Did he kill the man because the man had spread slanderous stories about him? Then there is mitigation. Did he kill the man in the heat of passion during a drunken fight? That is mitigation."

Crowe seemed to gather strength from all the heads nodding agreement in court. He turned to stare at the two killers seated alongside a glowering Darrow. "But this is a cold-blooded murder without a defense. The attempt, on a plea of guilty, to introduce an insanity defense before Your Honor is unprecedented. The statute says that it is a matter that must be tried by a jury."

Darrow responded forcibly. "The statute in this state provides that the court may listen to anything, either in mitigation of the penalty or in aggravation. The legislature has given wide latitude to either court or jury."

The verbal embellishments, and the legal intricacies, bounced back and forth in a marathon of courtroom jousting that stretched deep into the next day. The trial hung in the balance as the opposing sides argued over the presence of one of America's foremost experts in mental illness. At least Dr. White himself looked the picture of sanity as, ignoring the frantic bickering, he opened his briefcase and pulled out a sheaf of reports relating to some completely different cases. Remaining in the witness box, while the attorneys argued, he picked up his fountain pen and edited his own work. He appeared resigned to a long wait.

It took two days for a vacillating Caverly to make a ruling. He appeared to swing from one side to the other as the rival teams battled. Darrow allowed Walter Bachrach to present the more complex points of law, but he delivered the crucial summation. He looked askance at the prosecution table as he began to talk in a low voice. "Now I understand that when everything has been said in this case, from beginning to end, the position of the state's attorney is that the universe will crumble unless these boys are hanged."

Darrow sighed. "I must say that I have never before seen the same passion and enthusiasm for the death penalty as I have in this case, and there have been thousands of killings before this, much more horrible in detail, where there was some motive for it. There have been thousands before and there will probably be thousands again, whether these boys are hanged or go to prison. If I thought that hanging them would prevent any future murders, I would probably be in favor of doing it. In fact I would consent to have anybody hanged, excepting myself, if I thought it would prevent all future murders."

He half smiled at his little joke, sparing himself the noose, before he pressed on. "But I have no such feeling. I know the world will go on about the same in the future as it has in the past—at least I think so. My clients are not so important to the economy of things, either in their life or their death, and if this case is like all other cases, it ought to be tried calmly and dispassionately upon the facts."

Darrow pointed a gnarled finger at Robert Crowe. "I cannot understand the glib, light-hearted carelessness of lawyers who talk of hanging two boys as if they were taking a holiday or visiting the races." Crowe rolled his eyes, but the gibe still struck as Darrow said that any request for a hanging should be voiced with profound regret. "That has not been done in this case," he said mournfully. "I have never seen a more deliberate effort to turn the human beings of a community into ravening wolves . . . and to take advantage of anything that might make a stage of hatred against these boys."

His gaze fixed on Caverly as he reached the crux of his plea. "I don't believe there is a judge in Cook County that would not take into consideration the mental status of any man before they sentence him to death. I am not speaking of this as a matter of law. I am speaking as a matter of humanity, as a matter of common justice. It is hard enough to sentence a man to die, and every humane judge seems to find a reason by which he can find life, instead of a reason for taking. Life is the greatest and highest concern, even though the life must be spent behind stone walls."

There was a hush in the steaming courtroom as Darrow sat down. Leopold looked up at him, his eyes shining. Loeb kept his own head down as if the magnitude of Caverly's decision suddenly resonated inside him.

Fighting Bob was not quite done. "If Your Honor please," Crowe almost jeered as he gestured sarcastically at the boys, "they are not cold-blooded murderers, egotistical and secure in their conceit that they are above and beyond the law on account of their wealth and influence."

Crowe shook his head in mock amusement. "No, they merely committed some boyish little prank and they are sitting here sobbing for mercy, crying their hearts out."

He pulled his squat figure up to its full height and thrust out his strapping chest. "We have a duty to perform," he told Caverly. "Mr. Darrow would not hang anybody. But Your Honor and I are not like Mr. Darrow, the paid advocate, who has no oath of office, or no duty to the public to perform. We have sworn that we will execute the laws as we find them. The laws in this case demand the extreme penalty."

It was time, finally, for Caverly to announce his ruling. He adjusted the upward tilt of the fan on his desk and cleared his throat.

"The court is of the opinion," he intoned gravely, "that it is his duty to hear any evidence that the defense may present and it is not for the court to determine in advance what it may be. The court will hear it and give it such weight that he thinks it is entitled to."

Darrow nodded respectfully but Crowe tossed his head angrily. "The motion of the state . . . ," Caverly said. He looked down at Crowe. "What is your motion?"

"It is an objection," Crowe muttered.

"The objection to the witness is over-ruled," Caverly confirmed as he glanced across at the psychiatrist in the dock. "The witness may proceed."

DR. WHITE folded away the diversionary work that had occupied him during the extended delay. He replaced it with the prepared statement he had written following his examination of Leopold and Loeb. As Crowe glared at him, White read in a quiet, compelling style. Calling the defendants "Babe" and "Dickie," which followed Darrow's policy of stressing their youth, he drew the court into their world. His psychiatric evidence provided a chilling insight into two cracked minds.

In addressing Dickie's mental imbalance, White revealed that "there is a tendency for the fantasy life, an abnormal fantasy life, to realize itself in reality." He explained that Dickie would sometimes begin to divulge a surreal fantasy to him by saying, "And, do you know, Teddy . . . ," in the same voice he had used as a child when describing a dreamy reverie to his small bear in bed. And yet Babe Leopold's own unrealistic obsession with Dickie meant that, despite this emotional immaturity, his feckless lover emerged in his mind as a Nietzschean superman. "He knew Dickie lied to him about his marks at school, was boastful and untruthful, [but] he nevertheless says that Dickie not only approached perfection but far surpassed it. In a scheme of the perfect man which he drew up, he gave Dickie a scoring of 90, himself 63, and various others of their mutual acquaintances marks ranging from 30 to 40."

According to White, the cruel psychological kink in each boy provided an eerie series of echoes. "Dickie needed an audience. In his fantasies the criminal gang was his audience. In reality, Babe was his audience. . . . Now as Dickie sometimes plays the part of the superior and sometimes the inferior, no understanding of the relationship can be complete without first understanding this fantasy of Babe's.

"He, Babe, is generally the slave in the situation, [but] he is the slave who makes Dickie the king, maintains him in his kingdom, like the premier who occupies the principal state office over a weak king. He maintains the kingdom for the king [but] he is really the strong man. He expresses both components of his makeup with a desire for subjugation on the one part and a desire for supremacy on the other."

White was convinced that Dickie had struck the deadly blows that killed Bobby Franks. Yet the identity of the actual killer mattered less than their entwined destiny. They were lovers who had brought doom down on each other. "I cannot see how Babe would have entered into it at all alone because he had no criminalistic tendencies [like] Dickie," White concluded. "And I don't think Dickie would have ever functioned to this extent all by himself. So these two boys, with their peculiarly inter-digited personalities, came into this emotional compact with the Franks homicide as a result. . . . I do not believe the homicide can be explained without an understanding of [their] relationship."

During Crowe's cross-examination, White was asked, "Did Richard Loeb on May 21, 1924, know the difference between right and wrong?"

"He knew intellectually that it was against the law," the doctor replied.

"Did he know it was morally wrong?"

"He did not have adequate feeling toward its moral wrongfulness."

"But did he have sufficient capacity to refrain from killing?

White hesitated. "I don't know," he eventually admitted.

Crowe's next question was pointed: "Is Richard Loeb, in your opinion, insane?"

Walter Bachrach jumped up to voice his objection—which the judge sustained.

Crowe posed a different query. "What distinction, doctor, is there between mental illness and insanity?"

"Well," Dr. White said, "mental illness is sickness of the mind. Insanity is something you gentlemen know about. I don't. It is purely a legal and sociological term. It is not a medical term at all."

Darrow grunted in satisfaction. His first witness had done all that could be expected of him.

THE CUNNING old lion knew better than to trust the courtroom in his elaborate bid for justice and mercy. Darrow had been at work for months in an attempt to gauge and shape public opinion of a sensational trial. Even before he had instructed Leopold and Loeb to change their plea, he had sent out men to mingle with the crowds in Chicago to find out what was being said on the street. The hostile reaction toward his clients had hardened his resolve for them to admit their guilt. And Darrow was gratified that, once public shock at their switch of tactics had dissipated, the percentage of people in the city who favored immediate execution of the boys dropped from 60 to 40 percent.

That swing in perception had also been swayed by Darrow's persuading of the two fathers to issue a letter, printed in all the Chicago newspapers, in which they reiterated that they would make no attempt to free their boys from prison should they receive life sentences. They also declared, in the same statement, that Darrow's fee would be decided by the Bar Association. But rumors still persisted, much to his irritation, that Darrow would be paid a million dollars for defending the killers.

Darrow resorted to more furtive tactics in preparing the public for the psychological foundations to his defense. On the last weekend in July, just before the prosecution case closed and the furor over his use of medical experts erupted, he was unsurprised when the sympathetic psychiatric report into Leopold and Loeb's behavior was apparently "stolen" from his secretary's desk by a reporter for the *Chicago Daily Tribune*. Darrow, of course, knew the creative hack who then splashed the report, compiled by the respected psychiatrists Harold Hulbert and Karl Bowman, across the front page of his newspaper. The attorney expressed his "outrage" at the supposed theft and, feeling "duty-bound" not to prejudice any of the *Tribune*'s rivals, immediately made the report available to all other Chicago newspapers.

Hulbert and Bowman's examination of the boys resulted in the most exhaustive psychiatric study in criminal history. It fitted Darrow's purposes perfectly that, in their efforts to uncover the truth of the murder as well as to determine the mental health of the accused, the doctors had also carried out a series of tests into each patient's endocrine glands—which release hormones into the bloodstream and affect psychological well-being. The psychiatrists had concluded that Leopold suffered from "a disorder of the vegetative nervous system and of the glands of internal secretion."

Rather than being swaggering supermen, Leopold and Loeb appeared lost and vulnerable, their moods skittering from one extreme to another. Darrow's assessment of the difference between their characters endorsed the psychiatrists' view. They warmed most to Leopold. Even if he was clumsy in public, Leopold approached the psychological examinations and legal briefings with an inquiring mind stimulated by his genuine interest in the work of Hulbert and Bowman and Darrow himself. Loeb brought a shallower perspective, and he had even fallen asleep during a joint psychiatric session with Leopold.

The Hulbert and Bowman report depicted Leopold and Loeb as victims—just like Bobby Franks. But there was a tragic contrast. Where the fourteen-year-old had fallen prey to a random murder, his killers had been ruined by illness and the unfortunate entanglement of their disturbed psyches. It helped Darrow that those central planks to his defense had already been aired publicly before his witnesses were admitted on the stand. The struggle, now, continued in the courtroom.

E ARLY ON MONDAY, August 4, William Healy, a psychiatrist from Boston, followed Dr. White in another canny strategy planned by Darrow. Less forgiving in his descriptions of the defendants, Healy resisted any sentimental overview of the boys' failings. He considered Loeb to be "untruthful, unscrupulous, [and] disloyal even to his friends." Leopold, meanwhile, had "reacted in a most abnormal way in regard to the whole crime. His main concern seems to be whether or not the reporters say the right thing about him." Healy also suggested that Leopold contemplated the murder of Bobby "the same as he would decide whether to have pie for supper."

The psychiatrist was ready to provide graphic evidence of their previously secret "compact" of sex and crime. But he and Darrow felt uneasy at the prospect that his words, raw and shocking in 1924, would fill newspapers across America the following morning. Darrow suggested that they should, in the company of the prosecution attorneys, approach the bench. In that way, through whispered conversation, Healy could explain to Judge Caverly the relationship between Leopold and Loeb. His words would be recorded in the official court transcripts, but the newspapers should not be allowed to print them.

As the men gathered together in a hushed circle around the judge's bench, the rest of the crowd in the courtroom leaned forward, straining in anticipation. But Healy and the lawyers spoke in such furtive tones that not even the most sharp-eared reporters could make out the words. Their pens and pencils twitched with impatience. It made for a curious scene as Darrow, bowed in concentration, turned his head slightly to scan the court for any other alert eavesdroppers. He smiled quietly at the dissatisfied faces of the frustrated spectators. It was plain that they could hear little of the discussion.

On a day of blistering heat Darrow's gaze moved to the rear and sides of the courtroom. Line upon line of straw hats and colorful bonnets hung from the brass hooks that covered the walls, as if the visitors were enjoying a light and summery picnic rather than being denied the more sordid details from a salacious trial of sex and murder.

Healy, in a discreet murmur, explained to Caverly that "this compact, as was told to me separately by each of the boys on different occasions, consisted of an agreement that Leopold, who had very definite homosexual tendencies, was to have the privilege of . . ."

The psychiatrist's soft voice trailed away as he looked up at Darrow for guidance.

"Do you want me to be very specific?" Healy whispered again.

"Absolutely," Crowe hissed before Darrow could reply. "This is very important."

Darrow nodded encouragingly at Healy. The psychiatrist spoke even more quietly, so that the attorneys and the judge had to form such a tight bond around him that their heads were almost touching. "Leopold was to have the privilege of inserting his penis between Loeb's legs at special dates," Healy explained. "At one time it was to be three times in two months—if they continued their criminalistic activities together. Then they had some of their quarrels, and it [became] once for each criminalistic deed. Now their other so-called perverse tendencies seemed to amount to very little. They only engaged in anything else, so far as I can ascertain, very seldom, but this particular thing was very definite and explicit."

Healy quaked again, as if he dared not continue. Benjamin Bachrach intervened with an urgent whisper: "So that it need not be repeated, make it clear what the compact was."

A few stray reporters, in their desperation to hear, had edged closer to the front of the court. Before Healy could say anything else, Darrow called out loudly to the judge: "I do not suppose this should be taken in the presence of newspapermen, Your Honor."

Caverly looked up in surprise. He had been so wrapped up in Healy's account that he had failed to notice the encroaching hacks. Darrow pointed them out. The newspapermen remained rooted to the spot, as if by not moving a muscle they might be ignored.

The judge reached for his gavel and banged it against the wooden block. "Gentlemen, will you go and sit down!" he shouted.

The reporters stayed perfectly still for a moment longer, as if the command was being addressed to some other pesky eavesdroppers. They could not fool the judge. "You newspapermen!" Caverly snapped as he pointed at the men who, finally, grinned in recognition of their guilt. They still did not retreat. Caverly waved them back. "Take your seats!"

Once they had obeyed him, and calm was restored, the judge looked sternly at the press. "This should not be published!" he said in a clear order from the court.

As if fearing their permanent exclusion from the trial, most of the journalists nodded dolefully at Caverly. The judge turned back to Healy and, in another conspiratorial whisper, instructed him to continue. And so, once more, the heads bent toward him, the oiled hair of each man shining in the bright morning sunshine.

"They experimented once or twice with each other," Healy muttered huskily. "They experimented with mouth perversions but they did not keep it up at all. They did not get anything out of it. Leopold has had for many years a great deal of fantasy life surrounding sexual activity. That is part of the whole story and has been for many years. He has fantasies of being with a man, and usually with Loeb himself, even when he has connections with girls, and the whole thing is an absurd situation, because there is nothing but just putting his penis between this fellow's legs and getting that sort of thrill. He says he gets a thrill out of anticipating it. Loeb would pretend to be drunk and then he would almost rape him and would be furiously passionate at the time, whereas with women he does not get that same thrill and passion."

Healy relived the moment when, on a Pullman train on the way to Charlevoix, the boys first began their sexual relationship. "It was the first time in a berth, and it was when Leopold had this first experience with his penis between Loeb's legs. Then he found it gave him more pleasure than anything else he had ever done. To go on further with this, even in jail here, a look at Loeb's body or a touch upon his shoulder thrills him, so he says, immeasurably."

Leopold had intimated he was helplessly in love with Loeb. That fleeting confession meant more to Darrow than Leopold's convoluted explanations that he still saw himself as a "slave" to Loeb's "king." Rather than those half-baked themes of submission and domination, Darrow saw the love and pathos in an awkward nineteen-year-old whose brilliance had been framed by loneliness and mental illness, a boy who had meandered from reading and bird-watching to murder.

The clandestine exchange had reached its conclusion. Judge Caverly invited Healy to return to the stand—where he would be able to continue his account in a normal voice. Healy, having recovered from his muffled embarrassment, focused on the strangely interlocked but conflicting minds of Leopold and Loeb. He was clearly beguiled more by Leopold than the "lazy" and "unambitious" Loeb. Describing Loeb's intelligence as "no better than average," the psychiatrist lamented the fact that "he is not at all interested in mental tasks . . . he can work with fair attention and persistence, but it is a good deal of grind for him." Leopold was very different.

With a semblance of awe, the psychiatrist told the court how Leopold had displayed the mental gymnastics of an unusually gifted intellect. With his prodigious memory Leopold had scored far higher than anyone Healy had ever tested before when taking the Monroe Silent Reading test. He could scan a list of twenty words and then, a minute later, recite their exact order, even if he was asked to start at the bottom and work his way back up to the top. Leopold could also be given, as a random example, the thirteenth word and, immediately, he would be able to tell Healy which words occupied positions twelve and fourteen in the list.

But Leopold's emotional maturity, his ability to tell right from wrong, was at the opposite end of the spectrum. "He achieved a score of 56.5 out of 100," Healy said of a test in practical judgment. "That is the average for a twelve-year-old. Stated another way, twenty-five per cent of ten-year-old boys do better than that."

Healy revealed that, in one of Leopold's fantasies, he imagined himself in prison as Napoleon, a great leader cast out into exile on the barren island of St. Helena.

Crowe had heard enough. "Is [Leopold] insane?" he asked Healy during an intense session of cross-examination.

"He has a paranoid personality," the doctor suggested. "Anything he wanted to do was right, even kidnapping and murder. There is nothing in the feelings of sympathy which would prevent him because of his disintegrated personality. There is no place for sympathy and feeling to play any part. In other words, he had an established pathological personality before he met Loeb, but probably his activities would have taken another direction except for this chance association."

If Darrow was satisfied with the mounting evidence of mental illness, he was less sure what Caverly would make of the increasingly disturbing portrayal of his clients. Darrow hoped the psychiatric accounts would engender compassion, but he realized that creeping revulsion and shock might just as easily elicit a cruel judgment. There were enough people, even in the modern world, who would prefer to do away with such ravaged souls.

Crowe was one of them, a man with the power to fight for a state execution. He rose to his feet angrily. "We have now got to a point in this hearing," he complained, "where it appears as if the defense here is insanity. There is only one thing to do under the law—and this is to call a jury."

Caverly hesitated, as if weighing his options, and then he spoke clearly. "The motion is denied."

T HE STRENGTH of Darrow's evidence became increasingly apparent. Dr. Bernard Glueck admitted that, while studying Loeb's reaction to the murder, "I was amazed at the absolute absence of any signs of normal feelings. He showed no remorse, no regret, no compassion for the people involved in the situation. . . . The whole thing became incomprehensible to me except on the basis of a disordered personality."

Stressing that Loeb had admitted to him that he had used the chisel to murder Bobby Franks with his own hand, Glueck shuddered. He had extensive experience as a prison psychiatrist at Sing Sing in

Ossining, New York, but his encounters with Loeb were more disconcerting. "I have examined a lot of hardened criminals, men awaiting execution. And the hardened criminal shows in every response a kind of crudity. Loeb is affable, polite, and shows an habitual kind of refinement, and yet seems to be incapable of responding to this situation with adequate emotions."

Leopold, instead, found immense gratification in describing the murder as an expression of seemingly pure Nietzschean philosophy. "And so," Glueck concluded, "I think the Franks case was perhaps the inevitable outcome of this curious coming together of two pathologically disordered personalities, each one of whom brought into the relationship a phase of their personality which made their contemplation and the execution of the crime possible."

Dr. Harold Hulbert provided the fourth and final pillar of Darrow's uniform defense. He repeated the belief that Leopold and Loeb were also bound together in a way as wretched as it was fated. "Each boy felt inadequate to carry out the life he most desired unless he had someone else in his life to complement him, to complete him. Leopold wanted a superior for a companion. Loeb, on the other hand, wanted someone to adulate him for a companion. . . . The psychiatric cause for this is not to be found in either boy alone, but in the interplay or interweaving of the two personalities, their two desires caused by their two constitutions and experiences. This friendship between the two boys was not altogether a pleasant one to either of them. The ideas that each proposed to the other were repulsive. Their friendship was not based so much on desire as need."

Crowe and the prosecution were allowed, in a right of rebuttal, to call their own group of psychiatric experts. Their findings stood in stark contrast to the earlier depictions of fatally damaged boys. "The young man was entirely orientated," Dr. Archibald Church said of Loeb. "His logical powers, as manifested during our interview, were normal and I saw no evidence of any mental disease." He was supported by Dr. Hugh Patrick, who cleared Loeb of any mitigating psychological affliction. "Unless we assume that every man who commits a deliberate, cold-blooded, planned murder must be mentally diseased, there was no evidence of mental illness."

Dr. Harold Singer highlighted the relatively normal reactions the accused had shown during recent days in the dock. Their earlier irreverence, which seemed unbalanced, had become much more naturally appropriate. "Their demeanor has been distinctly different. There has been much less laughing . . . [and] they have occasionally, particularly Leopold, shaken their heads as if in dissent from various things that have been said. The laughing on the part of Loeb has changed frequently and quite abruptly to a very serious expression. I have observed him show signs of feeling and of clear thinking and normal emotional reactions—but no evidence of mental disease."

Darrow took charge of the cross-examination. He produced a masterly example of the art, weaving together consistently smart and tough interrogations that probed each witness. His strategy was twofold—initially to put forward questions that encouraged the prosecution psychiatrist to support, in broad terms, the views already expressed by his counterparts for the defense. When an inevitable divergence emerged, Darrow pushed hard to uncover inconsistencies in the testimony. He harried the witnesses remorselessly, with questions piling up hour upon hour. Darrow could also be savage.

"Do you want to hang these boys?" he asked Dr. Church bluntly, to howls of protest from the prosecution.

He was usually far more subtle in his searching appraisal. "Doctor Church, what do you mean by an emotion?" he asked when the psychiatrist said that he had found no evidence of mental illness in the emotional reaction of the defendants.

"Emotion is a play of feeling," Church suggested.

"Isn't there a difference between the part of the human anatomy which produces emotion and the mind, which is supposed to be the seat of reason?"

"No one has emotion unless he intellectually perceives it," Church parried.

"You are assuming the mind is the product of brain action?"

"Yes."

"The mind is probably a product of the whole organism, isn't it?" Darrow said.

"No, I don't think so."

"Is there anything in the mind excepting the manifestation of the physical organization?"

"Practically not."

"When is the most trying age in a young man?" Darrow asked in a sudden switch of tack.

"At the age of puberty and adolescence," Church replied.

"There comes a change of emotional life, doesn't it, as a rule?"

"Yes sir."

"They are then most apt to leave off habits that have been inculcated in them?" Darrow asked. "On account of new feelings or emotions?"

"They are," Church agreed.

"And it is the most prolific time for insanity with youth—is it not?"

"Yes."

Apart from wringing that concession from Church, Darrow also emphasized that none of the psychiatrists for the prosecution had made more than a cursory examination of the defendants. Unlike Drs. White and Healy, and especially Hulbert and Bowman, who had spent weeks analyzing both their physical and mental health, the prosecution experts had often confined their study to a brief conversation—sometimes lasting less than thirty minutes. Hulbert and Bowman's investigation of the endocrine glands had, Darrow claimed, produced far more telling insights into the reasons surrounding the murder than anything outlined in the less empirical observations of the rival doctors.

He also hammered away at any contradiction in their accounts. Darrow would ask a witness if he believed the defendants to be insane. The answer was a predictable "no"—which endorsed the defense claim that the case should be considered in this judicial hearing rather than in a trial by jury. Did the witness, however, believe that Leopold and Loeb might be mentally ill? If the answer remained in the negative, Darrow would raise his unkempt eyebrows in disbelief. Did the bizarre actions of Leopold and Loeb reflect the behavior of mentally healthy young men? As the psychiatrist on the stand began his complicated justification, Darrow cut in with a withering instruction that they answer the question with a simple yes or no. It was almost impossible to say yes—and every no reinforced Darrow's argument.

Yet the last witness summoned by State Attorney Crowe was ready to stand his ground against Darrow. Dr. William Krohn was practiced in courtroom testimony, and he refused to be manipulated. In a long and unstoppable speech, Krohn brushed aside Darrow's attempted interjections and concentrated on his own findings. He argued that the conditions under which he had examined the defendants had been "ideal" and that his investigation had been "thorough" and "complete." Talking doggedly over Darrow's shouted questions, he insisted that the two men "were not suffering from any mental disease when I examined them."

Darrow objected loudly to Judge Caverly. He had the right to cross-examination, without the witness resorting to "any introductory hot air."

"What did you say about hot air?" Krohn yelled.

Caverly tried to placate the witness. "Go ahead, doctor," he encouraged Krohn.

"Mr. Darrow is an authority on hot air," Crowe snickered as the courtroom rocked appreciatively.

Darrow responded fiercely and for the next few hours he pursued Krohn—attempting every trick he knew in an attempt to outwit the psychiatrist. But Krohn was immovable. Leopold and Loeb, he insisted, were completely sane. They were without any psychological defects. Even when Darrow threw up his arms in disgust and exclaimed his disbelief that any normal man could have acted in such a way, Krohn remained implacable. He had much experience in the matter, he said, and he had long since learned that entirely sane people could commit acts of unspeakable cruelty and depravity. Some murderers were simply evil and deserving of neither compassion nor pity.

Darrow and Crowe, meanwhile, hammered each other endlessly. After Crowe had dismissed the psychiatric findings of the defense as "laughable," Darrow retorted bitterly: "Yes, you would laugh at anything. . . . I think you would laugh at the hanging of these boys."

Crowe snorted. "We have heard a considerable amount about split personalities in this case. I was somewhat surprised to find that my old friend, who has acted as counsel and nursemaid in this case for two babes who were wandering in dreamland, also was possessed of a split personality. I had heard so much of the milk of human kindness that ran out in streams from his larger heart that I was surprised to know he also had so much poison in his system."

He quoted the address Darrow had made during a recent prison visit to Joliet: "I do not believe in the least in crime. I do not believe people are in jail because they deserve to be. They are in jail simply because they cannot avoid it on account of circumstances which are entirely beyond their control and for which they are in no way responsible."

Fighting Bob smacked a fist into his open palm as he claimed Darrow's "dangerous philosophy of life" was also on trial. He turned away from his brooding rival and looked up at the judge. It was Caverly's duty not to give in to that anarchic vision. He should feel compelled, instead, to protect the state and sentence the accused to death.

When Krohn stepped down at last, with Darrow having to concede defeat, it appeared as if the prosecution had clawed back its ascendancy. But the outcome of the case was still shrouded in uncertainty on the afternoon of August 19, 1924, when the State confirmed it had called its final witness. Only the closing arguments remained before Caverly would announce his sentence.

LEOPOLD AND LOEB crumbled a little during the prosecution's frenzied summary. Their haughty cool had been replaced by troubled expressions. "Your Honor," the assistant state attorney Joseph Savage seethed, "you have before you one of the most cold-blooded, cruel, cowardly, dastardly murders that was ever tried in the history of any court." He relived the gruesome moment when Bobby Franks was killed. "The blow was struck from behind," he reminded the judge, "that cowardly blow. What mercy did they show to him? Why, after striking four blows, they pulled him to the rear of the car and gouged his life out."

As people began to sob softly, and Bobby's father, Jacob Franks, left his seat in distress, Savage shouted out: "Mercy? Why, Your Honor, it is an insult in a case of this kind to come before the bar of justice and beg for mercy. I know Your Honor will be just as merciful to these two defendants sitting here as they were to Bobby Franks."

Savage roared on, speaking all through a muggy afternoon and on into the next morning, before he reached his last damning instruction. "Hang them!" he yelped to the packed courtroom. "Hang these heartless supermen!"

During a brief recess Leopold whispered to his older brother: "My God, Mike, do you think we'll swing after that?"

CHAPTER 7

THE BOOK OF LOVE

A WAY FROM the hysteria and the menace of Chicago, it was difficult for Mary to gauge the shifting momentum of the trial. Most
days she went to the local library in the big house on the hill in Palisades to read the newspapers in an attempt to keep pace with events
in court. Margaret tailed along behind her, eventually settling into a
sweet hush among the silent books and rustling papers as she lost herself in L. Frank Baum's *The Wonderful Wizard of Oz*, which had first
been published in 1900—in Chicago. She had once asked her mother
if Darrow was a kind of wizard, a question that made Mary laugh. But
there was wizardry in his words, in his use of language and emotion,
and so Margaret's perceptiveness had again surprised her.

They were still different, mother and daughter. Margaret liked
it best when the rented summer house was full of visitors, whereas Mary was happiest when the last person left and she could get
back to the quiet and her work. She had sent her Leopold and Loeb
articles to the *Nation* and waited in hope. But she cared more that
she had found something of her old self in writing. Mary had been
asked to edit the autobiography of Mother Jones, the labor activist

still described by some as "the most dangerous woman in America." She had already gathered most of the material in the preceding years when Mother Jones dictated her life story—and now she agreed to weave those memories into a book. A return to union politics thrilled her. Darrow, she knew, respected Mother Jones's steely courage and would also be pleased.

Mary had transformed her relationship with Margaret. It had taken her nine years, but now she understood that Margaret "bubbles with joy . . . she gains with praise and stiffens and becomes ugly with blame." So, four weeks after Lem's latest departure, as the trial entered its second month, Mary filled her initially "inexpressible loneliness" with love. She came to exult in her little girl, to see previously concealed depths and heightened pleasures in their relationship. Mary wrote of Margaret as being "abundant" and "gay" and, in the middle of a beautifully hazy August, of the moment when "she put a flower in her hair and looked a rich and voluptuous and beautiful little girl and, oh, such a darling companion."

If she had stayed in Chicago with Darrow, her secluded but strangely magical summer with Margaret would have never unfolded. Sometimes out of pain, she knew, beauty is born.

North Dearborn Street, Chicago, August 21, 1924

Just over two months had passed since Darrow and Mary Field stood together in the courtroom. Sixty odd days, in which he had been consumed by the fight to save Leopold and Loeb, had disappeared. As they had agreed, Darrow had not written to her during his difficult battle against Crowe and Savage. But on the draining summer night before he made his final address, and the most important speech of his life, Darrow paused for thought.

In his office, clogged with cigarette smoke despite the windows being flung open in a futile search for a cool gust of air, Darrow sat at his old desk. Crumpled balls of papers were strewn across its surface and on the floor around him as he drafted chunks of the marathon speech with which he would try to extend the lives of two boys. Darrow would speak, as always, without notes. But the familiar illusion of conversational ease was crafted by laborious preparation. Although he did not write out his actual speech, he polished and burnished individual passages around which he would weave his flowing pattern of words that he expected to last at least two days.

Many of his larger ideas had been expressed repeatedly during his public speeches and debates against capital punishment. This time, however, they needed to be bound tightly to the specifics of a painful and harrowing case. But Darrow needed to summon a poetic eloquence that could transcend depravity and murder with his sincere belief in compassion and understanding.

On his desk, alongside his almost illegible scrawl, Darrow reached for two books of poetry he hoped would infuse the spirit of his appeal. He had bought both for Mary. In the midst of their first affair, he had given her a copy of Omar Khayyám's *The Rubaiyat*, the translated series of Persian quatrains he found unbearably beautiful. He often recited aloud to her:

> *So I be written in the Book of Love*
> *I do not care about that Book above;*
> *Erase my name or write it as you will,*
> *So I be written in the Book of Love*

Khayyám's words had been written in Persia more than seven hundred years before, but Darrow felt a kinship with the poet, mathematician, and philosopher whose pessimistic view of man's weakness was tempered by his tender words. He also loved the way an old man could celebrate, as Khayyám did all those centuries ago, a life of freedom and pleasure: "Enjoy wine and women, and don't be afraid / God has compassion."

Darrow had always made Mary laugh when he read those words

with particular relish. On the day of their courtroom parting he had also given her, along with his own book on crime, a copy of A. E. Housman's *Last Poems*, published in 1922. They were a wistful companion to the English poet's famous collection, *A Shropshire Lad*, which evoked the tragedy of doomed youth. Their relevance to Leopold and Loeb was made poignant by Housman's own tortured homosexuality.

He flicked through the pages until he fell again on the poem he planned to recite in front of Judge Caverly. Housman's "The Culprit," an elegy spoken by a young English boy about to be hanged, had made Mary cry when Darrow read it to her. He hoped he might somehow do the same to all those who listened to him in the more formal territory of the court.

Darrow would not easily bear the hanging of his clients. All the stupidity and cruelty of the world seemed bound up in their fate. Babe and Dickie, whose childish names sounded like a forlorn wail for their squandered youth, had killed Bobby Franks. They had ripped the heart from another family without thought for anyone beyond themselves. They had ruined their own families. And, still, they had yet to suffer remorse or even regret.

But what would change if they were taken up the steps of the scaffold? Whose humanity would be enriched if a noose was placed around their necks? What hope could be drawn from the sight of a boy hanging from a rope with a broken neck and dangling feet?

Crowe and Savage hailed the death sentence as a deterrent. Yet it had not restrained Leopold and Loeb in their unhinged scheming. They had not thought once of the gallows while they cruised the streets of Kenwood on May 21, precisely three months before that very day. They had feared neither a trial nor death row. They had just been themselves, bound to the moment, senseless and hopeless. Killing them, in turn, seemed to Darrow a crime of even more somber resonance.

It was late that Thursday night in the Loop. Darrow bent to his task again. He pushed aside the collections of poetry and lit another cigarette. As the smoke drifted from his nose he picked up a black fountain pen and reached for a clean sheet of paper. This time his words would cut to the quick. As his left hand curled around the pen and raced across the page, he was hopeful that each slanting sentence might bring them all a little closer to salvation.

T HE NEXT AFTERNOON, just before two o'clock on Friday, August 22, 1924, a frantic crowd swarmed the Criminal Courts Building. Every seat inside the actual courtroom had already been filled but, in their desperation to hear Darrow speak, the mob pushed forward. The line of bailiffs guarding the main entrance held for a few minutes and then, when the pressure became intolerable, they split apart. People poured through the broken chain and ran up the six flights of stairs.

The heat inside the room was ferocious, but Judge Caverly, aware of the encroaching skirmish, ordered the doors to be sealed shut. A bailiff called out: "Judge, four of your friends are outside."

"Let them stay outside," Caverly shouted.

The fighting and screaming escalated, and a woman fainted. Men trampled over her to get closer to the court, and, amid the shoving, another bailiff's arm was broken. He was led away through a rear entrance, moaning in agony. Caverly called for more police assistance, and then, at ten past two, they managed to force back the crowd.

At an impatient nod from the judge, Darrow stood up. He wore a soft white suit and matching tie, and nervously brushed the hair from his sweaty forehead.

Darrow began to talk—but the words were lost in the noise from outside. He raised his hands in a muted appeal for quiet. "I think we had better wait," Darrow eventually suggested.

Caverly motioned him to return to his seat. Threatening to clear the entire building, he retreated to his chambers to telephone Chicago's chief of police. Describing the scenes as "disgraceful" and the "worst riot" he had ever witnessed in court, Caverly warned that people could be trampled to death if order was not restored.

Darrow tried again ten minutes later, at 2:24, but still he could not be heard. Caverly jumped up and, yelling above the din, ordered all six floors of the criminal courts to be vacated—bar his own court-room. It took twenty-five police officers, wielding wooden clubs, to drive the crowd down the stairs and into the street outside. Peace, at last, followed the bruised flesh and cracked heads that had been forced from the building. The stage was set.

Holding on to his suspenders in a comforting gesture, as he used his thumbs to stretch and push them away from his sweat-stained cotton shirt, Darrow took in a deep breath. He could smell the hot stench of humanity pressed up close against him. "Your Honor," he said, "it has been almost three months since the great responsibility of this case was assumed by my associates and myself. I am willing to confess that it has been three months of great anxiety—a burden which I would have been spared excepting for my feelings of affec-tion toward some of the members of one of these unfortunate fami-lies. This responsibility is almost too great for anyone to assume, but we lawyers can no more choose than the court can choose."

Darrow began to walk, as if it was easier to think and talk while he crisscrossed the courtroom. The absorbed hush was broken only by the sound of Judge Caverly's rotating metal fan and the more whis-pery paper equivalents that many women flapped in an effort to cool themselves.

"Our anxiety over this case has not been due to the facts that are connected with this most unfortunate affair, but to the almost unheard-of publicity it has received. Day after day the people of Chi-cago have been regaled with stories of all sorts about it, until almost every person has formed an opinion. And when the public is inter-ested and demands a punishment, no matter what the offense, great or small, it thinks of only one punishment, and that is death."

Darrow allowed that last word a little time to breathe. Judge Caverly took the thoughtful cue and rested his chin in the cupped palm of his hand.

"It was announced that there were millions of dollars to be spent on this case. Wild and extravagant stories were freely published as though they were facts. Here was to be effort to save the lives of two boys by the use of money in fabulous amounts, amounts such as these families never even had.

"We announced to the public that no excessive use of money would be made in this case, neither for lawyers nor for psychiatrists, nor in any other way. We have faithfully kept that promise. The psychiatrists are receiving a per diem, and only a per diem, which is the same as that paid by the State. The attorneys, at their own request, have agreed to take such amount as the officers of the Chicago Bar Association may think is proper in this case.

"If we fail in this defense it will not be for lack of money. It will be on account of money. Money has been the most serious handicap that we have met. There are times when poverty is fortunate."

A murmur rippled through the court. Darrow waited for stillness to return, noticing that Leopold and Loeb had hardly stirred as they drank in every word. Leopold, in particular, had almost come to worship Darrow as the greatest man he had ever met. More significantly, he and Loeb had been shocked out of their callous nonchalance.

"I told Your Honor in the beginning that never had there been a case in Chicago where, on a plea of guilty, a boy under twenty-one had been sentenced to death. I will raise that age and say, never has there been a case where a human being under the age of twenty-three has been sentenced to death."

Darrow pointed angrily at Crowe and his team. "I have heard in the last six weeks nothing but the cry for blood. I have heard from the office of the state's attorney only ugly hate. I have heard precedents quoted which would be a disgrace to a savage race. I have seen a court urged almost to the point of threat to hang two boys, in the face of science, in the face of philosophy, in the face of humanity, in the face of experience, in the face of all the better and more humane thought of the age."

Picking out Crowe's assistant, Joseph Savage, Darrow shook his head. "Mr. Savage . . . ," he sighed. "Did you pick him for his name or his ability or his learning?"

Darrow's voice rose above the tittering. "My friend Mr. Savage, in as cruel a speech as he knew how to make, said that we pleaded guilty because we were afraid to do anything else. Your Honor, that is true . . . I can tell Your Honor why. I have found that years and experience with life tempers one's emotions and make him more understanding of his fellow man. Why, when my friend Savage is my age, or even yours, he will read his address to this court with horror.

"I am aware that as one grows older he is less critical. He is not so sure. He is inclined to make allowance for his fellow-man. I am aware that a court has more experience, more judgment and more kindliness than a jury. Your Honor, it may be hardly fair to the court. I am aware that I have helped to place a serious burden upon your shoulders. And I have always meant to be your friend. But this was not an act of friendship.

"I know perfectly well that where responsibility is divided by twelve, it is easy to say, 'Away with him.' But, Your Honor, if these boys hang, you must do it. There can be no division of responsibility here. You can never explain that the rest overpowered you. It must be by your deliberate, cool, premeditated act, without a chance to shift responsibility. It was not a kindness to you. We placed this responsibility on your shoulders because we were mindful of the rights of our clients, and we were mindful of the unhappy families who have done no wrong."

Judge Caverly attempted to maintain the same impassive expression but, at Darrow's careful reminder, he shifted in his chair, as if in subtle acknowledgment of those onerous duties.

For the next hour Darrow set about undermining the prosecution. He ridiculed Crowe's main contention that the murder of Bobby Franks had been motivated primarily by a desire to prize a $10,000 ransom from his family. "The boys had been reared in luxury. They had never been denied anything, no want or desire left unsatisfied. No debts, no need of money, nothing. And yet they murdered a little boy against whom they had nothing in the world, without malice, without reason, to get $5,000 each?"

Darrow laughed dryly. "All right. All right, Your Honor, if the court believes it, if anyone believes it—I can't help it."

He stretched out his arm toward Crowe and Savage. "That is what this case rests on. It could not stand up a minute without motive. Without it, it was the senseless act of immature and diseased children, wandering around in the dark and moved by some emotion that we still perhaps have not the knowledge to understand thoroughly."

Darrow turned back to Leopold and Loeb. It seemed as the eyes of the accused had begun to glisten as Darrow, with aching voice, asked the question that had troubled them all for so long: "Why did they kill little Bobby Franks?"

He allowed his question to settle in the mind. And then, with pain and incomprehension framing his words, he answered: "Not for money, not for spite, not for hate. They killed him as they might kill a spider or a fly—for the experience. They killed him because they were made that way. Because somewhere in the infinite processes that go to the making up of the boy or the man something slipped, and those unfortunate lads sit here hated, despised, outcasts, with the community shouting for their blood.

"Are they to blame for it? There is no man on earth who can mention any purpose for it all or any reason at all. It is one of those things that happened; and it calls not for hate but kindness, for charity, for consideration.

"Mr. Savage, with the immaturity of youth and inexperience, says that if we hang them there will be no more killing. But this world has been one long slaughterhouse from the beginning until today, and killing goes on and on, and will forever. Why not read something, why not study something, why not think instead of blindly shouting for death? *Kill them!* Will that prevent other senseless boys or other vicious men or vicious women from killing? No! It will simply call upon every weak-minded person to do as they have done."

As if to chide Savage for his lack of reading, and his crude use of the misery felt by the parents of Bobby Franks, Darrow picked up a sheaf of notes and waved them at the young prosecutor in reminder that all three families affected by the murder had suffered. "I remember a little poem that gives the soliloquy of a boy about to be hanged, a soliloquy such as these boys might make," he said as he stretched out his hand to Leopold and Loeb.

Darrow glanced at Housman's poem, "The Culprit," the same poem he had read to his old lover, Mary Field. His voice rang out:

> *My mother and my father*
> *Out of the light they lie;*
> *The warrant would not find them,*
> *And here 'tis only I*
> *Shall hang so high*
> *O let not man remember*
> *The soul that God forgot,*
> *But fetch the county kerchief*
> *And noose me in the knot*
> *And I will rot*
> *For so the game is ended*
> *That should not have begun.*
> *My father and my mother*
> *They had a likely son*
> *And I have none.*

Darrow looked weary, his piercing eyes sunk deep into the back of his head as he scanned the silent courtroom. "This weary old world goes on," he said, "begetting, with birth and with living and with death: and all of it is blind from the beginning to the end. I do not know what made these boys do this mad act, but I do know there is reason for it. I know they did not beget themselves. I know that any one of an infinite number of causes reaching back from the beginning might be working out in these boys' minds, whom you are asked to hang in malice and in hatred and injustice, because someone in the past has sinned against them."

He shook his head at the endless cycle of human pain. And then he moved quickly across the court, as if harried by an urgent question. "Do you think you can cure the hatreds and the maladjustments of the world by hanging them? You simply show your ignorance and your hate when you say it."

His finger jabbed again toward Savage as he approached the last few minutes of that sapping Friday afternoon in court. "What is my friend's idea of justice? He says to this court and Your Honor, whom he says he respects, 'Give them the same mercy they gave to Bobby Franks.' Is that the law? Is that justice? Is this what a court should do? If the State in which I live is not kinder, more humane, more considerate, more intelligent than the mad act of these two boys, I am sorry I have lived so long."

Darrow's work was almost done for the day, and he would rise again in the morning to speak even more powerfully in his bid to save two boys who looked at him in wonder. The old man spoke softly, almost as if he were talking to himself: "I feel sorry for all fathers and mothers. The mother who looks into the blue eyes of her little babe cannot help musing over the end of the child, whether it will be crowned with the greatest promises which her mind can imagine or whether he may meet death upon the scaffold."

He looked up again at Judge Caverly. "All she can do is rear him with love and care," he murmured, his voice echoing above the hum of the fan, "to watch over him tenderly, to meet life with hope and trust and confidence, and to leave the rest to fate."

THE NEXT MORNING Darrow appeared rejuvenated as sunlight streamed in through the twelve-foot-high courtroom windows. "Your Honor," he began cheerfully, "last night I spoke about what is perfectly obvious in this case—that no human being could have done what these boys did, excepting through the operation of a diseased brain. I do not propose to go through each step of this terrible deed—it would take too long. But I do want to call the attention of this court to some of the other acts in this distressing and weird homicide; acts which show conclusively that there could be no reason for their conduct."

Darrow re-created the murder, picking out one scene after another to prove his point. "Dick and Nathan are in the car. They see Bobby Franks on the street, and they call him to get into the car. It is about five o'clock in the afternoon, in the long summer days, on a thickly-settled street, built up with homes, the houses of their friends and companions, known to everybody, automobiles appearing and disappearing, and they take him in the car—for nothing.

"Without any motive or any reason they pick up this little boy right in sight of their own homes, and surrounded by their neighbors. They drive a little way, on a populous street, where everybody can see, where eyes may be at every window as they pass by. They hit him over the head with a chisel and kill him, and go on about their business, driving this car within half a block of Loeb's home, within the same distance of the Franks home, drive it past the neighbors that they know, in the open highway, in the broad daylight.

"I say again, whatever madness and hate and frenzy may do to the human mind, there is not a single person who reasons who can believe that one of these acts was the act of men—of brains—that were not diseased. There is no other explanation for it. . . .

"They pull the dead boy into the back seat, and wrap him in a blanket, and this funeral car starts on its route . . . the car is driven for twenty miles . . . the slightest accident, the slightest misfortune, a bit of curiosity, an arrest for speeding—anything would bring destruction. For what? For nothing!"

Darrow's voice cried out at that "nothing" and then, almost in scholarly despair, he scratched his head. "The mad acting of the Fool in *King Lear* is the only thing I know of that compares with it. And yet doctors will swear that it is a sane act. They know better."

He had a faraway look in his eyes. Darrow was still in the speeding blue car.

"They go down a thickly populated street, and then for three miles take the longest street to go through this city—solid with business buildings, filled with automobiles backed upon the street, with streetcars on the track, with thousands of peering eyes, one boy driving and the other on the back seat, with the corpse of little Bobby Franks, the blood streaming from him, wetting everything in the car. And yet they tell me that this is sanity. They tell me that the brains of these boys are not diseased. You need no experts. You need no X-rays. You need no study of the endocrines. Their conduct shows exactly what it was, and shows that this Court has before him two young men who should be examined in a psychopathic hospital and treated kindly and with care.

"They get through south Chicago, and they take the regular automobile road down toward Hammond. There is the same situation—hundreds of machines; any accident might encompass their ruin. They stop at the fork of the road, and leave little Bobby Franks, soaked with blood, in the machine, and get their dinner, and eat it without an emotion or a qualm. . . .

"My friend Savage pictured to you the putting of this dead boy into a culvert. Well, no-one can minutely describe any killing and not make it shocking. It is shocking. It is shocking because we love life and because we instinctively draw back from death. It is shocking . . . but, Your Honor, I can think of a scene that makes this pale into insignificance."

The dread in court almost matched the reek, seeping in through the open windows, from the fish market below. "I can think," Darrow said, "and only think, Your Honor, of taking two boys, irresponsible, weak, diseased, penning them in a cell, checking off the days and the hours and the minutes until they will be taken out and hanged. Wouldn't it be a glorious day for Chicago?"

Darrow threw back his head, as if to breathe in the fishy tang more deeply. "Wouldn't it be a glorious triumph for the state's attorney? Wouldn't it be a glorious triumph of justice in this land? Wouldn't it be a glorious illustration of Christianity and kindness and charity?"

His voice lowered. The sneering irony was gone and so too, it seemed, was all hope.

"I can picture them," Darrow said softly, "wakened in the gray light of morning, furnished a suit of clothes by the State, led to the scaffold, their feet tied, black caps drawn over their heads, stood on a trap door, the hangman pressing a spring so that it gives way under them. I can see them fall through space—and—stopped by the rope around their necks."

A sob broke out near the front of the court, but Darrow pressed on. "That would surely expiate placing Bobby Franks in the culvert after he was dead. This would doubtless bring immense satisfaction to some people. It would bring a greater satisfaction because it would be done in the name of justice. I am always suspicious of righteous indignation. Nothing is more cruel than righteous indignation. To hear young men talk of justice—well, it would make me smile if it did not make me sad.

"And must I ask that these boys get mercy by spending the rest of their lives in prison, year following year, month following month, and day following day, with nothing to look forward to but hostile guards and stone walls. It ought not to be hard to get that much mercy in any court in the year 1924."

Darrow recovered the following day, with Sunday a day of rest even for an agnostic attorney acting in defense of the lost and the damned. He barely spoke to an agitated Ruby as he lay on his bed before, eventually, he rose and returned to the North Dearborn Street office to sharpen the final third of his speech. It was as if he were in a trance from which he would only emerge when the final words were said. He wrote fitfully, and read a little from *The Book of Love*.

THE SAME OLD crush enveloped the Criminal Court on Monday, August 25, 1924. Darrow could feel, in the crowd, a lust for blood. It was said that they came to hear him speak for the last time, and that was true of many, but their hunger for a hanging steeled him. He looked out at the blurred faces and shook his head. "Many may say now that they want to hang these boys; but I know that giving the people blood is something like giving them dinner. When they get it they go to sleep. They may for the time being have an emotion, but they will bitterly regret it. And I undertake to say that if these two boys are sentenced to death, and are hanged, on that day a pall will settle over the people of this land that will be dark and deep, and cover every humane and intelligent person with its gloom."

Despite the blue brilliance of the day outside, Darrow plunged further into the perennial darkness. "You can trace it all down through the history of man. You can trace the burnings, the boilings, the drawing and quartering, the hanging of people in England at the crossroads, carving them up and hanging them as examples for all to see."

Darrow had his own childhood in mind, as well as his father's, as he crisscrossed through history. In preparing himself the previous night, he had remembered how, when he had been a boy of eight or nine, his father had told him a haunting story. Amirus Darrow had attended a public hanging when he was still a young man. In his mindless rush to join the crowd in rural Ohio, and witness the punishment of a murderer, Amirus had pushed his way to the front of the throng. His view was clear when they led the man toward the gallows. He watched as they lowered the rope so that its noose could be placed around the man's head, which they then covered with a black hood. Amirus could stand it no longer. He turned away so that he might not witness the actual hanging. But the sound the crowd made, as the hangman's work was done, filled him with shame and revulsion.

Amirus Darrow never again attended a public killing, and his description of the scene marked his son. Even when Clarence, as a boy, visited his father's furniture shop in Kinsman, Ohio, the coffins near the back filled his head with grisly images of that hanging.

Almost sixty years later those same feelings coursed through him as he addressed the court. "You can read the stories of the hangings on a high hill, and the populace for miles around coming out to the scene, that everybody might be awed into goodness. Hanging for picking pockets—and more pockets were picked in the crowd that went to the hanging than had been known before. Hangings for murder—and men were murdered on the way there and on the way home. Hangings for poaching, hanging for everything, and hangings in public, not shut up cruelly and brutally in a jail, out of the light of day, wakened in the night-time and led forth and killed, but taken to the shire town, in the presence of a multitude, so that all might see that the wages of sin were death."

Darrow then drew back from the abyss. He suggested that perhaps man had changed a little in more recent years. There were fewer hangings, and in some countries they had even been abolished. Perhaps there was still hope, for he knew that "every step in the progress of the world has come from the human feelings of man. It has come from that deep well of sympathy which, in spite of all our training and all our conventions and all our teaching, still lives on in the human breast. Without it there could be no human life in this weary old world.

"Gradually the laws have modified, and men look back with horror at the hangings and the killings of the past. What did they find in England? That as they got rid of these barbarous statutes, crimes decreased instead of increased; as the criminal law was modified and humanized, there was less crime instead of more. I will undertake to say, Your Honor, that you scarcely find a single book written by a student of criminology that has not made the statement over and over again: as the penal code was made less terrible, crimes grew less frequent."

Darrow's voice again became more urgent as he stressed, "I am not pleading so much for these boys as I am for the infinite number of others to follow, those who perhaps cannot be as well defended as these have been, those may go down in the storm and the tempest without aid. It is of them I am thinking, and for them I am begging this court not to turn backward toward the barbarous past."

But he did not forget the boys and, with a detailed summary of their psychological disease, he took the court into the fractured minds of Dickie Loeb and Babe Leopold—before he underlined again their glaring youth and the pitiful life that awaited them behind bars. "I would be the last person on earth to close the door to any human being that lives, and least of all to my clients. But what have they to look forward to?"

He quoted another stanza from Housman, his voice echoing of "hollow fires burnt out to black" before ending with a last desolate line of poetry that captured the future of the defendants should they be saved: "There's nothing but the night." Darrow looked up meaningfully at Caverly and prepared to repeat those last few words again: "I care not, Your Honor, whether the march begins at the gallows or when the gates of Joliet close upon them, there is nothing but the night."

It was almost four o'clock that afternoon when Darrow reached his conclusion. "I am pleading for the future," he said huskily. "I am pleading for a time when hatred and cruelty will not control the hearts of men, when we can learn by reason and judgment and understanding and faith that all life is worth saving, and that mercy is the highest attribute of man."

Darrow felt his own eyes moisten when he saw that Caverly was crying. They were silent tears, but powerful enough to alter the shape of the judge's twitching mouth.

"I was reading last night of the aspiration of the old Persian poet, Omar Khayyám," Darrow murmured, as his own tears began to roll down his crevassed face. His voice, however, remained firm. "It appealed to me as the highest that I can vision. I wish it was in my heart, and I wish it was in the heart of all:

> *So I be written in the Book of Love*
> *I do not care about that Book above*
> *Erase my name or write it as you will*
> *So I be written in the Book of Love."*

Darrow's head was bowed, and his eyes were filled. And then, after ten long seconds, he looked up again and nodded to the judge. Slowly, Darrow returned to his seat, the silence following him with gathering force. The quiet held, as if no one dared break the spell.

Hunched mutely next to him, Leopold and Loeb were both deathly pale. When the judge finally found his voice and closed proceedings for the day, they rose unsteadily to their feet. As the prison guards approached, Loeb wiped the tears from his cheek. Leopold stumbled as he was led from the dock. He stretched out his arms, to stop himself from falling, and brushed past the stunned bailiffs as he headed blindly back to his cell.

THE NEXT MORNING, in the cool tranquillity of the library on the hill in Palisades, with Margaret sitting across a table from her as she savored the last few chapters of *The Wonderful Wizard of Oz*, Mary lost herself in a different kind of wonder. The *New York Times*, like most major papers across America, had reprinted Darrow's closing address. Ream upon ream of tightly packed sentences unfolded majestically before her. The tender beauty and noble virtue of his plea washed everything else away, and she glowed with pride. Her old lover was hailed as a poet, she wrote later, and a philosopher: "They compared him to Socrates and his old alpaca coat to a Greek robe." Darrow, Mary Field had always believed, belonged among the giants.

Robert Crowe spoke for two days but, in comparison to the emotive force of Darrow, he struggled to engage the court. In desperation he labored over the possibility that Leopold and Loeb had sexually assaulted Bobby Franks after they had killed him. "Apart from the fact that the pants were taken off, and the fact that they are perverts, I have a right to argue that they committed an act of perversion."

Judge Caverly ordered the "ladies of the court" to take their leave. "I will have the bailiffs escort you into the hallway," he thundered when many women resisted his demand. "There is nothing left here but a lot of stuff that is not fit for you to hear!"

The following morning, on Wednesday, August 27, Fighting Bob argued, "If the state had only half of the evidence that it did have, or a quarter of the evidence that it had, we would have had a jury in the box and a plea of not guilty. But trapped like a couple of rats, and with no place to escape except through an insanity defense, they proceeded to build it up. A weird, uncanny crime? The crime is not half as weird or uncanny as the defense that is put in here."

Crowe worked hard, and the sweat flew from him. But he made a mistake near the close when he sneered, "If the defendant, Leopold, did not say that he would plead guilty before a friendly judge, his actions demonstrated that he thinks he has got one."

Caverly was irate. "The court will order stricken from the record the closing remarks of the state's attorney as being a cowardly and dastardly assault upon the integrity of this court."

"It was not so intended, Your Honor," Crowe groveled.

"This court will not be intimidated by anybody at any time or place as long as he occupies this position," Caverly retorted before, a little more calmly, confirming that he would announce his judgment at 9:30 A.M. on September 10, 1924.

M ARY FELT the onset of an early fall as she noticed how some leaves had begun to turn in color and even spiral down to the ground. The nights drew in more quickly, and the days were crisper, as she and Margaret prepared for their return to New York and the start of a new school year. Lem was still away; on September 6, exactly seven weeks had passed since he set sail on *The Sylvia*. Mary wrote of how she was "longing for my mate"—but she heard nothing from her husband, or Darrow. She spoiled her little girl instead, taking Margaret to the Waldorf for lunch, while she waited for Lem and for the verdict from Chicago.

Margaret was radiant, and Mary could not be gloomy in her company. "This has been such a happy summer, Mother," Margaret sighed on a bittersweet September afternoon.

T HE TWO WEEKS of waiting for the judge's decision dragged by in slow agony for Darrow. Even though the last few days had gone well for the defense, all the old doubts and fears rose up in him. It seemed, despite the praise he had received, that the desire to hang Leopold and Loeb still raged. The boys had received countless letters in jail, promising that they would be killed should Caverly spare them the rope. Darrow knew that the judge himself had been sent death threats while he considered his sentence. All the eloquence of his summation suddenly seemed to matter little in the face of such venom.

And so, early on the morning of sentencing, Darrow sat in his office on North Dearborn Street. The blinds were shut and the air was thick with smoke as one cigarette followed another in a long bleak chain. He had caught sight of himself in the mirror when shaving that morning, and his face looked limp and haggard. His expression was as stricken and helpless as any man about to be condemned to the scaffold.

Later, in court, while Darrow rocked to himself on a tilted chair, Judge Caverly delivered his judgment. "In view of the profound interest that this case . . . ," he began, before looking up in irritation at the interruption of clattering cameras. He gave them a minute to take their photographs and then, demanding order, started to read again.

"In view of the profound interest that this case has aroused, not only in this community but in the entire country and even beyond its boundaries, the court feels it is his duty to state the reasons which have led him to the determination he has reached."

The words came and they did not suit Darrow. "The plea of guilty," Caverly stressed, "does not make a special case in favor of the defendants." He was equally emphatic that the case for insanity could not be accepted. "It is beyond the province of this court, as it is beyond the province of human science in its present state of development, to predicate ultimate responsibility for human acts . . . the court feels strongly that similar analysis made of other persons accused of crime would reveal similar or different abnormalities."

The rope had begun to tighten. Loeb twitched nervously, but Leopold's stare was blank, his body motionless. "The testimony in this case reveals a crime of singular atrocity. It is, in a sense, inexplicable. But it is not thereby rendered less inhuman or repulsive. It was deliberately planned and prepared for during a considerable period of time. It was executed with every feature of callousness and cruelty."

The judge paused for emphasis.

"The court is satisfied that neither in the act itself, nor in its motive or lack of motive, nor in the antecedents of the offenders, can he find any mitigating circumstances."

Darrow barely dared to breathe. They were in trouble.

Caverly pointed out that, in some states, three judges were joined together to share the burden of sentencing in such a case. "Nevertheless, the court is willing to meet his responsibilities."

As they waited, the unspoken word of *hanging* burned into Darrow. He decided to look up and face Caverly. "It would have been the path of least resistance to impose the extreme penalty of the law," the judge said quietly.

Darrow felt his body jerk forward as if he could not quite believe that last sentence.

"In choosing imprisonment instead of death," Caverly said, "the court is moved chiefly by the consideration of the age of the defendants."

Leopold gazed in astonishment as Darrow hung his head in relief. "Life imprisonment may not, at the moment, strike the public imagination as forcibly as would death by hanging," Caverly suggested. "But to the offenders, particularly of the type they are, the prolonged suffering on years of confinement may well be the severer form of retribution and expiation."

Darrow tried to control himself as Judge Caverly announced the formal sentences.

"For the crime of murder, confinement at the penitentiary at Joliet for the term of their natural lives."

"For the crime of kidnapping for ransom, similar confinement for the term of ninety-nine years."

The sentence was received in near silence. And for a long moment it seemed as if no one knew what he should do next. Judge Caverly made a show of shuffling his papers neatly together, and one of the more alert bailiffs tugged at both boys. Loeb stood up, still looking confused and frightened at the prospect of life plus ninety-nine years. And then Leopold followed him, jumping to his feet. He turned to Darrow in sudden delight. The old man who had saved him stretched out his hand, a sad little smile flitting across his seamed face.

Bedlam broke out then. Reporters raced to their phones, and

people rushed toward Darrow. They shouted and hollered, mostly at Darrow. Above the crowd and the din, he watched two boys being led back to the Bridge of Sighs—and away from the gallows.

Darrow closed his fist in silent triumph. He had won. Compassion had overcome vengeance. Leopold and Loeb had escaped death. He had won.

CHAPTER 8

THE PINCH HITTER
AND THE POLITICIAN

27 East Eleventh Street, New York, May 18, 1925

MARY FIELD PARTON slipped the shimmering new blue dress over her head, being careful not to muss up her delicately arranged hair, and pulled it down over her petticoat. As she smoothed the material over her hips in front of the bedroom mirror, she glanced at her silent daughter who, standing behind her, was reflected in the glass. Margaret was only four days away from her tenth birthday, but she looked almost unbearably solemn. Mary sighed, wishing for just a little more privacy, as she checked the pins holding her hair in place. It was the style Darrow liked most, with the long thick strands piled around her head, so that her wide and open face could be seen clearly when it was caught in thoughtful repose or, more usually, laughter.

She hesitated and then, as if choosing her mood for the evening, added a touch of blusher to her cheeks and a tracer of lipstick to the outline of her lower lip. This moment always fascinated Margaret, and because it was so rare for Mary to wear makeup she made a virtue of the ritual. Her top lip curled down and gently pressed itself against its glossy partner. Mary, feeling suddenly cheerful, winked at her girl as her two lips did their little shimmy of an embrace so that the lipstick might be spread evenly.

Mary would never think of herself as beautiful, but there were times when she relished the transformative value of spending a few minutes in front of the mirror. She was far too interested in words and ideas to attribute any lasting significance to her appearance. But, on an evening like this, she liked the sensation of making herself look more pretty than plain. Darrow, she knew, would notice the lipstick and the dress amid the small group of dinner party guests.

His occasional visits to the apartment always sent Mary mildly crazy at least a day before he arrived. From the moment she started cleaning each room to setting the table with the best silver, Mary was agog with the work that needed to be done. She would tidy and sweep, dust and polish, cook and yell, in between deciding what to wear and making sure Darrow's favorite, if illegal, whiskey was in the cabinet. And then, with an hour left before he punched the doorbell, Mary would dress and make herself up in a concentrated daze. She'd then go in hunt of the special glasses she used for the pineapple and gin cocktails with which she would welcome Darrow.

If domestic duties had once threatened to wear her down, Mary now looked at ease in the kitchen. It helped that she was about to see Darrow again. She might have felt less enthusiastic about washing and drying the best china dishes, or cutting and chopping and assembling the canapés, had their dinner been in honor of anyone else that hazy Monday evening. But Darrow's imminent arrival changed everything. She worked with efficiency and zeal.

In their circle of writers and artists Mary was loved for her mischievous humor. And so she smiled while preparing a symbolic large side salad—one of Darrow's pet aversions. He loathed all vegetables and any food that was "green." The fussy old attorney reserved a special loathing for spinach, cabbage, green beans, celery, lettuce, tomatoes, onions, and radishes. He also refused to eat chicken or lamb or veal. Though she wanted to keep him sweet, and avoided the meat dishes he disliked, Mary cackled as she tossed together a salad of fat green lettuce leaves and plump cherry tomatoes, which she garnished with strips of cucumber, celery, and radish. She could imagine Darrow's cry of disapproval, and the amusement of everyone else, as she walked out with her huge salad bowl on a wooden tray.

Lemuel, who had returned eight months before from his expedition to Greenland and Labrador, was notably quieter in such moments. He still praised the attorney's vision and compassion, but he seemed to shrink a little whenever Darrow was around—whether the huge and shambling lawyer was actually in their home or just a powerful name that hovered over them as they read the latest newspaper story about him or Mary revealed the next time she would travel alone to see him make a public speech or participate in a debate at some packed hall in the city.

As much as she loved Lem, he still sometimes seemed impossibly vague and dreamy when, in her more disgruntled moments, she made a stark comparison between her husband and Darrow. Lem was not out saving lives or standing up to injustice in a tumultuous court, or meeting George Bernard Shaw or H. G. Wells as Darrow did. But then she caught herself. She had not forgotten her disillusionment in Chicago the previous summer. Lem would never have hurt her, or let her down, as Darrow had done. He might not be a great man like Darrow, at least not in terms of his public stature or the quality of his work, but there was a goodness in Lem that bound her tightly to him. Mary knew she had missed Lem more than Darrow when she had been alone with Margaret in Sneden's Landing. And on the previous afternoon, when they whiled away a lazy Sunday at the summer house, she had felt remarkably happy as she watched Lem plant corn and beans.

But, still, Mary confronted a conundrum. She and Darrow were intellectually and philosophically suited, and emotionally and psychologically akin, yet separated by marriage to two different people.

When it had just been a triangle, with Ruby a distant point above their heads, it had seemed far easier. Mary, with her disdain for Ruby, felt untrammeled by guilt. And Darrow, being Darrow, did not indulge in self-recriminating accusations of deceit and hypocrisy. But, for Mary, everything was different now. She had Lem; and she loved him. Lem was, in the words she often used for him, her "precious chum."

Yet Mary understood that parts of their being remained hidden from each other and, often, even from themselves. She described Lem and herself in opposite terms to the immense icebergs that fascinated him on his Arctic travels. Their visible personalities were warm and appealing. The smaller and more mysterious shape of their secret selves lurked beneath the married surface, pinned down by a portion of loss and yearning that bore little relation to the overwhelming cheeriness that everyone saw in their life together. Yet Mary recognized that the invisible layer to their characters also made them most human.

Whether as faded as lust or impractical as fantasy, those emotions lurked somewhere in the depths, cut adrift from the rooted personae they presented in public. Lem existed in numerous overlapping worlds that stretched beyond his roles as husband and father. He, like Mary, was a writer and an adventurer. But the difference in their characters was most evident in the way he explored the physical world while she entered the sphere of ideas and feelings. Mary's hinterland stood in silent contrast to the faraway locations Lem visited whenever he needed excitement or regeneration.

She was more in tune with Darrow, for literature and philosophy offered terrain as challenging as it was beautiful, as soothing as it could be difficult. Even if he had been forty years younger, Darrow would not have joined a trek around the Arctic Circle. He had never understood the compulsion to conquer frozen new territories when there was still so much to set right in the immediate world raging around them. And so it seemed fitting that Darrow, with his history of defending the labor movement and his passion for books, had just written the foreword Mary would use in the autobiography of Mother Jones.

Mary had planned a low-key evening, with just four other guests besides Darrow and themselves, to celebrate the imminent publication of her first book. She would have opted for something grander, but she knew how preoccupied Darrow had become over the last few months. The redemptive glory of the Leopold and Loeb trial had not lasted. His delight had curdled the longer he waited for the families to settle their debt to him. And yet for weeks, and then months, he had heard nothing. It was almost as if he had done them a small favor—akin to sorting out a minor traffic offense—rather than unleashing the legal majesty that reversed an otherwise certain hanging.

Darrow's polite note to the Loebs, in late November 1924, suggesting that they might meet to discuss his fee, was ignored. Harry Fisher, his friend and an officer of the Bar Association, argued that his organization should both name and collect an amount that they estimated to be worth $200,000. Darrow declined, saying he did not want any public trouble between him and his clients. He wrote again in early December and, finally, a few weeks later, Jacob Loeb, Dick's uncle, turned up unannounced at Darrow's office.

Loeb had sneered: "You know, Clarence, the world is full of eminent lawyers who would have paid a fortune for the chance to distinguish themselves in this case."

The great old attorney remembered how, six months earlier, a tearful Loeb had pleaded for his help. And now that the seemingly impossible had been achieved, and the boys were still alive, Loeb looked very different. Telling Darrow that they could afford only $100,000, he deducted the $10,000 he had already paid for initial expenses. He then produced three checks of $30,000 each. The first was made out to Darrow while the two others were to be paid to the Bachrach brothers, who had assisted him. Darrow eyed his $30,000, thinking how far removed it was from the million dollars for which he was rumored to have sold his soul. And then, feeling "sick at heart," he signed the acceptance form spread across his desk. When he looked up, he saw that Albert Loeb's face was creased in a knowing smile.

The fading appeal of the law had waned still further for Darrow. "I had exhausted all the strength I could summon," he wrote. Although he would maintain his legal practice, Darrow believed he could supplement his comfortable bank balance with lectures and debates on his four favorite topics: "Is Life Worth Living?," "Prohibition," "Capital Punishment," and "Science v. Religion: The Merits of Evolutionary Theory." Darrow had forthright and unshakable views on each. Life was generally not worth living, he argued, although it would be made more bearable by the right to enjoy a drink and an immediate end to state hangings. The teaching of evolution, meanwhile, was played out on the front line of a fierce battle between religious fundamentalism and modernist thinking.

Darrow was despised by the religious right because he was such an eloquent opponent of puritanical certainty. On a pre-Christmas visit to New York, combining time with Mary and another debate against Prohibition, at the Opera House, Darrow asked on December 23, 1924: "What kind of poem can you get out of a glass of ice water? Take out of this world the men who have drunk, down through the past, you take away all the poetry and literature, practically all the work of genius that the world has produced. If you could gradually kill off everybody who ever drunk, or wanted to, and leave the world to prohibitionists—my God, would any of us want to live in it?"

He was equally scathing toward the pinch-faced zealots who hoped to ban the teaching of evolution. "Give them a drink," Darrow snorted, "and a book by Darwin. If they gave both half-a-chance their world would open up with the light of knowledge and the warmth of whiskey."

In more private company Darrow loved to tell the story of how Mary had once outwitted the Prohibition laws when she sailed from San Francisco to New York. Under her loose dress she had worn a rope harness from which she carefully dangled two bottles of brandy. It worked as a kind of pulley so that when the Prohibition agents at the docks patted her down to check for any contraband liquor, she could, in an astonishingly risky and delicate operation, raise and lower the bottles so that they avoided detection. Mary somehow got away with it, and when she arrived at her friends' apartment in Greenwich Village, she caused a whooping sensation by stripping off her dress to reveal the harness, her petticoat, and the two precious bottles.

But even such a riotous anecdote could set off another of Darrow's brooding jags on the rise of the evangelical right and the masters of Prohibition.

On March 30, 1925, after they had shared a "wonderful dinner with [Theodore] Dreiser and his gal," Mary compared the novelist and the attorney. "Dreiser has a beautiful credulity . . . he is cocksure of nothing, ever the country boy seeing the city for the first time!" She wrote of Darrow as "the old lion with slouching shoulders and arch wit, sad as an old world . . . C.D. has 'come to that state where the horror of the universe and its smallness are both visible at the same time . . . the twilight of the double vision.' A Passage to India."

E. M. Forster's novel had been published the year before and both Mary and Darrow considered it a masterpiece. Just like T. S. Eliot's The Wasteland, written three years earlier, Forster's novel probed the darkening spiritual condition of twentieth-century man. Darrow felt an instinctive empathy with Eliot and Forster, which Mary, in turn, found almost unbearably moving. Convinced by Darrow's greatness, she felt privileged to be back alongside him.

There was little lust, now, to cloud her emotions. It felt enough for Mary merely to be with Darrow. The day after their dinner with Dreiser they had gone to the Belmont Hotel in New York and seen Ed Nockles, who had been the only labor leader to defend Darrow throughout all his trouble with the bribery trials of 1912 and 1913. Yet the physical decline of Nockles came as a shock and seemed eerily symbolic of the collapse of labor—a movement they now hoped to celebrate with her book about Mother Jones.

The sudden sound of the doorbell echoed through her. *Darrow!* She tore off her apron and hurriedly checked her hair in the mirror. Lem beat her to the door and opened it to Darrow.

He swept into the apartment, half turning to shake Lem's hand and then to embrace a glowing Mary. Darrow complemented her on the blue dress and asked Lem about his next planned trip. He invariably made time to talk to Lemuel, and he was welcomed with exceeding politeness in return. But, as always, Darrow dominated attention and he reveled in the fact as everyone in the room hung on his words.

The old lion grumbled about the decline of America and the banality of life as if he secretly enjoyed such trials and tribulations. And then, eventually, he spotted Margaret staring at him from a far corner. "Good evening, young lady," Darrow muttered as he waved away Mary's entreaties that the little girl should come over and say a proper hello to him. He knew she did not like him much for he often made some wisecrack she didn't quite grasp while the adults roared with laughter. He did not mean to poke fun at her, but neither could he feign any enduring interest in a girl far too young for him to flirt with or engage in proper conversation.

His friends were more comfortable with children. Carl Sandburg, the poet and writer, then in the midst of his epic series of books about Abraham Lincoln, often brought his battered old guitar and sang syrupy ballads to Margaret as if his presence at the Partons was just an excuse to see her. On his last visit he had given her two of his books. Her favorite, *Rootabaga Stories*, was a collection of children's stories published five years before; but he had also, in an act of high seriousness, presented her with a book of poetry. On the inner sleeve of *Smoke and Steel*, an anthology of his Chicago poems, he had written: "Margaret Parton—May all the white moons of life be good to her little feet." Darrow just shook his head. What was that all about?

Fremont Older, Darrow's great ally and an esteemed newspaper editor from San Francisco, spoke just as earnestly to Margaret. He loved swimming and challenged her to a "winner-take-all race" later that summer, as her parents had just paid $5,000 to buy a small and dilapidated eighteenth-century farmhouse in Sneden's Landing. They would swim against each other in the warm lake; Older teased Margaret, saying that to meet her challenge he had already gone into training. Darrow wiggled an eyebrow at him as if he were crazy. Surely that kind of sweet talk would be better served on a twenty-year-old girl?

And then there was Sinclair Lewis, the sensation of American publishing. Margaret could not quite believe she knew a man who had recently written a novel, *Main Street*, that had sold two million copies. Her mother said his new book, *Arrowsmith*, about a noble young physician and his innovative scientific research, had just been awarded the Pulitzer Prize. But Lewis turned it down because he did not believe in cheap baubles. He was a communist, Mary whispered admiringly, as well as a wonderful writer.

Darrow liked Lewis but he was struck more by the fact that, during the Partons' long lunches or dinner parties, the author would often eat his meal with Margaret at the small table that had been set apart from the adults. While they ate together he would talk to her about cats—making Margaret laugh when he slipped a surreptitious tidbit of food to Patsy, her own tabby cat.

Darrow amused himself in a different way. He complained about the stupidity of the world and how little life mattered. He also read poetry in a deep and resonant voice that made everyone gaze adoringly at him—while Margaret sat alone in a corner of the room. She was in the same place now and, whether or not he was pricked by his conscience, Darrow directed a rare question her way. He asked her if she had any interesting news for him.

Margaret was surprised, but she responded eagerly. Yes, she said, there was something she wanted to show him. Darrow nodded sagely, and everyone else tittered as Margaret ran to her bedroom. "What can it be?" Darrow mused dramatically, "what can it be?"

She raced back into the room, clutching a piece of paper on which

Babe Ruth had written his name—after her father had recently met the great slugger on a journalistic assignment. Having often heard how Darrow loved baseball, Margaret hoped that she would, at last, win the affection of her mother's special friend.

Margaret showed Darrow the autograph and asked if he would sign his own name on the paper. Darrow squinted and grunted at her and then, with a small smile, he accepted her pen.

The room was hushed as everyone watched him. "Fame at last," he sighed, "alongside Babe . . ."

Margaret thanked him and then, barely looking down at the fluttering paper, she rushed over to show her mother the prize. Mary, beaming and radiant, read aloud the untidy scrawl: "Clarence Darrow, Pinch Hitter." The adults all hooted with laughter while Margaret, confused by the joke, flushed in embarrassment. She slipped away unnoticed, hating Darrow for his refusal to soften toward her and, especially, for the hold he exerted over her mother.

Mary did not notice how her daughter felt and later that night she wrote blithely of "Food, booze, friends: a holy trinity. M helpful and self-eliminating."

She was more concerned to note that Darrow initially "seemed older, leaden-hearted." He made some misogynistic asides that she wrote down: "'Women's one concern is the child—for that she gets a man. She is conservative, preoccupied with that one life-long function.'" His mood had clearly not been improved by the sight of Mary and Lem happily hosting their family party, and a streak of jealousy appeared to run through his gibes.

Mary, typically, forgave him, suggesting in her journal that "humanity's ignorance and stupidity oppress him." The whole mood of the party had changed for the better, anyway, in the moment when Darrow stunned them all. Leaning forward in his chair, with gleaming eye and jutting jaw, he had told them, with a sense of drama, how he had resolved to contest a trial that threatened to outstrip even Leopold and Loeb for controversy and nationwide scrutiny.

He did not have to say "John Scopes" or "Dayton"—for over the last week those previously unknown names had become familiar to everyone. Newspapers across the country had carried front-page stories about the likely fate of the schoolteacher charged by the State of Tennessee with breaking a new law. Scopes, an innocent and unremarkable-looking young man in his glasses and crisp white shirt, was about to face a dramatic trial in the small town of Dayton—where he had been declared guilty of teaching the theory of evolution to his biology students.

"America," Darrow growled, "has taken another step back to the Dark Ages." But he spoke with such relish that Mary could not doubt his appetite in volunteering to defend Scopes—for free.

They all knew what it meant, both to Darrow and the entire country, for ranged against him in the opposite corner was a man bent on bringing down divine judgment on Scopes. He was a fiery and formidable evangelist, a former politician who had stood for the presidency of the United States in three separate elections, a man who considered Darrow his enemy and a defender of all that was debased in the modern world.

Darrow smiled, as if he could already feel the fire that had been lit inside him. And then he laughed softly when he was asked if it was really true, that he would face the political and religious leader he had once supported, but with whom he had since engaged in bitter discord. He nodded in delight. William Jennings Bryan, the baleful messenger of God, had promised to fight him to the death.

THE DESTINIES of Darrow and William Jennings Bryan had been locked together three decades earlier when they came together as a seemingly united force at the 1896 Democratic Party convention in Chicago. Darrow, at thirty-nine, was just three years older than Bryan, and they appeared similar in both their showboating personalities and combative political ambitions. Each man was a brilliant orator, pledging soaring empathy for the underprivileged masses. Bryan, like Darrow, had established his renown as a radical attorney who fought the mighty railroad companies all the way to the Supreme Court. But, beyond their public pledge for justice and equality, both lawyers had a taste for fame and personal power.

While Bryan craved an unlikely presidential nomination in 1896, Darrow stood for Congress. Despite his protestations that he had no real interest in a political career, it had not taken long for Darrow to be persuaded by his mentor, John Peter Altgeld, the governor of Illinois, that he should contest the congressional election. His district was traditionally such a safe Democratic seat that Darrow was considered a certainty for Congress. Bryan's path to the presidency was less obvious.

Altgeld doubted Bryan's intellect and also believed that the congressman from Nebraska would struggle to draw together the more sophisticated urban wing of the party and the overwhelming rural base of his support. Darrow was less intransigent and, with his preference for the underdog, noticed that Bryan booked himself into the distinctly modest Windsor-Clifton Hotel, where rooms ranged in price from 75 cents to $2 a night. Unlike many of the other leading Democrats, Bryan did not then have the funds to finance a stay at a flashier convention hotel.

In 1896 America was in the grip of depression, and the country's farmers, particularly in the Midwest, insisted that they were being made to suffer by bankers on the East Coast. The farming and labor communities had become convinced that a simple solution would alleviate their economic woes. They wanted silver to become legal tender again so that the price of gold dropped and the national banks increased the circulation of money. In this way, they would find it easier to pay off their debts and return to more prosperous times. Darrow echoed both Altgeld and Bryan in his support of "the silver ticket" by stressing that he always felt affinity for "the debtor rather than the creditor."

Bryan was more ornate in the defining speech of his career. "The humblest citizen in all the land, when clad in the armor of a righteous cause, is stronger than all the hosts of error," he told an entranced convention. While his voice rolled and cracked and his face shone with sweat, Bryan described it as "a cause as holy as the cause of liberty—the cause of humanity."

Darrow saw something of himself in Bryan as the congressman worked the crowd of twenty thousand Democrats. Bryan instinctively knew how to control his audience, adjusting his pitch and fervor with masterly technique. Marveling at "his voice, his personality, his knowledge of mob psychology, his aptness for forming rhythmical sentences," Darrow listened admiringly to Bryan's convention speech.

Bryan reached a biblical crescendo: "We will answer demands for a gold standard by saying to them: 'You shall not press down upon the brow of labor this crown of thorns; you shall not crucify mankind upon a cross of gold.'"

Darrow was stunned. "I have enjoyed a great many addresses, some of which I have delivered myself," he wrote wryly, "but I never listened to one that [so] affected and moved an audience. Men and women cheered and laughed and cried. They listened with desires and hopes, and finally with absolute confidence and trust. Here was a political Messiah who was to lift the burdens that the oppressed had borne so long. When he had finished his speech, amidst the greatest ovation I had ever witnessed, there was no longer any doubt to the name of the nominee."

Only Altgeld had not been persuaded. "What did he say, anyhow?" he asked Darrow. But there was no stopping Bryan, the barrel-chested man from Nebraska, who now styled himself as "The Great Commoner." Darrow agreed to endorse his nomination.

He campaigned vigorously on behalf of Bryan, and Altgeld, while neglecting his own election prospects. This was less generosity on Darrow's behalf than an arrogant belief that he did not even need to fight for votes himself. He was so sure of his own victory that he poured his energy into helping Bryan to become president and Altgeld to secure another term as governor. That error sealed one of the crushing moments of hubris in Darrow's eventful career.

All three men narrowly lost their respective elections. Darrow was beaten by a mere six hundred votes, a result that would have been avoided had he not been so overconfident; he had declined to campaign for even a single day in his own district. His defeat wounded him—even if he claimed later that election failure had come as a relief. "The scheming and dickering and trading for political place never appealed to me, and I concluded early in life that if one entered a political course he must leave his independence behind, and this I could never abide."

If there was some truth in that assertion, it was also plain that Darrow resented the time he had given to Bryan at his own expense. His secret hope of forging a new life for himself in Congress had been ended abruptly by his foolish support of an even bigger fool in Bryan. Darrow's future, from that moment, became wedded to the law. He and Altgeld also believed that the expected presidential victory for the Democrats had been ruined by the Commoner's impoverished political vision. They regarded him increasingly as a one-dimensional and uncultured thinker who was not fit to run for the highest office.

Darrow's sudden bitterness toward Bryan was made apparent when they met at a political banquet early the following year. As Bryan appeared to have learned no lessons from his defeat, Darrow sniped angrily at him. "You'd better study science, history, philosophy, and read Flaubert's *Madame Bovary*, and quit this village religious stuff. You're head of the party before you are ready and a leader should lead with thought."

Standing alongside another of Darrow's future enemies, the poet Edgar Lee Masters, Bryan puffed out his chest: "Darrow's the only man who looks down on me for believing in God."

Darrow defended Nathan Leopold, Jr. *(seated, left)* and Richard Loeb *(far right)* in
[Tria]l of the Century after the nineteen-year-old boys admitted to the world's first "thrill
[killin]g" in Chicago on May 21, 1924. (© Bettmann/CORBIS)

Richard "Dickie" Loeb *(left)* and Nathan "Babe" Leopold, lovers and the sons of
millionaires, confessed to the kidnapping and murder of fourteen-year-old Robert
Franks—a crime that shocked America but that the university students described as "an
intellectual exercise." Leopold said coolly that "the killing was an experiment. It is just as
easy to justify such a death as it is to justify an entomologist killing a beetle on a pin."
(© Bettmann/CORBIS)

Darrow with a relaxed and smiling Loeb, who predicted that "this thing will be the making of me." The old attorney spoke of his fear that the defendants would be "wakened in the gray light of morning…led to the scaffold, their feet tied, black caps drawn over their heads, stood on a trap door, the hangman pressing a spring so that it gives way under them." (© Bettmann/CORBIS)

Loeb, in his typically fashionable bow tie, being led from the Cook County jail into the courthouse, with Leopold following near the rear of the procession. Like his accomplice, Leopold's black hair is slicked back in the "Valentino sheik" style that was so popular in America in 1924. (© Bettmann/CORBIS)

The sixty-eight-year-old Darrow outside court in Dayton. The Scopes trial sealed his reputation as "America's greatest one-man stage draw." (© Bettmann/CORBIS)

July 1925, Dayton, Tennessee. The Scopes Monkey Trial pitted two of the most famous men in America against each other in a heated duel. William Jennings Bryan *(seated, left)* cools himself with a fan while Clarence Darrow *(standing with arms folded)* observes the proceedings with his customary restraint. (© Bettmann/CORBIS)

The main street of Dayton, the tiny Tennessee town that became internationally notorious as the scene of the Scopes Monkey Trial. Some newspaper headlines turned Dayton into a dramatic setting—"Scopes Case a 'Duel to Death'" thundered the *Chicago Daily Tribune*—but the *New Republic* suggested that "In Dayton, religion takes the place of golf, bridge, music, art, literature, the theater, dancing, clubs. Take religion away, and the desolation and distress would be pitiable to contemplate." (© Underwood & Underwood/CORBIS)

During the trial of John T. Scopes, the "man-killing" heat became so oppressive that a temporary court was set up outside on July 20, 1925. After Darrow demanded the removal of a READ YOUR BIBLE banner, still visible here, he called Bryan to the witness stand, where, in front of a transfixed crowd of three thousand, he destroyed his opponent's credibility with a brutal cross-examination. (© Bettmann/CORBIS)

John Scopes *(center)*, the twenty-four-year-old who admitted his "guilt" in teaching evolutionary theory at Rhea County High School, was defended by Darrow *(right)* and Dudley Field Malone *(left)*. In a dramatic speech, Malone, who had once worked for Bryan, argued that "the least this generation can do is to give the next generation all the facts, all the available data, all the theories, all the information that learning, that study, has produced. Give it to the children in the hope of heaven that they will make it a better world of this than we have been able to make it....For God's sake let the children have their minds kept open." (© Bettmann/CORBIS)

Moments before the Scopes trial began on July 10, 1925, two former political allies and now bitter rivals sat together in court. Darrow *(left)* sat with his hands clasped in front of him. But he was not about to pray for divine assistance against a religious fundamentalist he denounced as the leader of "the cult of Morondom." Bryan *(right)*, a three-time Democratic candidate for president, regarded the Scopes case as a personal crusade against evolutionary theory. He warned that Christianity would "crumble" if Darrow was victorious. (© Bettmann/CORBIS)

Darrow and his wife, Ruby, sail to Europe and Britain, where the attorney was friendly with writers like George Bernard Shaw and H.G. Wells. Although they never divorced, Ruby was painfully aware of Darrow's affair with Mary Field Parton. In later years, while preparing Darrow's authorized biography after his death, she warned author Irving Stone that he should not include Mary: "Why don't you ignore her, forget her?...I urge you not to glorify her. I shudder at her indecency and cannot work with you under such circumstances....Let us draw a veil over M.F. and say no more."
(© CORBIS)

Clarence Darrow, with his typically disheveled hair, familiar suspenders, and a cigarette jutting from his mouth, reads his mail during the Scopes trial.
(© Underwood & Underwood/CORBIS)

Gladys Sweet, the young wife of Ossian and mother of baby Iva, showed courage in the face of terrible racism in Detroit. Even when the prosecution offered to drop all charges against Gladys in November 1925, the twenty-three-year-old "light-skinned Negro housewife" refused to leave her husband and the ten other defendants in the dock. She insisted that she would also face the charge of murder. (Walter P. Reuther Library, Wayne State University)

Dr. Ossian Sweet went on trial for murder in Detroit on October 30, 1925, the day he turned thirty. The previous month a lynch mob, angered at a black family's move into a white neighborhood, had gathered threateningly around the doctor's new home. The Sweets tried to defend themselves and, in the aftermath of the confrontation, a white man lay dead in the street. Sweet provided riveting testimony in court and inspired Darrow to produce arguably his greatest closing summary in April 1926. Despite his eventual acquittal, Dr. Sweet's life would still be blighted by tragedy. (Walter P. Reuther Library, Wayne State University)

2905 Garland Avenue, Detroit. The day after they moved in, the house that the Sweet family bought in September 1925 was surrounded by a mob of white neighbors led by the sinister Waterworks Improvement Association, with a spokesman urging them to "put the niggers out." Ossian Sweet said in court that "when I opened the door and saw the mob I realized I was facing the same mob that had hounded my people through its entire history....I was filled with a peculiar fear, the fear of one who knows the history of my race." (Walter P. Reuther Library, Wayne State University)

Clarence Darrow said Mary Field Parton, his lover and friend of thirty years, was "the cleverest woman I ever knew...." Mary and Darrow were involved in an affair between 1908 and 1912, and they resumed their relationship in the summer of 1924 just before Darrow plunged into his last great trilogy of trials. Mary, a writer and activist, died in 1969 at the age of ninety-one. (*Untitled*, Margaret Parton Collection, Coll 36, Special Collections and University Archives, University of Oregon Libraries)

"Your kind of God," Darrow snarled as he turned away in disdain. The roots of a hostile conflict had been sown.

Bryan made two more unsuccessful tilts at the White House—winning the nomination but losing the presidential elections of 1900 and 1908. As his political star faded, after an unhappy stint as secretary of state in Woodrow Wilson's government, Bryan announced his withdrawal from government when America entered World War I. Unlike Darrow, who passionately supported the war effort against German nationalism, he believed that the United States should stand apart from international affairs. Bryan turned instead to religious fundamentalism in a move that widened the divide between him and Darrow—who regarded his switch of focus as a desperate pitch to regain power. The attorney also declared that the accompanying biblical rhetoric was offensive to intelligent thought.

Darrow had long been skeptical of religious dogma. His own father, though born a Methodist, had lost his faith as a young man while immersing himself in philosophy and literature. Amirus Darrow repeatedly told his son, "The end of wisdom is the fear of God; the beginning of wisdom is doubt." Even if he thought the name his father had given him, Clarence, was a terrible cross to bear, Darrow endorsed that essentially questioning philosophy of life. Amirus Darrow had been an early disciple of Darwin and had introduced Clarence to concepts of evolution, as well as the freethinking of Voltaire and Thomas Paine, when he was not much more than a child. He came to share his father's pride in their place as "the village infidels" in Kinsman, Ohio.

In his later years he defined himself as an agnostic rather than an atheist—for he recognized the impossibility of scientifically proving that God did not exist. He found that doubt and uncertainty, as his father promised, made him more thoughtful. It also elicited a curious irony, for Darrow was often exalted as a man of great Christian compassion, a Christian by example rather than faith. Darrow argued that he had no quarrel with God but merely with his more prejudiced followers—of whom Bryan was a strident example.

His disdain deepened when the Great Commoner became a millionaire after he moved to Florida and cashed in on the property boom. But the greatest source of their mutual antipathy could be found in their conflicting perceptions of evolution. Bryan's views were sparked by little-known writers, like James Leuba, whose book, *The Belief in God and Immortality*, stated that 45 percent of American college students had turned away from "fundamental" Christian doctrines. Shocked by the statistics, Bryan campaigned relentlessly against evolution in speeches like "The Menace of Darwinism." He dismissed all empirical evidence to reach the nub of his reductive argument against evolutionary theory: "Man is infinitely more than science; science, as well as the Sabbath, was made for man."

Bryan also drew a direct link between Darwin's writing and the philosophies of Nietzsche. Darwin's premise of "the survival of the fittest" seemed to correspond with the myth of the Nietzschean "superman" that had beguiled German nationalists—as well as a wayward Jewish boy, Nathan Leopold, in Chicago. "Nietzsche carried Darwinism to its logical conclusion," Bryan wrote, "and denied the existence of God, denounced Christianity as the doctrine of the degenerate, and democracy as the refuge of the weakling; he overthrew all standards of morality and eulogized war as necessary to man's development."

Darrow's defense of Leopold and Loeb had disgusted Bryan. He saw such "amoral and degenerate prattling" as part of the wider "evil and lawless campaign to undermine Christian values—seen so clearly in the teaching of evolution." And now, in his shaping and support of a draconian new law in Tennessee, which outlawed the teaching of evolution, Bryan had the vehicle with which he planned to drive a wedge of fundamentalist dogma into the already constrained borders of American learning.

In contrast, Darrow endorsed "the evolutionists," a loose grouping of scientific thinkers and writers whom the *Nashville Tennessean* newspaper linked to "the liberals, the feminists, the radicals of all degrees and shades, the 'birth controlists,' the psycho-analysts, the agnostics . . . the socialists, social service workers and professional 'causers'" who supposedly threatened the traditional fabric of American society.

The most extreme faction on the right was represented by the Ku Klux Klan. After the original Klan had been destroyed by the 1871 Civil Rights Act, an even more sinister force had emerged after an absence of forty-four years. The new KKK had risen up in 1915, and its violent racism toward Negroes was accompanied by a loathing for "urban ethnics," from Jews to Catholics, "evolutionists" who read Darwin, and "immoral" women epitomized by the sexually confident flappers of the Jazz Age. For the Klan, and their more respectable fundamentalist peers, the supremacy of white male Protestantism was sacrosanct.

Darrow, having just defended two homosexual Jewish intellectuals, and being so vocal in his support of black America, stood for everything the Klan feared and hated. To the fundamentalist right, he was America's plain-speaking nightmare. "Purity of the Anglo-Saxon race?" he snorted when asked about the sanctity of "ordinary" America and why he defended different racial and ethnic groupings. "That's the greatest race of sons-of-bitches that ever invested the earth. Mind you, if there is such a race then I am one of them—because my ancestors lived in this country for three hundred years. But I do not brag about it. I apologize for it."

Fundamentalists clung instead to a self-righteous certainty underpinned by literal interpretations of the Bible. Bryan, their new leader, insisted, "When the Bible ceases to be an authority—a divine authority—the Word of God can be accepted, rejected, or mutilated according to the whim or mood of the reader." Anarchy would then result, whether it took the form of heavy petting with wanton flappers or the teaching of Darwin's theory of evolution. Christian society, in consequence, would crumble and die.

The limitations of Bryan's knowledge and intelligence infuriated Darrow. Driven by an ambition to usurp such "fundamentalist idiocy," he was galvanized by the prospect of outwitting the Great Commoner in another trial that would capture national attention. The battle lines had been drawn—with Darrow and all freethinking people on one side and, across the divide, Bryan and his evangelical army bearing down on them.

CHAPTER 9

HELLFIRE PREACHERS
AND BIOLOGY TEACHERS

T WO SUMMERS before, in June 1923, Bryan and Darrow had clashed over evolution for the first time. Bryan sparked the acrimony when he wrote to the *Chicago Daily Tribune* to complain that his beliefs had been misrepresented. At the end of a rambling diatribe he warned that "it is improper and subversive . . . to teach atheism or agnosticism or true Darwinism or any other hypothesis that links man in a blood relationship to any other form of life."

The *Tribune* responded with a stinging editorial on June 29: "It would be difficult to conceive a more clumsy statement. . . . The interchangeable use of atheism and agnosticism in itself indicates that its author was not only ignorant of evolution but of words in common usage. It is perhaps possible to teach atheism, but it is impossible to give a course of instruction in agnosticism. The agnostic professes merely not to know. All scholarship involves doubt. Indeed, thinking is doubting."

For an avowed agnostic and thinker like Darrow, the temptation to enter the fray was irresistible. On July 4, in a front-page scoop, the *Tribune* published Darrow's reply, which was dominated by his demand that Bryan should answer a long list of questions to clarify his views on evolution. There were fifty-five queries in total, ranging from the serious to the comic and back again: "Is the account of the creation of the earth and all life in Genesis literally true, or was it an allegory? Was the earth made in six literal days, measured by the revolution of the earth on its axis? Did God create man on the sixth day? Did he rest on the seventh day? Was Eve literally made from the rib of Adam? Did God curse the serpent for tempting Eve and decree that thereafter it should go on his belly? How did he travel before that time? Did God tell Eve that thereafter he would multiply the sorrows of all women and that their husbands should rule over them? Under the biblical chronology was not the earth created less than 6,000 years ago? Has not man probably been on earth 500,000 years? Does not geology show that by fossil remains, by the cutting away of rock for river beds, by deposit of sorts, that the earth is much more than a million years, and probably many million years, old? Can one not be a Christian without believing in the literal truth of the narrations of the Bible here mentioned?"

On the day Darrow's questions were published, Bryan preached to a crowd of ten thousand believers at Billy Sunday's tabernacle in Winona Lake, Indiana. Sunday, who had played professional baseball for the Chicago Cubs before becoming an evangelical preacher, was another fervent opponent of evolution. He formed a potent combination with Bryan, whose address marked the climax of the Fourth of July celebrations organized by the Christian Citizenship.

Bryan was presented with a copy of Darrow's questions. "Mr. Darrow is one of two atheists with whom I am acquainted," he smiled. "I am not worried about an atheist who admits he is one. The man who denies the existence of God is not likely to have much influence. My controversy is not with atheists like Mr. Darrow, but with those who claim to be Christians and who substitute the guesses of evolution for the word of God. Evolution has no place for the miracle and the supernatural. That leaves the Bible as merely a man-made book shorn of its divine authority. The unproven hypothesis of evolution is the root cause of nearly all the dissension in the church . . . the teaching of evolution as a fact instead of a theory has caused students to lose faith in the Bible."

His warnings were echoed by Sunday and T. T. Martin, who had just published his booklet *Hell and the High Schools: Christ or Evolution: Which?* Martin insisted that "the teaching of evolution is being drilled into our boys and girls in our High Schools during the most susceptible, dangerous age of their lives."

A robust attempt by the Kentucky General Assembly to ban such teaching had been narrowly outvoted in 1922, and so the hopes of Bryan and his biblical brigade had switched to neighboring Tennessee. John W. Butler, a farmer and Primitive Baptist from rural Macon County, had run for the Tennessee House of Representatives on an antievolution ticket—motivated by the fact that a local preacher had told him how a young woman in his community had moved away from the church after completing a university course that included the study of evolution.

When he decided to stand for election again in 1924, Butler resolved to turn his general policy into a specific Tennessee law. Butler revealed that, on the day he turned forty-nine, "I was thinking 'what will I do on my birthday?' and I said to myself, 'well, first thing, I'll get that law off my mind.' So I wrote it out after breakfast at home just like I wanted it. I had the stenographer up at the Capitol type it for me."

The legal words were as arid as they were destructive. "Be it enacted by the General Assembly of the State of Tennessee that it shall be unlawful for any teacher in any of the universities, normals and all other public schools of the State which are supported in whole or in part by the public school funds of the State, to teach any theory that denies the story of Divine creation of man as taught in the Bible, and to teach instead that man has descended from a lower order of animals."

On January 21, 1925, without even the pretense of a debate, But-ler's birthday scribble was passed as a legal bill in the Tennessee House of Representatives by seventy-one votes to five. Each member of the House had been presented, before voting, with the text of Bryan's tirade against the teaching of evolution, "Is the Bible True?" Butler proposed that anyone found guilty of breaking the new law would be punished by a fine of up to $500.

There was a measure of dissent in Tennessee. "Perhaps if there is any other being entitled to share Mr. Bryan's satisfaction at this Tennessee legislature it is the monkey," argued the *Chattanooga Times*. "Surely if the human race is accurately represented by that portion of it in the Tennessee House of Representatives, the monkey has a right to rejoice that the human race is no kin to the monkey race." The *Rockwood Times* was less amused: "The quicker this jackass measure is booted into a waste basket, the better for the cause of enlightenment and progress in Tennessee."

The State Senate hesitated over both Butler and an accompanying motion, introduced by Senator John Shelton, that the teaching of evolution would become an official felony. After its Judiciary Committee voted against Shelton's proposal, on the basis that legislators "should not make laws that even remotely affect the question of religious belief," the Senate agreed to take a formal vote on the Butler bill.

In February 1925, in the buildup to that Senate vote, Billy Sunday drew almost two hundred thousand people to his eighteen-day crusade in Memphis. On a wild opening night he offered "a star of glory to the Tennessee legislature for its action against that God-forsaken gang of evolutionary cut-throats." Denouncing Darwin and Darrow as "infidels," he warned that "education today is chained to the devil's throne." His voice rearing and tripping over the words, as if he was on the verge of speaking in tongues, Sunday fired off a typical evangelical riff:

"Teaching evolution!" he cried. "Teaching about pre-historic man!" Sunday's sob had become a scalded yelp. "No such thing as pre-historic man. . . . Pre-historic man! Pre-historic man!" As his eyes rolled and his body shook in revulsion, the *Commercial Appeal*, a Nashville newspaper, noted primly that "Mr. Sunday gagged as if he was about to vomit."

Bryan was more conciliatory in advising members of the Senate to remove any penalty from an antievolution bill. He believed that it would divert attention from the core issue of eroding faith in the Bible, and he also pointed out, "We are dealing with an educated class that is supposed to respect the law." But his guarded suggestion was ignored. On March 21, 1925, the Senate passed the Butler bill, without amendment, by twenty-four votes to six. Governor Austin Peay, who then signed his official ratification of the new law, was hailed by Bryan, who promised that "other states North and South will now follow the example of Tennessee."

If Governor Peay and William Jennings Bryan both doubted the Tennessee law would ever be challenged, a select group in New York thought differently. In his office at the American Civil Liberties Union (ACLU), the organization's founder, Roger Baldwin, read out details of the Butler Act to his incensed colleagues. Baldwin emphasized that the new law violated both the freedom of speech and the rights of teachers. The ACLU resolved to financially and legally support any Tennessee teacher willing to defy the act by openly instructing his or her students in evolutionary theory.

They duly printed an advertisement in newspapers across Tennessee—asking for a volunteer who would put himself forward as an ACLU-sponsored defendant in a "test case" against the dubious constitutional validity of the Butler Act. Tennessee initially ignored the call from New York with almost blanket disdain. There was one exception. On May 4, 1925, George Rappleyea, a small and determined thirty-one-year-old man in glasses, rushed to Robinson's Drugstore in the mining town of Dayton in Rhea County. Nestling prettily but otherwise anonymously in the Tennessee River Valley, about forty miles upstream from Chattanooga and the same distance again, in the opposite direction, from Knoxville, Dayton's population was limited to 1,800. Rappleyea, a young New Yorker who had grown up on Third Avenue before moving south to work as a chemical engineer in Dayton, saw an opportunity to publicize his little-known new home.

"Have you seen the morning paper?" Rappleyea asked the drugstore owner, Frank E. Robinson, as he waved his copy of the *Chattanooga Times*. Rappleyea knew that Robinson, as the chairman of the Rhea County School Board, was always dreaming up unfulfilled publicity schemes for Dayton. He pointed to the ACLU ad and claimed to have strong contacts with the New York organization. This was their chance to finally put Dayton on the map.

A small group of men gathered around them. The school superintendent, Walter White, who had been a Republican state senator, supported the Butler Act; but he was just as keen to bring national attention to the town. Robinson was quickly convinced, and they sent word to three local attorneys to join them. John Godsey agreed to lead the defense of a potential challenger of the law while two brothers, Herbert and Sue Hicks, declared themselves ready to prosecute any teacher who taught evolution in a Dayton public school. Sue Hicks had been named after his unfortunate mother who died in the midst of giving birth to him—but he was an otherwise ordinary fellow who had become friends with John Scopes, the local high school's science teacher and part-time football coach.

Scopes would be the perfect candidate for their publicity scam. With his horn-rimmed glasses and blond hair he was scholarly but approachable, as sociable as he was thoughtful. There was nothing radical or offensive about John T. Scopes. He had been born in Kentucky, in the small city of Paducah, in 1900. His father was a railroad mechanic whose work had moved the family to Salem, Illinois, which just happened to be the birthplace of William Jennings Bryan. When Scopes was at the University of Kentucky, vaguely pursuing a law career, he had heard Bryan make a campaign speech against the teaching of evolution in 1922. He disagreed with Bryan's argument, but

Scopes was not involved in any overt protest against fundamentalist Christianity. He laughed instead, with his friends, at the whistling sound the gap-toothed preacher made whenever he said a word dominated by the letter "s."

Soon after graduating from Kentucky in 1924, Scopes accepted a position as a science teacher at Rhea County High School in Dayton. He was not too brokenhearted by his failure to become a lawyer. Scopes settled instead into a routine working life where he taught science classes and coached the football team—whose best player was a deadly runner called Punk Cunningham. Scopes's father, like Darrow, might have been an agnostic and a socialist, but Scopes occasionally attended the Methodist church as a way of meeting people. He was a well-liked teacher, even if some of the more conservative members of Dayton society noticed that he also enjoyed smoking and dancing with some of his girl students at Saturday night shindigs. But it was widely accepted that Scopes was a relatively typical young man of twenty-four.

And so, true to his popular image, Scopes was engaged in an energetic game of tennis on the high school courts that May morning when he received a message. Doc Robinson wanted to see him urgently. A perspiring Scopes dutifully and politely hurried to the drugstore. Robinson offered him a chair, and the boy who worked as a soda jerk brought the young teacher a glass of cold lemonade. "John, we've been arguing," Rappleyea murmured artfully, "and I said nobody could teach biology without teaching evolution." Scopes nodded his agreement, while wondering at the real reason for their sudden need to see him.

Scopes had recently taken over the teaching of biology from the school principal, who had fallen ill during the last weeks of term. He reached out for a copy of Hunter's *Civic Biology*, which Robinson sold from his shelves as a school textbook, and turned to the section on evolution. Scopes admitted that he had encouraged his students to use the book as they studied for their final exams. "Then, you've been violating the law," Robinson laughed, conceding that he, too, was guilty of dealing in illegal goods.

The drugstore owner paused meaningfully, and then, looking intently at Scopes, he asked the question that would change the young teacher's life: " 'John, would you be willing to let your name be used in this test case for Dayton?' "

Scopes, being an essentially agreeable sort, nodded innocently. He drank his lemonade while watching Robinson pick up the store telephone and call the *Chattanooga Times*. "This is F. E. Robinson in Dayton," he said. "I'm chairman of the school board here. We've just arrested a man for teaching evolution."

Robinson then called the *Nashville Banner* and repeated the news with unconcealed relish. Scopes sucked up the last of his drink with a straw and then, after shaking hands with a beaming Robinson and all the others in an excited group, he waved good-bye. Robinson was about to wire confirmation to the ACLU in New York, but Scopes had a tennis match to complete. His detour to the drugstore had been interesting, but he was anxious to get back to the serious business of playing.

T HE NEXT MORNING Scopes made front-page news in the *Nashville Banner*. He was bemused to read that his academic status had been elevated to a professorship. "J.T. Scopes, head of the science department of the Rhea County High School, was charged with violating the recently enacted law prohibiting the teaching of evolution in the public schools of Tennessee. Professor Scopes is being prosecuted by George W. Rappleyea, manager of the Cumberland Coal and Iron Company, who is represented in the prosecution by S.K. Hicks. The defendant will attack the new law on constitutional grounds. The case is brought as a test of the new law."

The Associated Press ran the story on its wires. Scopes and Dayton were soon spread across almost every major newspaper in America. The juggernaut of the "Scopes Monkey Trial" had begun to roll with unstoppable momentum.

A week later, on May 12, 1925, both Bryan and Darrow were at work, raising hell onstage. At a fundamentalist gathering in Pennsylvania, Bryan focused on the Scopes story. In an inflammatory speech he denounced scientists who had defended Scopes as "dishonest scoundrels," Bryan sounded triumphant: "We've got them now where they've got to come up and fight. In Tennessee there is a law that they can't teach man is descended from any lower form of life. Now that law has been deliberately violated."

Bryan announced that he would happily assist the prosecution in punishing the violator. A preliminary hearing had already been held in which three justices of the peace decided that Scopes should answer the charge before a grand jury on May 25, 1925. "The fundamentalists are so interested in the case that I said I'd be one of their counsel if the law department of the state of Tennessee doesn't object," Bryan enthused. "This is one of the greatest questions ever raised. . . . Can a handful of scientists rob your children of religion and turn them out atheists? We'll find 109 million of the other side. For the first time in my life I'm on the side of the majority."

The Dayton self-publicists were heartened by the extent and intensity of this sudden fascination with their town, and Sue Hicks declared it "a great honor" to have the active support of a charismatic religious leader like Bryan.

Darrow heard of his rival's intentions early the following morning. He had spent the previous evening lecturing on "The Sane Treatment of Crime" at the annual conference of the American Psychiatric Association in Richmond, Virginia. Darrow had made history by becoming the first lawyer to present a keynote address in the association's eighty-year existence—and he achieved another innovation by sneaking in numerous comic references to the standoff between science and religion in Tennessee.

Yet his humorous asides were quickly forgotten when, on arriving in New York, Darrow read reports of Bryan's move against Scopes. "At once I wanted to go," he wrote later. "My object, and my only object, was to focus the attention of the country on the program of Mr. Bryan and other fundamentalists in America. I knew that education was in danger from the source that has always hampered it—religious fanaticism. To me, it was perfectly clear that the proceedings bore little semblance to a court case, but I realized that there was no limit to the mischief that might be accomplished unless the country was roused to the evil at hand."

The week before, in New York, Darrow had discussed the Scopes case with Arthur Garfield Hays, the ACLU's legal representative, and Dudley Field Malone, an eminent lawyer. All three had concluded then that Scopes would be defended best by local Dayton lawyers. But Bryan's entry changed everything. Darrow went back to see Malone so that, together, they might offer their services to the ACLU and take on Bryan.

In Dayton, George Rappleyea, anxious in the face of rumors that Chattanooga was on the verge of launching an audacious bid to stage the trial itself, had no idea of Darrow's plan. He had an even fancier name in mind to defend Scopes—H. G. Wells, regarded as the "Father of Science Fiction" after the publication of his late-nineteenth-century novels *The Time Machine*, *The War of the Worlds*, and *The Invisible Man*. Reminding reporters that Wells's 1896 novel *The Island of Doctor Moreau* had examined issues of society, community, religion, Darwinism, and eugenics, Rappleyea claimed, "I am sure in the interests of science Mr. Wells will consent."

Apart from the fact that he lived on the other side of the Atlantic, in London, the fifty-eight-year-old writer was not a lawyer. He dismissed the offer from Dayton with much amusement, but he was eventually delighted to learn that his friend Darrow would assume the task of defending Scopes and the theory of evolution in which Wells believed so forcibly.

When Wells had been asked by an interviewer how he had enjoyed his visit to America in 1918, he conjured up his enthusiasm in one sentence: "Well, I met Clarence Darrow!" He later wrote that Darrow represented "a fine flower of American insurrection . . . he believes superstitiously in the individual unorganized free common man, that is to say he is a sentimental anarchist. He is for an imaginary 'little man'—against monopoly, against rule, against law, any law."

Darrow, of course, was opposed vehemently to the Butler law. On May 14, 1925, he and Malone wired Scopes's lawyer, John Neal, who had driven from Knoxville to offer his own legal expertise to the young teacher: "We are certain you need no assistance in your defense of Professor Scopes, who is to be prosecuted for teaching evolution, but we have read the report that Mr. William Jennings Bryan has volunteered to aid the prosecution. In view of the fact that scientists are so much interested in the pursuit of knowledge that they cannot make the money that lecturers and Florida real estate agents command, in case you should need us, we are willing, without fees or expense, to help the defense of Professor Scopes in any way you may suggest or direct."

Although Rappleyea and his colleagues were elated, for a heavy-weight slugfest between Bryan and Darrow would offer Dayton untold publicity, the ACLU executive was concerned. They had planned a sober legal debate in an effort to overturn an unjust law. Darrow, whom they described as "too radical," would overshadow that hope. Even though Neal had wired back his acceptance of Darrow's offer, the ACLU resolved to overturn the decision at a meeting the following month. Scopes, Neal, and Rappleyea were all sent rail tickets to New York so that they might publicize their case. But it was made clear that they would also meet with the executive to decide whether Darrow should appear in Dayton.

DARROW'S AGITATED state of mind reflected the chaos at the ACLU meeting. He and Malone sat in as observers while the vitriol poured forth—mostly against Darrow's involvement. The arguments weighed heavily in favor of ousting him as, despite shouted objections from individual members, ACLU leaders said that the attorney was "a headline-chaser" whose celebrity would obscure the serious issues at stake. They argued that his standoff against Bryan threatened to turn the battle for civil liberties into an undignified and personality-driven carnival.

Malone angrily rebuffed an attempt to employ his services alone. He would not offer any help to the defense unless Darrow, the greatest lawyer in America, was at his side. In desperation Roger Baldwin, the ACLU chairman, turned to Scopes. They needed to hear the views of the defendant in a case that would be splashed across the front pages of every major newspaper in the United States and the world beyond.

Scopes spoke quietly but pointedly. Ever since Bryan had joined the prosecution, he said, Dayton had begun to fill with "screwballs, con-men, and characters. It's going to be a gouging, roughhouse battle. And if it's going to be a gutter fight, I'd rather have a good gutter fighter."

The room fell silent. Scopes glanced shyly at Darrow. He had not meant any offense with his words. Darrow took none and, instead, simply nodded when Scopes reiterated that he would be happy with any other lawyer the ACLU might suggest could help them. But he was utterly sure of one thing. He wanted Darrow to lead his defense in Dayton.

THE NATURAL order appeared obvious out in the country under a sunlit sky. For Mary, alone again with Margaret in the summer house, with the heat pressing down on her, thoughts of Darwin and Darrow, of evolution and the old law of the jungle, were hard to avoid. Man might be evolving, despite evidence to the contrary in Tennessee, but the rhythms of the world maintained an ancient pattern over their two acres of land. "Death in a cruel form visits our home," Mary wrote when recording an innocent encounter. "M witnesses death daily. Leaves fall, wood decays, blossoms wilt, ripened fruit falls, grass dies and trees fall, vegetables and fruit are eaten by parasites—and yet when Patsy, the cat, sprang at a little song bird and caught it by its singing throat, M cried piteously and went upstairs to weep alone at so 'cruel a world.' I reminded her that she had eaten meat for dinner, that all life lived on life. She must not hate Patsy."

The little girl's distress did not last, and a few nights later, beneath the soaring temperatures, Mary wrote, "M barefoot and stripped to near nakedness does not seem to feel the heat. A cool bath, white sheets, a white cool moon—then sleep for our little girlie."

Mary slept less well, "with all the bug bites—creatures eating creatures." She and Darrow had already planned their next meeting, back in New York, and Lem had disappeared again. Her lovely husband sometimes seemed to melt away with the heat. They would be in a taxi together, on the way to catch the little boat that crossed the Hudson at Dobbs Ferry, and Lem would suddenly tell her that he could not come with her and Margaret. And then, even when he was with them at the country house, he sometimes left early for the city in the morning and did not return that night. Mary had grown used to his elusiveness, and she found a rhythm of her own.

The farmhouse was often full of visitors—"good stories, good drinks, good friends" she wrote after Carl Sandburg and a small number of their circle came for dinner early that June. Margaret, who loved her mother's writer-friends most of all, was in her element. This time there was no Darrow to divert Mary, or shun a ten-year-old girl. And so Margaret did not even mind that the adult conversation soon drifted to Dayton, and Darrow's confrontation with Bryan, while Sandburg railed against the evangelical right. He despaired of all the sentimental phrases being spouted, ridiculing the meaningless banality of sayings like "Give my heart to Jesus," as if serious debate about religion and scientific knowledge would be sullied to the point where it resembled a cheap love song. They yearned for Darrow's imperious victory over Bryan.

On June 9, after a short spell at home, Lem left unexpectedly for Canada. He said he would be back in a couple of days, which Mary assumed meant at least a week. She was seeing Darrow in less than forty-eight hours but the country, at night, felt lonesome to her. It was fine in the day, as she worked and Margaret gamboled around with the boy from next door, "but at night one needs the cafés and lights of the city." All her "books, magazines, papers measure the depth of loneliness. Lovers need not read. Nor children. Nor simple people."

Darrow, instead, was an ardent reader and the most complicated man she knew. She took to brooding on his enduring if hapless marriage. He would never leave Ruby, no matter how much he cheated on her. The old devil was in a prison of his own making; and it was a fate he openly acknowledged. "Darrow says of his actions in getting divorces: He saved others; himself he cannot save."

Mary and Darrow spent the afternoon of June 11, 1925, at their favorite hotel in the city, the Belmont on Park Avenue, whiling away the hours with what she described later as "visit and nonsense." And then, after a leisurely supper at the hotel, she accompanied him in the private car the ACLU had arranged to take him and John Scopes to the Civic Club on East Thirty-fourth Street—where the young schoolteacher would be the guest of honor at a fund-raising dinner held for his defense. Scopes, bewildered by the noise and the crowds of New York, had failed to arrive at the hotel on time. The car left without him and as Darrow and Mary sat together in the backseat, talking softly, Scopes had to find his own way by foot. He got hopelessly lost and arrived at the Civic almost forty minutes late, flustered and red in the face with embarrassment.

At least Scopes could recover while the main speeches were made. As the *New York Times* reported, "He smiled through a pessimistic speech by Mr. Darrow, who used wit to disguise his pokes at the failure of man to improve the human animal, and laughed heartily when his counsel, Dr. John H. Neal, invited everybody to Tennessee for the trial, assuring them that the climate was fine in the summer and that there were no mosquitoes.

"Mr. Darrow seemed down-hearted about the condition of man because so many people had tried to improve the world and had only succeeded in making it worse. 'It is best to leave everyone free to work out things for himself,' he said. 'Nature is doing it in a big, broad way and doing it pretty successfully. If we try to help her we only muddy it up. I have tried most of my life to improve this world a little and I don't think it has changed much. Man has been on this earth for half a million years and he will probably stick around another half a million and then give way to something else. The best way is to try to interfere as little as possible with other people, to cultivate a broad sense of charity, and let people live their own lives."

John Neal stood up and gestured toward Darrow, his imminent partner in a case already tagged as another Trial of the Century. "In the beginning," Neal said, "there were only two clear voices, two individuals, who realized what this case was. One was Mr. Bryan. He saw this law was his child and rushed to its defense. I am glad he did, for there were some people who thought the law would not be properly defended. Another man saw immediately what was involved. Clarence Darrow. Despite the despair sometimes in his words you can see in him a great charity for mankind."

Mary smiled at Darrow, glazed pride shining in her eyes. He rocked slowly in his chair, his gaze holding hers as the club erupted in applause. Mary was the first to look away. She loved Darrow but, still, she was married and, once more, he would soon be gone. He carried the same expression that she remembered seeing in Chicago almost a year earlier, on the morning when they had said good-bye before the Leopold and Loeb trial. It was almost time to let him go again.

After the Scopes dinner, they attended an interminable meeting on Russian politics, where Darrow sat on the stage, looking again to Mary "like a prisoner." When it was his turn to stand and talk, he mentioned Christian Rudovitz, the Russian dissident he had helped save seventeen years earlier, echoing her memory of the night they had first met in Chicago, at a rally to protest his extradition back to Moscow. Darrow had since been affected by the news that, in his freedom, Rudovitz had eventually wound up lost and homeless on the streets of Chicago. Even the happy stories, it seemed, became muddied and deflated with time.

They returned later that night to the Belmont, but Darrow's mood had darkened. He had barely smiled when Mary tried to entertain him with the tales she'd been told by the night watchman at the Russian meeting—who had whispered that, behind the stage, in the depths of a hall where opera divas had once sung, ghosts walked when the building fell black and empty.

The only ghost Darrow could imagine was the looming shadow of William Jennings Bryan. He seemed weighed down and marked by all he had seen in sixty-eight years of life. The brutal ordeal awaiting him in Dayton would, once more, test his conviction and endurance. It was incredible to think that, so soon after Leopold and Loeb, he should face another colossal trial.

Mary agreed that she would leave him to sleep alone. With Lem still in Canada she had made arrangements for Margaret to spend the night with friends, so that she might stay at the Belmont. Now she looked lost and a little pitiful at the change of plan. The apartment on East Eleventh Street was being rented out for the summer, and she could not get back to the country at such a late hour. In a pinch she could stay with her friend Freddie O'Brien and his new wife, Margaret, but they lived all the way across town. Mary wrote down Darrow's reaction in her diary the next day. "D. gave me a little silver—not enough—for taxi fare. Spent night with F. O'Brien."

Darrow could sometimes seem so mean, as much with money as in love, that Mary had to remind herself of his contrasting generosity in less expected moments. She thought of this paradox when final preparations were being made for the publication of *The Autobiography of Mother Jones*. Years before, when Darrow had seen the old activist wander out of his office in Chicago in the ragged clothes that she always wore, he had reached into his pocket and drawn out a crumpled wad of bills. His bunched fist held almost $100, which he wanted to give to Mother Jones as a gift. He knew she would not accept it in person, so Darrow gave the money to Mary and asked her to spend it on some new clothes for Mother Jones. Winter was coming in Chicago, and he did not like the thought of her shivering in the same worn dress and tattered shawl.

And so Mary forgave Darrow, again, and met him the following morning. On that sunny Friday, June 12, with her battered old lover looking more cheerful, they walked up to Central Park. They strolled through the vast green oasis in the heart of the city and "talked of many things, particularly the inroads the 'good' people made. Requires constant weeding, says Darrow, that the fragile flowers in the little garden of man's desert may appear. The danger to liberty is from the good people. The 'good' do not let law of natural selection and survival operate."

CHAPTER 10

MONKEY BUSINESS

Dayton, Tennessee, July 8, 1925

DARROW ARRIVED in Dayton looking as unkempt and weary as ever. He was different on the inside. There, hidden from the world but surging through him with gathering force, Darrow felt an intense resolve. He stepped off the train and was gaped at by a crowd half the size of the thousand fevered souls who had turned out to welcome Bryan the previous day. Shedding his dark wool jacket for shirtsleeves as soon as he reached Dayton, Bryan had worn a white pith helmet against the blistering sun. The politician, like a tongue-flicking lizard, basked in the heat.

"This is the day I have been waiting for!" Bryan exclaimed as he waved to his adoring followers.

Darrow also longed for the day. Here was his chance to bring down Bryan and all his reactionary claptrap. The national newspapers were caught up in the mood. That morning, in a typical front-page spread, the *Chicago Daily Tribune* produced a banner headline: "Scopes Case a 'Duel to Death.'" The subtitle below was hardly less sensational— "Bryan Arrives, Breathing Fire, Bible on Trial!"

With most churches in the South portraying Darrow as the Devil and Bryan as Jesus, the *Tribune* suggested that "Mr. Bryan seems to feel so sure of the outcome of this case that he is willing to risk the whole Bible and the entire Christian religion on the outcome. He has not tried a case in twenty-eight years, and he faces Clarence Darrow and a battery of first-class lawyers, but he offers no compromise, no hope for religion or for a divine authority if evolution wins."

Bryan, in his rallying call, warned that "if evolution wins, Christianity goes." Dayton, galvanized by the Great Commoner, appeared to believe him. The main road into town was festooned with banners that barely fluttered in the suffocating heat. Darrow saw them all, but he said little as he read the words with a sardonic eye. They came in the form of blunt questions (WHERE WILL YOU SPEND ETERNITY?), commercial guidance (YOU NEED GOD IN YOUR BUSINESS), and gentle beseeching (SWEETHEARTS, COME TO JESUS).

Dayton evoked much of what Darrow remembered from the small dusty town of his childhood in Kinsman, Ohio. Only Kentucky separated Ohio from Tennessee, and the same rituals of religious zeal and fearful ignorance rose up—just as they had done when he was a boy sixty years before. His passion for learning and knowledge, now symbolized by the modest figure of John Scopes, instilled a determination not only to defeat Bryan and the fundamentalists but to crush them.

Beyond the fact that he had not worked in a court of law for almost three decades, Bryan was a superficial and lazy thinker. He had become used to preaching to congregations that shouted out "Hallelujah!" or "Amen!" in echoing choruses of approval. Darrow sensed that Bryan had forgotten how to plot a methodical argument that might compel a nonbeliever to change his or her mind. He had grown soft and bloated. Darrow, in contrast, was hardened to courtroom jousting and intellectual battle. But he did not underestimate his messianic opponent.

He recognized the cold fury that still lurked deep within Bryan.

Darrow saw it reflected in the malice of Bryan's expression, when his face set against his opponents and his eyes glittered. He had the look of a man who believed in the utter certainty of hell for his enemies, while closing himself to forgiveness or understanding. Bryan did not trade in compassion or subtlety of thought. He preferred to deal in iron certainties and biblical retribution for those who questioned his belief.

Yet Bryan, certain of victory against Darrow, had been in an unusually jocular mood on his arrival. As he posed cheerfully for photographers he'd joshed: "No monkey-business, boys!" In fact, a carnival atmosphere swept through the whole town. Apart from the brightly colored flags and slogans draped over Main Street, most of the local stores featured images of monkeys and coconuts. In between the lemonade counters and hot-dog stands, erected especially for the trial, vendors sold "Your Old Man's a Monkey" buttons and *Hell and the High Schools* booklets. The diminutive author of that tract, T. T. Martin, bustled around Dayton, distributing leaflets mocking "Mass Meetings for Infidels, Scoffers, Atheists, Communists, Evolutionists and Others" or making impromptu street-corner sermons on the literal truth of the Bible.

Martin, who doubled as the field secretary of the Anti-Evolution League of America, had been interviewed by H. L. Mencken, the great columnist who championed Darrow and loathed Bryan. In describing Martin as the only Christian in Dayton "who did not perspire hate," Mencken showed rare benevolence toward a fundamentalist preacher. He chose instead to make gentle fun of Martin wearing, despite the weather, a long-tailed black coat and a gleaming white dog collar so large "that he could pull his head into it like a turtle." Martin was also an enthusiastic seller of his earlier publications—offering dog-eared copies of *God or Gorilla?* to curious customers.

J. R. Darwin's Everything-to-Wear store in the middle of town allowed the Anti-Evolution League to set up its temporary headquarters in the same building. A huge banner had been set up outside the store: DARWIN IS RIGHT—INSIDE. Martin had, in turn, helped spread the even larger slogan hanging across the courthouse wall—READ YOUR BIBLE DAILY.

A reporter for the *New Republic* suggested that "in Dayton, religion takes the place of golf, bridge, music, art, literature, the theater, dancing, clubs. Take religion away, and the desolation and distress would be pitiable to contemplate." Yet Mencken was almost captivated. "The town, I confess, greatly surprised me. I expected to find a squalid Southern village, with darkies snoozing on the horseblocks, pigs rooting under the houses and the inhabitants full of hookworm and malaria. What I found was a country town full of charm and even beauty."

Darrow soon recognized that, away from the blinkered minds, a simple kindness could characterize people in Dayton. Part of his own easy charm resided in the same lack of pretension and an abundance of plain-speaking generosity. He also knew not to take himself too seriously and so, when the Progressive Club asked him to speak on the night of his arrival, Darrow remembered his small-town roots with folksy humor. "For a while I was practicing law and playing poker on the side and I almost starved. But then I started playing poker and practicing law on the side, and I made enough money to go to Chicago to open an office."

The laughter confirmed that the joke had been good enough to allow his mention of the otherwise evil practice of poker playing. As long as there was a glimmer of wit in Dayton, Darrow retained his hope. He became more serious and suggested that "Scopes isn't on trial. Civilization is on trial. The prosecution is opening the doors for a reign of bigotry equal to anything in the Middle Ages. No man's belief will be safe if they win."

RUBY DARROW'S appearance in Dayton a day later was expected. She was always at the side of her famous husband during his most high-profile trials. And, despite the cracks running through their marriage, Ruby usually knew what was best for Darrow. She was horrified to discover that he had spent the night before, with the rest of his defense team, at the Mansion—as a large and normally deserted house was known in Dayton.

George Rappleyea, who had dreamed up the case, arranged for Darrow's boys to base themselves in the secluded Mansion. It was a picturesque setting that would appeal to the color writers and feature photographers who had already descended on Dayton. The Mansion would also give the defense the necessary space to prepare their courtroom strategy in lengthy nighttime meetings. Rappleyea let slip to a batch of reporters, as casually as he could, that local people believed the house was haunted. He would do anything to squeeze out just a little more publicity for the town.

Apart from some newly installed beds, a table, and chairs, the Mansion did not feature a fully functioning plumbing system. It did not take Ruby long to persuade a banker in Dayton that he should give up his home so that Darrow might be made comfortable for the duration of the trial. The banker's son, Howard Morgan, just happened to be one of the students who studied evolution in Scopes's classroom. Howard would be summoned by the prosecution as one of their witnesses against the young teacher. And so, whether out of guilt or simply because he had been bulldozed by Ruby, Howard's father found alternative accommodation for his family.

On Friday morning, July 10, Darrow and Ruby left the Morgan house and made the short walk to Rhea County Court. Dayton had been awake for hours, ever since a loudly clanging cowbell had been rung at 6:30 to rouse all the newspapermen staying at the Acqua Hotel. Over two hundred reporters had settled in town, with the most distant quartet hailing from Britain and the Continent—two from London and a journalist each from France and Germany. Dayton's pride in having sparked international interest was tempered only by the derision aimed at it by celebrated writers like H. G. Wells and George Bernard Shaw—two friends of Darrow's.

Having just won the Nobel Prize for Literature, Shaw wrote an article, "Where Darwin Is Taboo," for the *New Leader*. He sneered, "It is not often that a single State can make a whole continent ridiculous, or a single man set Europe asking whether America has ever really been civilized. Tennessee and Mr. Bryan have brought off the double event . . . [but] Tennessee and Mr. Bryan have had a nasty jolt. They have come up against a modern idea. Not a new idea, of course; only the idea of evolution."

Shaw's friendship with Darrow had begun twenty-two years before. On his honeymoon in Europe with Ruby, in August 1903, Darrow had sought out Shaw. They met in London, at the Langham Hotel, where the Darrows stayed, and Shaw wrote that, "for some instinctive reason I like [Darrow], perhaps because he is a genuine noble savage—with that cheekbone he only wants a few feathers and a streak of ochre to be a Mohican."

Darrow, a touch sadly, had seen the creased folds of his flesh and the bags under his eyes shroud those once memorable cheekbones. But, still a warrior, he was set on claiming the bald head of Bryan. He would do his scalping slowly and quietly. In keeping with his deceptive image, he ambled into court as if he had just completed a leisurely stroll down a lovely country road. Darrow did not mind that the attention of the packed room was focused less on him than the imminent appearance of his rival—for he knew Bryan was the undoubted star in Dayton.

As people had begun to cram the old brick courthouse since seven, the morning crispness had already given way to a stifling atmosphere in the second-floor courtroom. Seven hundred spectators were seated while another three hundred had found places to stand and watch as Darrow coolly removed his straw boater and then his jacket to reveal a dazzling combination of blue suspenders and a cream tie over his striped pastel cotton shirt topped by a gleaming white collar. But the applause was less for his dashing look than the court's first sight of the Great Commoner.

Bryan emerged at two minutes before nine. At sixty-five, he still cut an imposing figure. His chest was deep and wide, as if he were an ancient wrestler who had grappled with far more formidable foes than Darrow. There was a bulk about Bryan, with his gut straining against his short-sleeved shirt, that made his son, William Jennings Jr., and the rest of his prosecution team look puny in comparison. He gave off the imperious air of a man long accustomed to power and the deference of others as he strode through court.

His rapturous reception gave way to even louder cheering when he turned to Darrow. As if to prove that he could be as Christian in character as anyone, Bryan stretched out his hand. Darrow took it graciously, and the two men sat down and spoke with some of their old affection. It would not last, but in that moment, two enemies could remember that they had once worked together for the same political party and held briefly similar ideals. As they spoke quietly together Darrow sat to the right of Bryan, his hands clasped in front of him but with his fingers noticeably not joined together. He was not about to pray for divine assistance. Instead, he looked calm, if mildly pensive, with an unlit cigarette jutting from his right hand. Judge John Raulston had warned him, and all the other attorneys, that smoking was banned in his courtroom.

Bryan's left hand held a palm-leaf fan, in readiness for the heat that had already made Darrow's short new haircut look damp against his brow. The Commoner had also prepared himself by removing his tie and tucking his collar inside his shirt. In this way, exposing his shoulder blades and the base of his neck to the open air, he looked as if he were wearing a collarless shirt. Bryan feigned a bluff exterior, but he revealed his nerves when he occasionally rubbed his bald head as if to reassure himself.

The exchange between the two men seemed studiously amiable. They did not appear ready to tear each other's hearts out. Darrow did most of the talking, his head inclined toward Bryan, and he caused his opponent to laugh. It was hard to detect his real feelings toward Bryan—a man he derided as the leader of a cult called "Morondom." There was no point, however, wasting any animosity during the pretrial niceties. Darrow needed to conserve himself, because he had taken a great risk in coming to Dayton, where the weight of opinion was stacked so solidly against him. He had managed to overturn the odds a year earlier, during his defense of Leopold and Loeb, but his restored reputation could be shredded by a humiliating defeat to the evangelists of Tennessee.

Judge Raulston was accompanied to his bench by a pair of tall policemen carrying palm fans that were far larger than that chosen by Bryan. Throughout the trial it would be the task of the two officers to twitch and swish the fans near Raulston's high-backed chair so that they could at least attempt to protect him from the lacerating heat. A bank of electric fans was also turned in the direction of the judge, and Bryan's prosecution table. Yet the defense team were afforded none of that cooling air as they sat in the sunbaked section of the room, beyond the limited reach of the fans. Darrow did not try to compensate for the disadvantage. He saw that everyone who pumped their papers as fans merely seemed to expend energy, which made them perspire still more.

His table, adjacent to Bryan and the prosecution, was on a stage about a foot lower than the platform holding both the jury box and the judge's bench. In contrast to most other courtrooms, Rhea County had altered the traditional seating plan. On entering the wooden double doors into the court, the judge's bench was found to the right—rather than at the front of the room. Raulston would face the lawyers and the public gallery to his left with row upon row of wooden chairs nailed to the floor and laced to one another by black wrought iron. Eighteen massive windows were set back deep enough in the foot-thick walls to offer up wide ledges already filled with eager spectators.

As Raulston allowed the photographers more time than they needed to take their pictures of him posing with his gavel, it looked to Darrow as if only John Scopes expected nothing for himself. The young defendant, sitting at Darrow's table, tried to avoid the cameras as best he could and sat with his head down as he waited for his trial to begin.

They were all asked to bow their heads. During the ensuing long prayer, Reverend Cartwright of Dayton cried out: "Oh God, our divine father, we recognize thee as the supreme ruler of the universe, in whose hands are the lives and destinies of all men, and of all the world." Darrow, in his role as a politely dubious agnostic, raised his gaze less to the heavens than the magnificent black roof that soared twenty feet above their heads. He listened patiently as the prayer filled the passing minutes. And when he looked down again, he saw that almost a thousand heads were still lowered in supplication—with the exceptions being a few newspapermen and his partners in defense. Arthur Garfield Hays of the ACLU and Dudley Field Malone, Darrow's friend and a lawyer more used to handling divorce litigation in Parisian courtrooms, stared steadily ahead.

Malone cut a debonair and handsome figure, in his elegantly formal suit, while the short and stocky Hays occupied a more concentrated presence. Hays, clearly, focused on the intricacies of a complicated legal hearing rather than any sense of individual style. His bustling grasp of the more arcane and tedious details of law meant that he was an astute choice to manage the defense strategy—while his keen literary sensibility and passion for civil liberties, which had led to his work with the ACLU, gelled with the interests of both Malone and Darrow.

With a reputation spanning international divorce law and an impassioned record of legal support for the women's suffrage movement, Malone himself brought an intriguing mix to the team. As a member of the Democratic Party he had also, in a curious irony, worked as an assistant to Bryan when the Great Commoner served as secretary of state in Woodrow Wilson's government from 1913 to 1915. Despite that relatively conventional backroom role, Malone had a real taste for grandstanding radical politics; five years later, in 1920, he had attempted to become the governor of New York on a combined farming and labor ticket. Malone had failed with that left-field electoral bid, but he was still determined to stand up to the reactionary right in America.

All three defense lawyers were soaked with perspiration as the court threatened to become unbearable during the midmorning heat. Judge Raulston, in recognition of the fact, had already confirmed a "coats off" policy in court. Darrow was more impressed by the single-mindedness of Malone, who had promised to wear his double-breasted jacket buttoned up throughout the trial—no matter how high the temperature climbed. The reporters, like Darrow, were less rigid and they sat in shirtsleeves, pencils tucked behind their ears as they waited. Their typewriters and telegraph machines, which had been clattering away noisily before nine, were silent. Darrow, with a wink of recognition, saw that Mencken, his old iconoclastic friend, had found himself a place not far from the defense. Mencken pulled an amused face as Reverend Cartwright's prayer thundered to a close with the words "God help us be loyal to God, and loyal to truth."

Using his gavel as an instrument of law rather than just a handy prop for the photographers, Raulston rapped on his desk and called for the case of "The State of Tennessee versus John Thomas Scopes." As a special grand jury had to confirm first that the defendant still had a case to answer, Raulston, a devout Baptist, announced that he would provide his instructions by reading out both the Butler Act and the opening thirty-one verses of the Bible.

Raulston read with portentous gravity from his well-thumbed Bible, as if the spirit of the Lord had entered him. He paused dramatically after the twenty-fourth verse of the first chapter of Genesis. The truth, as he and the State of Tennessee believed, would unfold in the next three verses:

And God made the beast of the earth after his kind, and cattle after their kind, and everything that creepeth upon the earth after his kind: and God saw that it was good.

And God said, Let us make man in our image, after our likeness; and let them have dominion over the fish of the sea, and over the fowl of the air, and over the cattle, and over all the earth, and over every creeping thing that creepeth upon the earth.

So God created man in His own image, in the image of God created He him; male and female created He them.

The indictment against Scopes was duly confirmed after an hour-long recess, and the grand jury stepped down. Yet before Darrow could question their prospective replacements, who were all hopeful of being chosen for the trial jury in order to win one of the twelve best seats in the house, the judge announced that they would break for lunch at 11:30.

Spectators, enjoying the holiday mood, spread picnic blankets on the neat square of green lawn outside the courthouse. In the welcome shade, offered by the surrounding oak and sweet-gum trees, they tucked into their hampers of food while listening to the passing preachers, one of whom kept moaning out loud, "For I say unto you, repent and be baptized!" A motionless blind singer, meanwhile, sang hymns in a throaty growl and accompanied himself on a small hand organ. For those who had not brought their own lunch, barbecues smoked and sizzled along the length of an adjoining pit that had been dug for the purpose. It was almost fifty feet in length, and four feet in width, and provided strong competition to the cheaper hot-dog stands.

Inside the courtroom that, according to Mencken, had "the atmosphere of a blast furnace," the typewriters clattered into action. Even the telegraph wires began to whir again as copy for the evening papers was prepared. Yet there was meager excitement during the afternoon session as Darrow quizzed the aspiring jurors about their attitude toward evolution. Most of them, despite their hopes of being selected, could not hide their antipathy.

J. R. Massingill, a preacher, was among the early candidates. "Did you ever preach on evolution?" Darrow asked.

"Yes," Massingill replied, "in connection with other subjects. I have referred to it in discussing them."

"Against it," Darrow asked bluntly, "or for it?"

"I'm strictly for the Bible."

"I am not talking about the Bible," Darrow sighed. "Did you preach for or against evolution?"

Massingill, with his hold on a precious place in the jury slipping away, looked to Raulston for help. "Is that a fair question, Judge?"

"Yes," the judge conceded. "Answer the question."

"Well, I preached against it, of course."

The crowd, rejuvenated by their lunch, applauded happily. Darrow shook his head. Massingill was struck off his list.

He was more sympathetic to Jim Riley. "Do you know anything about evolution," Darrow asked the farmer.

"No, not particularly," Riley said.

"Heard about it?"

"Yes," Riley nodded. "I've heard about it."

"Ever heard anyone preach any sermons on it?"

"No, sir."

"Ever read anything about it?"

"No, sir," Riley repeated before, after a brief pause, adding, "I can't read."

"Is that due to your eyes?" Darrow asked.

"No," Riley said quietly. "I am uneducated."

After Riley had been accepted as a jury member, Hays suggested that his admission to illiteracy "was said with such plain, simple dignity that we felt we had at least one honest man."

Riley and Massingill's diverse encounters with Darrow were the muted points of interest in a tedious afternoon—at the end of which, if nothing else, the dozen men to sit in judgment of Scopes had been chosen. Nine of the twelve jurors were farmers. Two of the remaining men, a teacher and a shipping clerk, worked as part-time farmers. The jury foreman, Jack R. Thompson, meanwhile, was a former U.S. marshal turned farm-owner. Almost all the jurors were bound up in the church. There were six Baptists, four Methodists, and one member of the Campbellite Disciples of Christ. Only W. F. Roberson, one of the full-time farmers, failed to announce his specific religious allegiance—although he attended church occasionally and planned to visit more often in the future.

"It was obvious after a few rounds that the jury would be unanimously hot for Genesis," Mencken wrote. "The most that Mr. Darrow could hope was to sneak in a few men bold enough to hear the evidence against Scopes before condemning him. Such a jury, in a legal sense, may be fair, but it would certainly be spitting in the eye of reason to call it impartial."

As the judge announced an early recess to a downbeat day, meaning the trial would only begin in earnest on Monday, Mencken found an alternative way to kick off the weekend. Approached at the Acqua Hotel a few hours later by Nellie Kenyon, a young and pretty reporter who wanted to interview him for the *Chattanooga Times*, Mencken asked her what he might do for entertainment on a Friday night in Dayton.

Kenyon told him that there was a small movie house, the drugstore, a baboon in a cage down the street, and a Holy Roller meeting on the outskirts of town.

"What's a Holy Roller meeting?" Mencken asked.

"A meeting of religious people," Kenyon said.

"How about taking me tonight?" Mencken suggested.

As the light faded from Dayton, Mencken and Henry Hyde, his colleague from the *Baltimore Evening Sun*, were driven down a bumpy dirt track in Kenyon's spluttering Model T Ford. After a couple of miles, on the edge of a wooded glade in the foothills of the Smoky Mountains, they parked the car and approached the kerosene lamps dangling from some distant trees. They could see sinister shapes in the gloom.

Mencken was transfixed by the sight of a tall, thin mountain man preaching to a small gathering. Dressed in dungarees, the preacher hollered out praise to God while men and women rolled on the ground, panting and gasping. The mountain man then shrieked, as if scalded, and dived into the prone heap of his informal congregation. A fascinated Mencken spent the next few hours watching the wild abandonment of the Holy Rollers.

He was hardly less startled when, on their return to town around eleven that evening, they discovered that the courthouse lawn was still packed with people listening reverently to the words of some visiting evangelists. It was unlike any Friday night Mencken had ever experienced in Baltimore or New York. The town was so soaked in religious mania it seemed as if not even Darrow, one of the few men Mencken truly admired, could stem the fervor. All the reason of science would surely be swamped in the backwoods of Tennessee.

THE WEEKEND passed in a blur. While Darrow and his team toiled in the Mansion, Scopes allowed himself the devious luxury of attending a Saturday night dance. He took to the floor with most of the young girls who caught his eye and then agreed to accompany one of his partners back to her hotel. She claimed to be too frightened to make the walk on her own. And there, in the dark, Scopes was caught in a modern honey trap. As the girl threw her arms around his neck, and started kissing him, Scopes was "momentarily paralyzed"— less by the kiss than the flaring pop of a flashlight as a photographer snapped his exclusive shot. The bespectacled teacher, an unwitting celebrity in a trial that captured the American imagination, had fallen for a newspaper setup.

Bryan, meanwhile, preached twice on Sunday, with the *Chicago Daily Tribune* reporting, "In the spirit of 'Onward, Christian Soldiers,' Dayton girded itself today for the war against evolution and liberalism which will begin in earnest with the resumption of the trial of John T. Scopes. The sun seemed to stand still in the heavens, as for Joshua of old, and to burn with holy wraith against the invaders of this fair Eden of fundamentalism. It was the most hectic Sabbath that Dayton has ever known."

Darrow, coolly and stealthily, confronted the forces of fundamentalism on Monday afternoon. After another monotonous morning as John Neal, his colleague, argued pedantically that the Butler Act violated the freedom of both religion and speech, Darrow opened with a smile. He had noted the sniping comments of Ben McKenzie, Bryan's assistant on the prosecution, against the big-city lawyers of the defense from New York and Chicago. But he was more amused that everyone in court referred to him as "Colonel Darrow"—or just plain old "Colonel."

After he had pushed back his chair and wiped the sweat from his brow, Darrow tugged on his lavender suspenders as he trudged across court. A stray dog had settled down near the front, to take a nap in the withering heat, and Darrow was careful not to tread on his tail. He gestured toward the prosecution as, charmingly, he began one of the great speeches of his career.

"I know my friend, McKenzie, whom I have learned not only to admire but to love in our short acquaintance, didn't mean anything in referring to us lawyers who come from out of town. For myself, I have been treated with the greatest courtesy by the attorneys and the community."

As the heaving room murmured appreciatively, the judge rapped his gavel. "No talking, please, in the courtroom."

Darrow did not mind the interruption. "I shall always remember that this court is the first one that ever gave me a great title of 'Colonel.' I hope it will stick to me when I get back north."

A solemn Judge Raulston leaned forward. "I want you to take it back home with you, Colonel," he urged.

"That is what I am trying to do," Darrow said with another small smile before he glanced at Bryan and his son who were, he was about to remind the court, based respectively in Miami and Los Angeles. "So far as coming from other cities is concerned, why, Your Honor, it is easy here. I came from Chicago, and my friends, Malone and Hays, came from New York. And on the other side we have a distinguished and very pleasant gentleman who came from California . . ."

He nodded to William Jennings Bryan Jr. and then, his face darkening, pointed to the young man's father. "And another who is prosecuting this case, and who is responsible for this foolish, mischievous, and wicked act, who comes from Florida."

Darrow shrugged at the stupidity of the man he needed to defeat. "The case we have to argue is a case at law, and hard as it is for me to bring my mind to conceive it, almost impossible as it is to put my mind back into the sixteenth century, I am going to argue as if it was serious, and as if it was a death-struggle between two civilizations."

Considering the statute itself, Darrow suggested that "it is full of weird, strange, impossible and imaginary provisions. It says you shan't teach any theory of the origin of man that is contrary to the divine theory contained in the Bible. Now let us pass up the word 'divine'! No legislature is strong enough in any state in the Union to characterize and pick any book as being divine. Let us take it as it is. What is the Bible?"

Even the wailing babies near the back of the court, and the chugging clatter of the passing trains, fell silent. It was as if all Dayton had paused to listen to the great Infidel. Darrow, however, looked less like the Devil than a kind and thoughtful grandfather, the sort of plain man they admired round these parts. "Your Honor," he said of the Bible, "I have read it myself. I might read it more or less wisely.

Others may understand it better. Others may think they understand it better when they do not. But in a general way I know what it is. I know there are millions of people who look on it as a divine book, and I have not the slightest objection to it. I know there are millions of people in the world who derive consolation in their times of trouble, and solace in times of distress, from the Bible. I would be pretty near the last one in the world to do anything to take it away."

The pain and the strife of life, as well as its rare pleasures and humors, were mapped across Darrow's weathered face. He did not look so different from a farmer or a factory worker. And so they listened attentively as Darrow continued. "I feel exactly the same toward the religious creed of every human being who lives. If anybody finds anything in this life that brings them consolation and health and happiness, I think they ought to have it. I haven't any fault to find with them at all.

"But what is it? The Bible is not one book. The Bible is made up of sixty-six books written over a period of about a thousand years, some very early and some of them comparatively late. It is a book primarily of religion and morals. It is not a book of science. Never was and was never meant to be. . . . It is not a book on geology. They know nothing about geology. It is not a book on biology. They know nothing about it. It is not a work on evolution; that is a mystery."

As Darrow changed tack, he noted, "Here, we find today as brazen and as bold an attempt to destroy learning as ever was made in the Middle Ages. The only difference is we have not provided that they shall be burned at the stake. . . . But there is time for that, Your Honor. We have to approach these things gradually."

His cynicism was biting, but Darrow remained in control. "There is nothing else, Your Honor," he suggested calmly, "that has caused the difference of opinion, of bitterness, of hatred, of war, of cruelty, that religion has caused. With that, of course, it has given consolation to millions. But it is one of those particular things that should be left solely between the individual and his Maker, or his God, or whatever takes expression with him, and it is no-one else's concern. . . . Your life and my life and the life of every American citizen depends after all upon the tolerance and the forbearance of his fellow man. If men are not tolerant, if men cannot respect each other's opinions, if men cannot live and let live, then no man's life is safe. No man's life is safe."

Darrow almost shuddered at his repetition before he returned to the crux of the trial. "Here is the state of Tennessee going along its own business, teaching evolution for years," he said quietly, before turning and pointing fiercely again at Bryan, "and along comes somebody who says, 'We have to believe it as I believe it. It is a crime to know more than I know.' And they publish a law inhibiting learning. That law makes the Bible the yard-stick to every man's intellect, to measure every man's intelligence and learning. Nothing was heard of all that until the fundamentalists got into Tennessee."

Mencken stood on a table on the side of the courtroom so that he could see Darrow more clearly above the throng. He had the look of a wise old bloodhound, sniffing out the scent of Darrow's prey, as he lifted his huge head above everyone else. In between using a big white handkerchief to mop his brow, Mencken kept his hands clasped behind his back as he rocked in solemn appreciation of his friend's work. For Mencken it was as if the truth resided in the world-weary but endlessly moving ache in Darrow's voice.

"You have but a dim notion of it, you who have only read it," Mencken wrote of Darrow's speech. "It was not designed for reading, but for hearing. The clang-tint of it was as important as the logic. It rose like the wind and ended like a flourish of bugles. The very judge on the bench, toward the end of it, began to look uneasy."

Darrow had fixed his eye on Raulston, and his head tilted toward the judge as if to close the distance between them, an agnostic from Chicago and a Baptist in Dayton.

"Sorry to interrupt you," Raulston said, "but it is adjourning time."

"If I may," Darrow retorted, "I can close in five minutes."

"Proceed tomorrow," Raulston snapped.

Darrow stared him down, and kept talking. "Your Honor knows the fires that have been lighted in America to kindle religious bigotry and hate," he warned as the judge sank back in defeat. "If today you can take a thing like evolution and make it a crime to teach it in the public school, tomorrow you can make it a crime to teach it in the private schools, and the next year you can make it a crime to teach it to the hustings or in the church. At the next session you may ban books and the newspapers. Soon you may set Catholic against Protestant and Protestant against Protestant, and try to foist your own religion upon the minds of men. If you can do one you can do the other.

"Ignorance and fanaticism is ever busy and needs feeding. Always it is feeding and gloating for more. Today it is the public school, tomorrow the private. The next day the preachers and the lecturers, the magazines, the books, the newspapers. After a while, Your Honor, it is the setting of man against man and creed against creed until with flying banners and beating drums we are marching backward to the glorious ages of the sixteenth century, when bigots lighted fagots to burn the men who dared to bring any intelligence and enlightenment and culture to the human mind."

When the final bugle had sounded, and the last word of a telling speech had been digested, Dayton had been floored by Darrow. He had faced them down and forced them to think. Ben McKenzie, one of his rivals at the prosecution table, was an obvious example. In court he wrapped his arm around Darrow and said huskily, "That was the greatest speech I ever heard in my life—on any subject."

For Marcet Haldeman-Julius, a writer from Kansas, and one of the few women reporting on the trial, Darrow now "stands a giant among mental pygmies . . . one of the few really great men in America."

Later that night, as if in confused celebration, a young chimp from a circus in Atlanta was paraded down Main Street in Dayton. His owner had dressed him in a suit and given him a cane. The small monkey sauntered down the street, sometimes stopping to scratch his head, as Dayton gave in just a little bit more to the madness of the Scopes trial.

The fundamentalists took more serious note of the fact that, an hour later, the town experienced a sudden power failure and was plunged into darkness. Water had already dried up from the taps but, in the pitch-black of night, the lack of light seemed symbolic to God-fearing folk. "The wrath of God has visited us!" a preacher yelped as the Holy Rollers lit their kerosene lamps.

Yet God had not lashed out against the monkey mayhem or even the speech of Darrow the Infidel. There was a simpler explanation. The axe of a careless worker had severed a water main and, to rectify the fault and avoid electrocuting the repairmen, the power supply had been turned off. The lights soon came back on—but the more sensible monkey from Atlanta had already disappeared into the night.

T HE FOLLOWING day, across the country, with the glowing newspaper reports from Dayton spread out around her, Mary Field Parton reached for her diary. There had been times during the trial when the comparative banality of her own forty-seven-year-old life appeared to smother her as she mutated, rather than evolved, into a more mechanical form from the unknown future. "I am a Robot: a washing, cooking, ironing machine, fed regularly, oiled a little," she had written a few weeks earlier. "All night to fight phantoms. Not a robot by night—but a defeated, broken lump. Oh, if only Margaret didn't need me."

She had been infuriated rather than dejected two days before, on Sunday afternoon, when with Lem back from Canada, they had entertained the Tonetti family, who had sold them the old farmhouse in Sneden's Landing. Mary Tonetti and her husband, the famous Italian sculptor Francois Tonetti, held on to most of the houses—renting them out to artists and writers they liked. But, on an apparent whim, the Tonettis had decided to sell to the Partons. Mary liked Mrs. Tonetti because she was a woman of formidable intelligence who had worked in the past with the great architect Stanford White. She was also fond of the Tonettis' daughter, Lydia, but enraged by their doltish son-in-law. He set Mary's frayed nerves on edge by revealing his utter ignorance of both Darwin and the compelling trial unfolding in Tennessee. "Mrs. Tonetti, an intellectual, explains what evolution means to him—a puckish, red-haired man with a high-flute voice as unmodulated as a tom-boy. He doesn't know anything of the trial in Dayton, of Clarence Darrow!"

Her disbelief had since been replaced by a renewed sense of awe on that sultry Tuesday morning of July 14, 1925. "Darrow made a wonderful plea for truth," she wrote, "for at least the scientific latitude which is freedom to seek truth. A haunting climax to a beautiful life—rising from a particular case to a defense of the abstract and the general. Defending not the little scientist but the laboratory; not the 'criminal' who taught, but the 'crime' of teaching in freedom. A great voice crying out . . ."

BRYAN WAITED for his moment to strike. Even when Darrow challenged Judge Raulston early that same morning, and protested against the Tennessee ritual of beginning each day in court with a prayer as prejudicial to the defense, he simply fanned his pinched face. Bryan's colleagues, Ben McKenzie and Tom Stewart, led the counterattack. The Great Commoner knew that Darrow was always going to be denied and so he reserved his strength.

All through Tuesday and much of the following morning, Bryan sat back as the arguments raged. He was unsurprised by Raulston's ruling in answer to Darrow's request that the charge be quashed on the grounds that the Butler Act was unconstitutional. As he swished his fan, Bryan smirked as he listened to the words. "I fail to see," Raulston said in his singsong whine, "how this act in any way interferes or, in the least, restrains, any person from worshipping God in the manner that best pleaseth him."

Bryan enjoyed the judge's biblical rhetoric—nodding intently at every "thou" or "pleaseth"—for he knew how it irked Darrow with his preference for ordinary English. The Commoner just about managed a sneer when Darrow said, "Of course, the weather is warm and we may all go a little further at times than we ought. But Mr. Stewart is perfectly justified in saying that I am an agnostic, for I am. And I do not consider it an insult but, rather, a compliment. I do not pretend to know where many ignorant men are sure. That is all agnosticism means."

And even when Malone, Darrow's right-hand man from New York, tried to goad Bryan by showing how he had fallen from the days when he advocated the search for truth, the politician turned preacher rebuffed him with a twitch of his fan. "When the proper time comes," Bryan said, "I shall be able to show the gentleman that I stand today just where I [once] did—but this has nothing to do with the case at the bar."

His supporters, cramming the court, whooped and applauded—and the judge did little to quell their delight. Bryan, "silent and grim since the trial began," the *New York Times* noted, "looked upward for an instant . . . his nose in the air, like a warhorse ready for battle."

After the prosecution had called Superintendent White, of Rhea County School, to confirm that Scopes had used Hunter's *Civic Biology* to teach the theory of evolution, two of his students were summoned. Howard Morgan, the eager freshman whose family home had been given up to Darrow, testified that Scopes had once discussed evolution with them in class. "Well," Darrow wondered, "did he tell you anything else that was wicked?" "No, not that I remember," the boy said to widespread laughter.

Morgan then explained the theory of man's evolution, as taught by Scopes. "The earth was once a hot, molten mass, too hot for plant or animal life to exist upon it. Then the sea cooled off the earth and there was a little germ of a one-cell organism which formed. And this organism kept evolving until it got to be a pretty good-sized animal and then come on to be a land animal and it kept on evolving—and from this came man, and that man was just another mammal."

Ben McKenzie, the joker in the prosecution pack, jumped up to say that, finally, he understood evolutionary theory: "God issued some sort of protoplasm or soft dishrag and put it in the ocean and said, 'Old boy, if you wait around six thousand years I will make something out of you.'"

A less cheery senior, Harry Shelton, confirmed that they had used Hunter as a textbook during biology revision under Scopes. Yet, as Darrow prompted him, Shelton admitted that his exposure to *Civic Biology* had not persuaded him to stop attending church every Sunday. His life had remained utterly unchanged.

Frank Robinson, the owner of the drugstore in which the plot to challenge the law had been developed, took the stand next. As he testified that Scopes himself admitted he had violated the law by teaching Hunter's book, Darrow pointed out the immediate contradiction. "You were a member of the school board," he reminded Robinson, "[but] you were selling [the book], were you not?" While the crowd in court chuckled, Darrow waved airily at Doc Robinson as if the entire charade against Scopes had been exposed again.

"You are not bound to answer these questions," Darrow grinned.

Even the prosecution enjoyed the joke. "The law says teach, not sell," Attorney General Stewart quipped.

There appeared to be little point in calling for further testimony and so Stewart, after reiterating that there were many others willing to make the same points against Scopes, rested the prosecution case. Less than an hour had passed since he had summoned his first witness that scalding Wednesday afternoon.

When Darrow called Maynard M. Metcalf, the eminent zoologist, as the defense's first expert witness, Stewart reminded his opponent that "we have a rule in this state that precludes the defendant from taking the stand if he does not [do so] first." But Darrow had no intention of calling John Scopes at any point during the trial. He would not testify because, as Darrow insisted, "Your Honor, every single word that was said against this defendant, everything was true."

The absurdity of the case against Scopes was heightened still further—prompting Mencken to be the first to tag it "The Monkey Trial."

John T. Scopes, the hapless defendant, the innocent young teacher guilty only of spreading knowledge to his students in Tennessee, cut an increasingly mute and isolated figure in court. He was not quite a monkey in a cage but still, as he wrote later, "I sat speechless, a ringside observer at my own trial, until the end of the circus."

CHAPTER II

A DUEL TO THE DEATH

Rhea County Court, Dayton, Tennessee, July 16, 1925

"I T WAS PLAIN to everyone, when Bryan came to Dayton, that his great days were behind him, that he was definitely old and headed at last for silence," H. L. Mencken wrote. "Hour after hour he grew more bitter. What the Christian Scientists call malicious animal magnetism seemed to radiate from him like heat from a stove. From my place in the courtroom, standing upon a table, I looked directly down upon him, sweating horribly and pumping his palm-leaf fan. His eyes fascinated me; I watched them all day long. They were blazing points of hatred. They glittered like occult and sinister gems. Now and then they wandered to me, and I got my share. It was like coming under fire."

Mencken and Darrow embraced a theory as to the source of that apparent hatred. It went beyond evangelical fury and reached back to the days of politics, the days when Bryan had waged three presidential campaigns and yet been denied a place in the White House each time. His power in America, once so close to being absolute, had diminished. Even his most recent attempts to find office had failed.

He had not been elected to the Senate by the people of Florida, nor to the position of moderator by the Presbyterian Church. Bryan had turned instead to fundamentalist religion as a way of reclaiming adulation and hitting back at those rivals who undermined him.

"One day it dawned on me," Mencken claimed, "that Bryan, after all, was an evangelical Christian only by a sort of afterthought—that his career in this world, and the glories thereof, had actually come to an end before he ever began whooping for Genesis. So I came to this conclusion: that what really moved him was a lust for revenge. The men of the cities had destroyed him and made a mock of him; now he would lead the yokels against them. . . . The hatred in the old man's burning eyes was not for the enemies of God; it was for the enemies of Bryan."

That scalding Thursday afternoon, having glowered silently for almost a week, Bryan declared himself ready to address the court. It was time to cut loose against Darrow. The previous afternoon, when Darrow questioned Maynard Metcalf, the zoologist, Bryan had flushed angrily. After Metcalf had provided, in Mencken's words, "one of the clearest, most succinct and most eloquent presentations of the case for the evolutionists that I have ever heard," Darrow asked him a simple question that probed the literal truth of the Bible: "Is the earth more than 6,000 years old?"

"Well, I think 600 million years would be a more moderate answer," the scientist answered laconically.

Judge Raulston had hardly been less shocked and called for a glass of water. The only consolation for Bryan had been the decision to banish the jury from court during Metcalf's cross-examination, for a final ruling had yet to be made whether scientific evidence would be admissible. As Raulston deliberated, Bryan prepared to intervene. Rumors that the Great Commoner was about to talk had swollen the courtroom with people. Those who were unlucky enough to be denied entry had been forced out across the hallway and down the stairs.

Bryan had dressed for the occasion. The deep purple of his shirt was accentuated by a black bow tie. A matching purple passion seemed to lift him, and anticipation rocked the court so powerfully that Raulston issued a warning: "The floor on which we are now assembled is under great weight. I do not know how well it is supported but sometimes buildings and floors give way when they are unduly burdened. So I suggest to you to be as quiet in the courtroom as you can. Have no more emotion than you can avoid. Especially have no applause, because it isn't proper in the courtroom."

There was little chance the delirious throng would heed that instruction. Determined to exult in Bryan's address, they were not disappointed. "This is that book," Bryan roared, waving a copy of Hunter's *Civic Biology* as if it was a deadly weapon he might smash against Darrow's head. "There is the book they were teaching your children that man was a mammal and so indistinguishable among the mammals that they leave him there with 3,499 other mammals—including elephants! Talk about putting Daniel in the lion's den! How dare those scientists put man in a little ring like that with lions and tigers and everything that is bad . . . shutting man up in a little circle like that with all these animals that have an odor that extends beyond the circumference of this circle!"

As people honked with laughter, Darrow wrinkled his nose in distaste. The stench of humanity, at least in Dayton, did not resemble the uplifting aroma of superior beings.

"The Christian believes man comes from above," Bryan insisted, "but the evolutionist believes he must come from below." Turning in derision to *The Descent of Man*, Bryan revealed that Darwin had charted the evolution of apes and monkeys, which he grouped together as Simiadae. "They then branched off into two great stems—the New World and the Old World monkeys. And from the latter, at a remote period, man, the wonder and glory of the universe, proceeded."

Bryan spread his arms wide so that dark rings of perspiration were visible against the shimmering purple of his shirt. Man had apparently evolved, he snorted, "not even from American monkeys, but from Old World monkeys!"

The creaking courtroom erupted again. Judge Raulston was too immersed in the jubilation to demand a return to order. Bryan, the ringmaster, assumed control. "Now we have our glorious pedigree! And each child is expected to copy the family tree and take it home to his family to be substituted for the Bible family tree. That is what Darwin says!"

Bryan turned toward Darrow. His esteemed rival understood the dangers. The previous summer, amid his desperate quest to save Leopold and Loeb, Darrow had alluded to the insidious way in which the killers had been warped by the ideas of Nietzsche and, by extension, Darwin. Darrow sighed as Bryan began to quote liberally from his address during the Chicago trial. With selective editing he presented a scenario in which Darrow blamed Nietzsche's philosophy, apparently sparked by Darwin's belief in the survival of the fittest, for the murder of Bobby Franks.

"Mr. Darrow is the greatest criminal lawyer in America today. His courtesy is noticeable—his ability is known—and it is a shame, in my mind, in the sight of a great God, that a mentality like his has strayed so far from the natural goal that it should follow. Great God, the good that a man of his ability could have done if he had aligned himself with the forces of right instead of aligning himself with that which strikes its fangs at the very bosom of Christianity."

The courtroom muttered its assent. "Now, my friends," Bryan said before glancing at the judge. "I beg your pardon, if the court please. I have been so in the habit of talking to an audience instead of a court that I will sometimes say 'my friends . . .'"

He gestured at Darrow and his colleagues. "Although I happen to know not all of them are my friends!" The crowd roared again and Bryan motioned for silence.

"When it comes to Bible experts," he said as he gazed at the prosecution, "do they think that they can bring them in here to instruct members of the jury? The one beauty about the Word of God is [that] it does not take an expert to understand it."

As cries of "Amen!" reverberated around him, Bryan's voice rose with conviction. "These people come in from outside the state and force upon the people of this state, and upon their children, a doctrine that refutes not only their belief in God but their belief in a Saviour and belief in heaven and takes from them every moral standard that the Bible gives us. . . . The facts are simple, the case is plain, and if these gentlemen want to enter upon the field of education . . . then convene a mock court for it will deserve the title of mock court if its purpose is to banish from the hearts of people the Word of God."

The court applauded as Bryan returned to the prosecution table. Individual people, some with tears in their eyes, offered Bryan their thanks and congratulations. Darrow murmured his disbelief to Arthur Garfield Hays: "Can it be possible this trial is taking place in the twentieth century?"

Two seats away from Darrow, Dudley Field Malone gathered himself in studied silence. With his arms crossed in defiance over his double-breasted jacket, Malone waited for the hubbub to quiet down. His intelligent, jowly features were set in fierce determination. Malone was ready to shock the court into a new way of thinking.

While Bryan fanned his sodden face and waved to admirers, Malone slowly unbuttoned his jacket. Courtroom spectators hushed, watching Malone loosen the clothes he had wrapped so tightly around himself for five days in a stifling court. He finally removed his jacket and folded it over the back of his chair. He had transfixed an entire courtroom without saying a single word.

Then, for the next thirty minutes, Malone spoke with such controlled fury that he made Bryan's speech seem forgettable. He began quietly—"Whether Mr. Bryan knows it or not, he is a mammal, he is an animal, and he is a man"—but his voice soon boomed. "Are we to have our children know nothing about science except what the church says they shall know? I have never seen harm in learning and understanding, in humility and open-mindedness. And I have never seen clearer the need of that learning when I see the attitude of the prosecution, who attack and refuse to accept the information and intelligence which witnesses will give them."

Malone's words were made even more compelling by his reminder that he had worked under Bryan in the State Department. "My old chief," he said with regret, "I never saw him back away from a great issue before. I feel that the prosecution here is filled with a needless fear. I believe that if they withdraw their objection and hear the evidence of our experts, their minds would not only be improved, but their souls would be purified."

There was no gasp of outrage because Malone, by now, had Dayton in his grip. "When he began the people were against him," one reporter suggested, simply because he came from New York. "They do not like his clothes nor his accent. But he had not been speaking more than ten minutes before there was a change."

Malone spoke such sense that it was impossible to resist the logic of his argument. "The least this generation can do," he argued, "is to give the next generation all the facts, all the available data, all the theories, all the information that learning, that study, has produced. Give it to the children in the hope of heaven that they will make it a better world of this than we have been able to make it. We have just had a war with twenty million dead. Civilization need not be so proud of what the grownups have done. For God's sake let the children have their minds kept open. Close no doors to their knowledge. Shut no door for them!"

His voice reverberated, as loud as it was certain. "We are not afraid! Where is the fear? We meet it! Where is the fear? We defy it! We ask Your Honor to admit the evidence as a matter of correct law, as a matter of sound procedure, and as a matter of justice to the defense in this case."

The reaction of the crowd made the applause for Bryan seem wan in comparison. "Women shrieked their approval," a reporter wrote from court. "Men, unmoved even by Darrow, could not restrain their cheers." A policeman brought in from Chattanooga to help quell the masses thumped so hard on a table that it cracked beneath his fists. One of his colleagues rushed across to assist him. Yet the second policeman was rebuffed by his table-thumping friend, who protested: "I'm not trying to get order. I'm cheering!"

Even John Washington Butler, the man responsible for drafting the Tennessee law against evolution, was swept along. From his unlikely position in the press box he hailed Malone's address as "the finest speech of the century." An ecstatic Mencken rushed over to embrace Malone. "Dudley," he yelled above the tumult, "that was the loudest speech I ever heard!"

In his filed copy from court Mencken suggested, with a smidgeon of abandonment, that Malone's delivery "roared out of the windows like the sound of artillery practice, and alarmed the moonshiners and catamounts on distant peaks. Trains thundering by on the nearby railroad sounded faint and far away. The yokels outside stuffed their Bibles into the loudspeaker horns and yielded themselves joyously to the impact of the original. In brief, Malone was in good voice. . . . At its end they gave it a tremendous cheer, a cheer at least four times as hearty as that given to Bryan. For these rustics delight in speechifying, and know when it is good. "

As the reporters rushed through their assessments, deciding unanimously in favor of Malone over Bryan, the two men sat quietly at their adjoining tables in a rapidly emptying courtroom. Only John Scopes separated them as, looking straight ahead, Bryan said softly, "Dudley, that was the greatest speech I have ever heard."

Malone, also gazing into the distance, responded even more gently. "Thank you Mr. Bryan," he said, remembering how they had once worked together and understanding that he had shredded his old boss. "I am just sorry it was I who had to make it."

Later that night, Darrow joined Malone and Hays in the Mansion on the edge of town. As they discussed the case in that rambling house, on the top of a small hill, Darrow looked up toward the black sky as thunder rolled across Dayton. "Boys," he said dryly, knowing that Bryan and the Holy Rollers would expect a swift retort from God, "if lightning strikes this house tonight . . ."

THE NEXT MORNING, with Dayton feeling unusually fresh following the heavy rain, the aftermath of a momentous day rippled across America. For the *New York Times*, the Scopes trial had produced "the greatest debate on science and religion in recent years." Ralph Perry, in the *Nashville Banner*, concluded grandly that "all prophesies that the trial of the Rhea County school teacher would rank with that of Galileo are well nigh fulfilled." Philip Kinsley, in the *Chicago Daily Tribune*, suggested that, in Dayton, "they were still talking of their disappointment in Bryan, the spectacle of this old gladiator of many battles tumbled in the dust by the shining spear of Dudley Field Malone's logic."

Bryan, however, had not lost his allies in court. Everything still hinged on Judge Raulston's ruling as to whether or not expert testimony would be allowed to influence the outcome of the trial. Raulston, who liked to describe himself to the hordes of newspapermen in an exaggerated southern drawl as "jers 'nother moun'in jerdge,"which in translation meant "just another mountain judge," had sided with Bryan and the prosecution throughout the first week. If Raulston remained intransigent and rejected scientific proof as a mere irrelevance, Darrow's hopes of a landmark legal victory would crumble.

He had faced the same crucial decision over expert witnesses with Leopold and Loeb eleven months earlier, but it seemed less a strange coincidence than just further evidence that man was increasingly being confronted with new ways of thinking that challenged the old certainties passed down through the ages. Judge Caverly in Chicago had taken days to reach a decision, but, when he did, he declared that psychiatric insights could set a precedent in a murder trial. In Dayton, with Judge Raulston, Darrow faced a far more un-yielding mind-set.

"We say 'keep your bible'," Malone had cried out, "but keep it where it belongs, in the world of your conscience. Do not try to tell an intelligent world that these books, written by men who knew none of the accepted fundamental facts of science, can be put into a course of science." It was possible, after all, to believe in both Christianity and evolution—and to read the Bible without accepting its every state-ment as the unequivocal truth when explaining man's existence in the universe.

It might take another hundred years, or even longer, but Dar-row believed that illuminated knowledge would eventually transcend blind obstinacy. In Dayton, however, he did not have the luxury of time. Darrow needed Raulston to surprise him and allow the scien-tific witnesses to be heard. His lone cause for optimism was that the judge occasionally appeared torn between "doing right for Tennes-see," whose lawmakers called for a swift verdict against Scopes, and not having his own reputation for impartiality tarnished during a trial of national significance. But a drastic shift of attitude was required for him to break ranks with local sentiment.

Judge Raulston sounded notably nervous when, after the usual morning prayer had opened the sixth day of the case, he turned to his notes and began to read in a halting voice. It took him a long time to get to the nub of his verdict: "In the final analysis this court, after a most earnest and careful consideration, has reached the conclusions that under the provisions of the act involved in this case, it is made unlawful thereby to teach in the public schools of the state of Tennes-see the theory that a man descended from a lower order of animals. If the court is correct in this, then the evidence of experts would shed no light on the issues. Therefore the court is content to sustain the motion of the attorney general to exclude the expert testimony."

In that stilted moment, Raulston had ruined any chance that Darrow might prove the innocence of Scopes in the face of an unjust law. The trial was all but over. There would be no justice for John Scopes, the ACLU, or, in Malone's words, the next generation of Tennessee children. Their minds would remain tightly shut.

They had expected Raulston's obdurate ruling, but, hearing its confirmation, Hays snorted dismissively. His sneering response was echoed by Darrow, who warned the judge, with biting disdain, that "the state of Tennessee don't rule the world yet." Tom Stewart, the chief prosecutor, defended Raulston. "I think it is a reflection upon the court," Stewart said of the defense's aggressive reaction.

"Well," Raulston murmured, smiling faintly at the suggestion, "it don't hurt this court."

Darrow could not contain his scorn. "There is no danger of it hurting us . . ."

"No," Stewart quipped, "you are already hurt as much as you can be hurt."

"Don't worry about us," Darrow said coldly.

All that remained, or so it seemed, was for Darrow to fight for the retention of expert testimony in a written record of the trial. If it would not be heard by a jury in Dayton, at least it should be considered on appeal to a higher court.

Bryan intervened with a sharply observed point: "If these witnesses are allowed to testify, I presume they would be subject to cross-examination."

Hays, the ACLU's lawyer, was anxious to avoid lengthy cross-examination of his witnesses. He knew that if religious miracles like the virgin birth were discussed at length by scientists, there was a great risk of alienating a whole swathe of churchgoing Americans who might have otherwise been willing to accept the basic tenets of evolution.

When Darrow, supporting Hays, began to protest, he was interrupted by the judge. "Colonel, what is the purpose of cross-examination?" Raulston asked. "Isn't it an effort to ascertain the truth?"

Darrow glared at the judge: "Has there ever been any effort to ascertain the truth in this case?"

The temperature in the courtroom had climbed just another notch. When Raulston finally relented, he ruled that written affidavits could be submitted as part of the trial record—as would prepared statements that were read aloud in a court empty of its twelve jurors.

Darrow requested an adjournment for the rest of the day so that the defense would have the necessary time to assemble the written documentation.

Raulston asked why the defense needed so long to compile their evidence. Darrow was furious: "I do not understand why every request of the state and every suggestion of the prosecution should meet with an endless waste of time, and a bare suggestion of anything that is perfectly competent on our part should be immediately overruled."

"I hope you do not mean to reflect upon the court."

"Well, Your Honor has the right to hope."

"I have the right to do something else, perhaps," Raulston snapped back.

Darrow waved his hand wearily. "All right," he sighed. "All right."

Raulston ordered that court would reconvene on Monday at 9:00 A.M. As Darrow drifted away, to work out his final strategy for a supposedly lost cause, he found an unlikely supporter in the author of the Butler Act. "I'd like to have heard the evidence," John W. Butler suggested outside court. "It would have been right smart of an education to hear those fellows who have studied the subject. Of course we got 'em licked anyhow, but I believe in being fair and square and American. Besides, I'd like to know what evolution is myself."

Mencken, along with dozens of other reporters, packed his suitcase. In his last biting report from Dayton that Friday afternoon, he insisted that "all that remains of the great cause of the State of Tennessee against the infidel Scopes is the final business of bumping off the defendant. There may be some legal jousting on Monday and some gaudy oratory on Tuesday, but the main battle is over, with Genesis completely triumphant."

THE PROSECUTION celebrated all weekend. Tom Stewart announced "a glorious victory" while Bryan suggested smugly that "the Tennessee case uncovered the conspiracy against Biblical Christianity." He pinpointed Darrow as the real enemy of religion. "He protested against opening court with prayer and has lost no opportunity to slur the intellect of those who believe in orthodox Christianity. Mr. Darrow's conduct during this case ought to inform the Christian world of the real animus we face." His son, William Jennings Bryan Jr., decided that, with their work done, he would return early to California.

Darrow had walked a doomed road many times before—most notably in Los Angeles, down and out on a charge of bribery, and again in Chicago when it appeared as if the whole of America wanted to hang Leopold and Loeb. And so, that weekend, Darrow hatched a plan to unseat Bryan and expose the ignorance and hypocrisy of fundamentalism. The need for secrecy was such that he made no attempt to stop Mencken's departure. Darrow could not afford to let slip to a single reporter what he had planned for Monday morning. It was essential that he catch the prosecution unawares, so he prepared himself surreptitiously. As the rest of his team checked the written statements provided by eight expert witnesses, Darrow and his friend Kirtley Mather, the Harvard geologist, worked together on Sunday evening.

The attorney whipped through the questions he would pose in the morning to his witness. They were drawn mostly from the list he had supplied to the *Chicago Daily Tribune* when, in the summer of 1923, he had tried to force Bryan to answer the obvious inconsistencies in any literal interpretation of the Bible. The Great Commoner had managed to dodge the bullets then, but now, in Dayton, Darrow was intent on gunning him down. Their duel was far from over.

Rhea County Court, Dayton, Tennessee, July 20, 1925

The mood in court seemed unusually tense. In contrast to the zestful days of the previous week, when even the police officers guarding the entrance had joined in the gaiety, there was little joking that Monday morning. An air of thrilling uncertainty could be felt after Judge Raulston arrived early with his face still etched in anger. He looked ready to make a dramatic statement amid speculation that he might even do the unthinkable and cite Clarence Darrow for contempt of court. And so, scenting blood, all Dayton turned out for the day that would end Scopes's monkey trial in total victory for Bryan and the prosecution.

Such expectation meant the fight for space was even more intense than usual, and the policemen holding back the mob outside were notably taciturn. And, inside, the normally cheery bailiff followed his usual chant of "Oh yes, oh yes, oh yes, this Honorable Circuit Court is now open pursuant to adjournment," with a surly rejoinder to the noisy crowd: "This ain't no circus!"

Judge Raulston, with darkened face and quavering voice, went to work immediately and upheld the rumors by charging Darrow with contempt. Referring to his outburst on Friday, Raulston said, "Men may become prominent but they should never feel themselves superior to the law or to justice. He who would hurl contempt into the records of my court insults and outrages the good people of one of the greatest states of the Union." He ordered that Darrow should appear before him the following morning to answer the charge on a fixed bail of $5,000.

It was the first time in his long career that Darrow had been cited for contempt. He was briefly indignant, but, since his overriding objective remained a showdown with Bryan, Darrow retained his discipline. The best way to get Bryan on the witness stand was to accept the charge and press ahead. Darrow politely asked Raulston for permission to have the prepared scientific statements read out aloud in court—even if the jury would remain excluded from any scientific testimony. It was one way of educating the masses.

The prosecution protested, and argument raged again. Darrow broke the impasse by suggesting that, rather than reading all sixty thousand words of testimony, Hays should offer up some extracts to the court. Raulston concurred, with the stipulation that the defense should take no more than an hour of his time.

Hays initially read quickly, as if determined to cram as many words into his allotted slot as possible, but he relaxed after an interruption from the prosecution. "Can I have time out?" Hays yelled, and even Raulston laughed as he gave his assent. From that point onward Hays read more clearly, and the brilliantly lucid and reasoned arguments of scientists and academics from Harvard, Rutgers, the University of Chicago, and Johns Hopkins flooded the court with logic and knowledge. People listened attentively, with Raulston looking as if he hardly dared breathe as he concentrated on each word. Hays read until noon when, calmly and thoughtfully, Raulston adjourned for lunch—confirming that the defense could continue after the recess.

As they entered the afternoon session Darrow smoothed the path for his next assault by openly expressing contrition to Raulston for his reaction three days before. "I have been practicing law for forty-seven years," he said, "and I have been most of the time in court. I have had many a case where I have had to do what I have been doing here, fighting the public opinion of the people in the community where I was trying the case. I never yet have in all my time had any criticism by the court for anything I have done."

Darrow paused and then, looking directly at the judge, he murmured, "I haven't the slightest fault to find with the court. Personally, I don't think it constitutes contempt, but I am quite certain that the remark should not have been made and the court could not help taking notice of it. I am sorry that I made it . . . and I want to apologize to the court for it."

As Darrow was applauded, Raulston held up his hand. "My friends, and Colonel Darrow," he said, his eyes shining, "the Man that I believe came into the world to save man from sin, the Man that died on the cross that man might be redeemed, taught me that it is godly to forgive, and were it not for the forgiving nature of Himself I would fear for man. The Saviour died on the cross pleading with God for the men who crucified Him. I believe in that Christ. I believe in these principles. I accept Colonel Darrow's apology."

With his spirit restored, Raulston decided to hold the afternoon session on the courthouse lawns. "The court should adjourn downstairs," he instructed. "I am afraid of the building. The court will convene down in the yard."

Despite rumors that cracks had appeared in the ceiling of the room immediately below the second-floor court, John Scopes suggested instead that "even the judge, who had an electric fan that the rest of us didn't share, couldn't stand the man-killing heat."

Raulston led them out to a platform, built for the weekend sermons, which had been erected in the shadow of giant oak and maple trees surrounding the courthouse square. The temperature was a little less oppressive and a crowd of three thousand covered every square inch of courthouse lawn. The *New York Times* noted that "the spectators, instead of being only men, were men, women, and children, and among them here and there was a Negro. Small boys went through the crowd selling bottled pop. Most of the men wore hats and smoked."

One of the most momentous cross-examinations in American legal history was preceded by a tranquil prelude as Hays completed his reading of the testimony supplied by the expert witnesses. But any temptation to nod off in the early afternoon shade was resisted as Darrow protested against the huge banner spread across an entire outside courtroom wall: READ YOUR BIBLE.

He demanded its removal, arguing that the message was prejudicial to the defense. When the prosecution objected, Darrow came up with another defiant proposal. "We might agree to get up a sign of equal size on the other side and in the same position, reading HUNTER'S BIOLOGY or READ YOUR EVOLUTION," he threatened.

Ben McKenzie of the prosecution jumped to his feet. "I have never seen the time in the history of this country when any man should be afraid to be reminded of the fact that he should read his Bible, and if they should represent a force that is aligned with the Devil and his satellites . . . when that time comes then there is time for us to tear up all Bibles, throw them in the fire and let the country go to hell!"

"I should like to move the court to expunge the last remarks," Malone roared.

Raulston nodded. "Yes, expunge that part of Mr. McKenzie's statement from the record—where he said you were satellites of the devil. Anybody else want to be heard?"

As Malone and Bryan shouted above the clamor, court officer Kelso Rice cried out: "People, this is no circus! There are no monkeys up here! This is a lawsuit. Let us have order."

Bryan backed down first. "I cannot see why 'Read Your Bible' would necessarily mean partiality toward our side . . . but if leaving that up there during the trial offends our brother, I would take it down during the trial."

Raulston agreed and ordered the removal of the biblical banner. He was more startled when, shortly after the workmen had completed the task, Hays announced that the defense would like to call one last witness.

"Hell is going to pop now," Malone whispered to an equally surprised Scopes.

"The defense desires to call Mr. Bryan as a witness," Hays said solemnly. "We should like to take Mr. Bryan's testimony for the purposes of our record."

As the prosecution lawyers objected, Judge Raulston appeared flummoxed. "Do you think you have a right to his testimony?" he asked Darrow tentatively.

McKenzie answered first. "I don't think it is necessary to call him," he yelled.

"If you ask him about any confidential matter," Raulston warned Darrow, "I will protect him."

Darrow responded coolly: "I do not intend to do that."

Bryan intervened to say that he was willing to take the stand—as long as he could also question Darrow, Malone, and Hays.

"Not at once?" Darrow smiled.

Bryan ignored him. "Where do you want me to sit?" he asked Hays.

"Mr. Bryan," the judge asked anxiously, "you are not objecting to going on the stand?"

"Not at all," Bryan smirked.

"Do you want Mr. Bryan sworn?" Raulston asked Darrow.

"No."

"I can make affirmation," Bryan insisted. "I can say, 'So help me God, I will tell the truth.'"

"No," Darrow said. "I take it you will tell the truth, Mr. Bryan."

He motioned his old rival to a seat on the platform. Once Bryan had settled himself, Darrow began quietly. "You have given considerable study to the Bible, haven't you, Mr. Bryan?"

"Yes, sir, I have tried to . . ."

"Well, we all know you have. We are not going to dispute that at all," Darrow murmured.

"I have studied the Bible for about fifty years," Bryan confirmed. "Of course, I have studied it more as I have become older."

Darrow took a step toward Bryan, closing the polite distance between them. He looked straight at Bryan, whose face was now only a few feet away from him. "Do you claim that everything in the Bible should be literally interpreted?"

"I believe everything in the Bible should be accepted as it is given there; [but] some of the Bible is given illustratively. For instance:

'Ye are the salt of the earth.' I would not insist that man was actually salt, or that he had flesh of salt, but it is used in the sense of salt as saving God's people."

"But when you read that Jonah swallowed the whale," Darrow said, "or the whale swallowed Jonah, how do you literally interpret that?"

"I read that a big fish swallowed Jonah," Bryan protested. "It does not say whale."

"Doesn't it?" Darrow said incredulously. "Are you sure?"

"That is my recollection of it. A big fish, and I believe it, and I believe in a God who can make a whale and can make a man and make both do what he pleases."

Darrow grunted while Bryan flapped his fan in his face. "Now you say the big fish swallowed Jonah," Darrow murmured, "and he remained there how long? Three days?"

Bryan twitched his head in agreement. Darrow's next query was blunt. Was the big fish made to swallow Jonah?

"I am not prepared to say that. The Bible merely says it was done."

"You don't know whether it was the ordinary run of fish—or made for that purpose?"

"You may guess," Bryan scowled. "You evolutionists guess."

"But when we do guess," Darrow said quietly, "we have a sense to guess right."

"But do not do it too often."

"Do you believe that he made such a fish and that it was big enough to swallow Jonah?"

"Yes, sir," Bryan said, encouraging a wave of applause with his next statement. "Let me add: One miracle is just as easy to believe as another."

"Or just as hard," Darrow countered.

"It is hard to believe for you, but easy for me. When you get beyond what man can do, you get within the realm of miracles. And it is just as easy to believe the miracle of Jonah as any other miracle in the Bible."

Bryan had the contented look of a man who believed he had done well. Yet he did not know that they had just begun, for him, an ordeal that would stretch over the next two punishing hours.

"Do you consider the story of Joshua and the sun a miracle?" Darrow continued.

"I think it is."

"Do you believe Joshua made the sun stand still?"

"I believe what the Bible says." Bryan hesitated for the first time. "I suppose you mean that the earth stood still?"

"I don't know," Darrow shrugged. "I am talking about the Bible now."

"I accept the Bible absolutely."

"The Bible says Joshua commanded the sun to stand still for the purpose of lengthening the day, doesn't it, and you believe it?"

"I do."

"Do you believe at that time the entire sun went around the earth?"

"No," Bryan said, and then paused. "I believe the earth goes around the sun."

"Do you believe that the men who wrote it thought that the day could be lengthened or that the sun could be stopped?"

"I don't know what they thought."

"You don't know?" Darrow said with a theatrical sigh.

Bryan flushed. "I think they wrote the fact without expressing their own thoughts."

For the next ten minutes the questions about Joshua and the sun bounced back and forth, Darrow probing relentlessly. The crowd tried to bolster Bryan, applauding and cheering him whenever possible, but he became increasingly uncertain. There was clearly something deeply, even cruelly, personal between two old men. Bryan might have been sixty-five, and Darrow three years older than him, but their enmity felt raw.

"Now, Mr. Bryan," Darrow said with feigned politeness, "have you ever pondered what would have happened to the earth if it had stood still?"

"No."

"You have not?"

"No. The God I believe in would have taken care of that, Mr. Darrow."

"I see," Darrow said as his hands curled around his suspenders. He looked thoughtfully at Bryan. "Have you ever pondered what would naturally happen to the earth if it stood suddenly still?"

"No."

"Don't you know it would have been converted into a molten mass of matter?"

Bryan jabbed his finger angrily at Darrow. "You testify to that when you get on the stand," he snapped. "I will give you a chance."

"Don't you believe it?" Darrow asked.

"I would want to hear expert testimony on that."

Darrow smiled. The memory of their struggle over "expert testimony" was too fresh for even Bryan's supporters to ignore. "You have never investigated the subject?"

"I don't think I have had the question asked," Bryan complained.

"Or ever thought of it?"

"I have been too busy on things that I thought were of more importance than that," Bryan said as he fanned himself furiously.

Darrow moved in mercilessly. "Mr. Bryan, could you tell me how old the earth is?"

"No, sir, I couldn't."

"Could you come anywhere near it?"

"I wouldn't attempt to," Bryan said. But, as the crowd muttered its recognition of his failure, he tried to hit back. "I could possibly come as near as the scientists do—but I would rather be more accurate before I give a guess."

Bryan, pushed by Darrow, eventually suggested that the date of man's creation might have started around 4004 B.C.

"That estimate is printed in the Bible?" Darrow said as he scratched his head.

"Everybody knows," Bryan stammered, "at least I think most of the people know. That was the estimate given."

"But what do you the think the Bible itself says?" Bryan could not answer, so, after a spiky silence, Darrow asked another question, sounding like a frustrated teacher confronting a recalcitrant boy who had refused to do his homework. "Don't you know how it was arrived at?"

"I never made a calculation," Bryan confessed sadly.

"A calculation from what?"

"I could not say."

"From the generations of man?" Darrow prompted.

Bryan had begun to sweat heavily under the fierce grilling. "I would not want to say that."

"What do *you* think?" Darrow said with mild encouragement.

Bryan, like the class dunce, shook his head helplessly. "I do not think about things I don't think about."

"Do you think about things you do think about?" Darrow asked in a derisory echo.

"Well," Bryan said hesitantly, "sometimes . . ."

There was such an outbreak of laughter, as much from Bryan's supporters as his opponents, that a demand for order had to be made. Bryan sounded like a buffoon, but, with his pride wounded, he refused all attempts from the prosecution to cut short his testimony. He hoped, somehow, that he could still defeat Darrow.

"Would you say that the earth was only 4,000 years old?" Darrow repeated.

"Oh no," Bryan said with a little more bluster, "I think it is much older than that."

"How much?"

"I couldn't say."

"Do you think the earth was made in six days?" Darrow asked, changing tack.

"Not six days of twenty-four hours."

"Doesn't it say so?" Darrow said in seeming surprise.

"No, sir."

Sensing another looming disaster for Bryan, Tom Stewart tried to intervene for the prosecution. "I want to interpose another objection," he shouted. "What is the purpose of this examination?"

Bryan spoke before the judge could respond. "The purpose is to cast ridicule on everybody who believes in the Bible," he said fiercely. "And I am perfectly willing that the world shall know that these gentlemen have no other purpose than ridiculing every Christian who believes in the Bible."

Darrow's voice rang out. "We have the purpose of preventing bigots and ignoramuses from controlling the education of the United States, and you know it."

Bryan was indignant. He glared at Darrow. "I am simply trying to protect the word of God against the greatest atheist or agnostic in the United States!"

Prolonged applause rolled across the courthouse lawn and, feeling emboldened, Bryan jabbed his finger at Darrow and yelled again. "I want the papers to know that I am not afraid to get on the stand in front of him and let him do his worst! I want the world to know."

Yet Darrow still wanted to know what the Bible meant when it spoke of a "day." Had it really only taken six days to create heaven and earth?

"I do not think those were twenty-four-hour days," Bryan conceded weakly. "My impression is that they were periods."

Darrow gazed at him. "Have you any idea of the length of these periods?" he eventually asked.

"No, I don't."

"Do you think the sun was made on the fourth day?"

"Yes."

"And they had evening and morning without the sun?"

"I am simply saying it is a period," Bryan said with a quiver in his voice.

"They had evening and morning for four periods without the sun, do you think?"

"I believe in creation as there told—and if I am not able to explain it I will accept it."

"Creation might have been going on for a very long time?" Darrow suggested casually.

Bryan, despite himself, nodded. "It might have continued for millions of years," he admitted. "But I think it would be just as easy for the kind of God we believe in to make the earth in six days or in six million years—or in six hundred million years. I do not think it is important whether we believe one or the other."

The whole case against Scopes, and the teaching of evolution, had crumbled to dust. "Yes," Darrow exclaimed softly. He looked over at the thousands in the now hushed crowd. They knew that he had won. Their silence sounded full of Bryan's defeat.

"All right," Darrow said. "Do you believe that the first woman was Eve?"

"Yes."

"Do you believe she was literally made out of Adam's rib?"

"I do."

"Did you ever discover where Cain got his wife?"

"No, sir," Bryan grimaced. "I leave the agnostics to hunt for her."

"Were there other people on the earth at that time?"

"I cannot say."

Darrow folded his arms across his chest. "You cannot say. Did it ever enter your consideration?"

"Never bothered me."

"There were no others recorded, but Cain got a wife."

"That is what the Bible says."

"All right," Darrow said with a dismissive wave of his hand. "Do you believe in the story of the temptation of Eve by the serpent?"

"I do," Bryan said, as if in a daze.

"Do you believe that after Eve ate the apple, or gave it to Adam, whichever way it was, that God cursed Eve and decreed that all womankind from thenceforth and forever should suffer the pains of childbirth in the reproduction of the earth?"

"I believe what it says."

Darrow turned to the Bible he now held in his hands. He read aloud from a place he had already marked: "And the Lord God said unto the serpent, 'Because thou hast done this, thou are cursed above all cattle, and above every beast of the field. Upon thy belly shalt thou go and dust shalt thou eat all the days of thy life.'"

The crowd almost shivered—whether at Darrow's dramatic reading of the wrath of God or simply because a cool wind, promising rain from the north, gusted suddenly across the courthouse lawns.

Darrow stood completely still, the Bible spread open in his hands. "Do you think that is why the serpent is compelled to crawl upon its belly?" he asked.

"I believe that," Bryan repeated, in parrot fashion.

"Have you any idea how the snake went before that time?"

"No, sir."

"Do you know whether he walked on his tail—or not?"

"No, sir. I have no way to know," Bryan said as the crowd laughed again.

In his humiliation, Bryan jumped up. "Your Honor," he said, his face a picture of rage, "I think I can shorten this testimony. The only purpose Mr. Darrow has is to slur the Bible, but I will answer his question. I will answer it all at once, and I have no objection in the world. I want the world to know that this man, who does not believe in a God, is trying to use a court in Tennessee—"

"I object to that!" Darrow shouted as he again approached Bryan.

Bryan continued the rest of his sentence, as to how Darrow was using a court in Tennessee, "to slur at [the Bible]. And while it will require time, I am willing to take it."

Darrow was now in his face, his battered old nose almost pressing against the cheek of Bryan. "I object to your statement," Darrow hissed. "I am examining you on your fool ideas that no intelligent Christian on earth believes!"

Judge Raulston, having lost control of his outdoor courtroom, had seen enough. He slammed the gavel against his bench. "Court is adjourned until nine o'clock tomorrow morning."

Raulston rose from his seat and hurried back to the courthouse. Darrow, with some disgust, turned sharply away from Bryan. His old enemy was left alone, slumped in his chair, as Darrow was embraced by his exultant supporters. The case against fundamentalism had been won with conviction—at least for a day.

TUESDAY, as if in heavenly sympathy for the vanquished Bryan, began with soft rain falling across Dayton. Judge Raulston ordered a return to normal proceedings and a proper courtroom and such was his haste to cut short an often uncontrollable trial that he began at 8:58 A.M. Darrow was still scrambling toward his seat inside when Raulston made a telling announcement. "I feel that the testimony of Mr. Bryan can shed no light upon any issue that will be pending before the higher court." The judge ruled that, apart from preventing Darrow's attempts to hound Bryan with further questions, the previous afternoon's dramatic exchange would be expunged from the record. "The only question we have now is whether or not this teacher, this accused, this defendant, taught that man descended from a lower order of animals. As I see it, after due deliberation, I feel Mr. Bryan's testimony cannot aid the higher court in determining that question."

Bryan was the first to respond—asking whether Darrow's questions as well as his own answers would be removed from the official court account of the trial.

"I expunged the whole proceedings," Raulston said sympathetically.

Bryan complained that he had been denied the opportunity to respond fully to Darrow's accusations of ignorance and bigotry. Without the prospect of placing Darrow on the stand, Bryan would have to trust the press to give him the space to put questions of his own to Darrow. They had reported his humiliation in such gory detail that he hoped they would now provide him a similar opportunity to test Darrow. "I think it is hardly fair for them to bring into the limelight my views on religion and stand behind a lantern that throws light on other people, but conceals themselves. I think it is only fair that the country should know the religious attitude of the people who come down here to deprive the people of Tennessee of the right to run their own schools."

"I object to that," Darrow sighed.

"I overrule the objection," Raulston insisted.

"That is all," Bryan muttered as he sat down again.

Some might have expected Darrow to be downcast, but the old battler was jubilant. His real war, against Bryan, had been settled in emphatic style. Victory had been confirmed joyously in the verbatim accounts of his cross-examination that had been printed in virtually every major newspaper in the United States. The truth of Bryan's crushing defeat was indisputable, with most editorials echoing the belief that Darrow had "brought about a striking revelation of the fundamentalist mind in all its shallow depth and narrow arrogance."

The judicial technicalities mattered far less than the sweeping moral triumph for Darrow and the defense. They would fight the prosecution again, on a more equal footing, in a higher court of appeal. But Darrow had one last devilishly clever trick to play on his demoralized rival. There was little point in perpetuating a legal charade and so, to crush Bryan still further, he decided to call an abrupt end to the monkey trial of Dayton.

"To save time," Darrow said briskly, "we will ask the court to bring in the jury and instruct them to find the defendant guilty."

A surprised murmur swept across the courtroom. Bryan, outflanked yet again by Darrow, looked on in disbelief. He had consoled himself through a torturous night with the thought that he still had a chance to redeem himself in his closing statement. That would be his moment of subdued absolution as he proved again that he was not the ignorant failure Darrow had depicted. He had worked for weeks on his final address, and its importance to him now was painfully apparent. But with one devastating gesture Darrow had stripped him of his last hope. By asking for a guilty verdict, the defense had eradicated the need for a closing address—from either side.

Bryan looked haunted as Attorney General Stewart accepted the offer. "We are pleased to accept the suggestion of Mr. Darrow," he said, purposely avoiding eye contact with his colleague.

The jury, much to their irritation, had been excluded from all but a few hours of the trial. Yet they appeared relieved that the circus was almost over and they soon could return to their crops. After they had filed back into place, Darrow motioned to the judge. "May I say a few words?"

Raulston waved his assent, and Darrow approached the twelve men. Bryan lifted his head, as if a grandstanding speech from Darrow might give him his chance after all.

"Gentlemen of the jury," Darrow said amiably, "we are sorry to have not had a chance to say anything to you. We will do it some other time. Now, we came down here to offer evidence in this case and the court has held under the law that the evidence we had is not admissible. So all we can do is take an exception and carry it to a higher court. As far as this case stands before the jury, the court has told you very plainly that if you think my client taught that man descended from a lower order of animals, you will find him guilty. . . . We cannot argue to you gentlemen under the instructions given by the court. We cannot even explain to you that we think you should return a verdict of not guilty. We do not see how you could. We do not ask it."

Darrow had again sunk the stiletto into Bryan. By refusing to plead for a "not guilty" verdict, and sitting down again so promptly, he shut down any last opportunity for Bryan to make his planned address. Darrow knew that Bryan considered it his masterpiece, for he had boasted for weeks how he'd honed his attack on the evolutionists to perfection. The old politician slumped in his seat. Darrow had piled yet more agony on him.

The Great Commoner waited silently while Darrow cracked jokes with Malone and Hays. A guilty verdict was so inevitable that the jury did not even retreat to a private room to debate the issues. The twelve farming men, instead, hung around in the corridor outside the main courtroom until, a mere nine minutes later, the foreman led them back inside.

"Mr. Foreman," Judge Raulston asked Jack R. Thompson, "will you tell us whether you have agreed on a verdict?"

"We have, Your Honor," Thompson replied.

"What do you find?"

"We have found for the State," Thompson said. "[We have] found the defendant guilty."

There was no cry of shock or despair, let alone any exclamation of joy. The Scopes trial was over and, despite the legal outcome, destined to go down in history as a great vindication of Darrow. After saving Leopold and Loeb ten and a half months earlier, Darrow had, in his own words, conquered another "summer for the gods."

The forgotten figure of John Scopes, who had just been fined the minimum penalty of $100, struggled to the front of court after John

Neal, for the defense, underlined the defendant's right to respond. Looking up at Judge Raulston, the young teacher spoke briefly, but with courage and passion. "Your Honor, I feel that I have been convicted of violating an unjust statute. I will continue in the future, as I have in the past, to oppose this law in any way I can. Any other action would be in violation of my ideal of academic freedom—that is, to teach the truth as guaranteed in our Constitution, of personal and religious freedom. I think the fine is unjust."

Bryan, unable to celebrate victory, suggested that "here has been fought out a little case of little consequence as a case, but the world is interested because it raises an issue, and that issue will soon be settled right, whether it is settled on our side or the other side."

Darrow argued instead that "this case will be remembered because it is the first case of this sort since we stopped trying people in America for witchcraft, because here we have done our best to turn back the tide that has sought to force itself upon this modern world, of testing every fact in science by a religion dictum."

The last word went to Arthur Garfield Hays, who looked up at the judge cheerily. "May I, as one of the counsel for the defense, ask Your Honor to allow me to send you *The Origin of Species* and *The Descent of Man* by Charles Darwin?"

"Yes," Raulston nodded enthusiastically as the court heaved again with laughter. "Yes . . ."

THE EULOGIES for Darrow and the lamentations for Bryan filled America's newspapers the next morning. Mencken, naturally, led the way in his condemnation of Bryan. "This three-time candidate for the Presidency came in a hero and he sat down in the end as one of the most tragic asses in American history," Mencken wrote in the *Baltimore Evening Sun*, which had already agreed to pay Scopes's fine in a gesture of solidarity.

Even though he was furious with himself for missing a seminal moment in his otherwise glittering journalistic career, Mencken could still relish both Bryan's downfall and the ascendancy of his friend. "I made up my mind to show the country what an ignoramus he was," Darrow wrote to Mencken, "and I succeeded."

Yet Mencken was too smart not to also shudder a little at all that he had witnessed in Dayton. As he confided to a friend, "I set out laughing and returned shivering. The fundamentalists are on us! They will sweep the south and middle west, Bryan or no Bryan."

He might have been depicted in many newspaper accounts as a broken man, but Bryan remained fanatical about his cause. He challenged Darrow to answer nine questions of his own—a task that the attorney completed in writing within an hour of receiving them. Starting with a simple "Do you believe in the existence of God as described in the Bible?," Bryan tried to pull the same stunt on Darrow that he had suffered on the courthouse lawn. Yet Darrow was so disarmingly frank in his confessions of uncertainty that he appeared the more reasonable man.

Bryan, having convinced the editor of the *Chattanooga Times* to print his undelivered closing address in its entirety, distilled the key points into the two sermons he delivered that Saturday. After speaking in Chattanooga and Winchester, Bryan led the prayers at a Methodist church in Dayton on Sunday morning. He then ate a heavy lunch, with Judge Raulston and their wives, and retired upstairs for an afternoon nap.

His gluttony had been documented for decades. In 1900, a journalist had written of all Bryan had devoured during a breakfast in Virginia: "An enormous melon, two quails, a formidable slice of ham with six eggs, batter cakes immersed in butter, accompanied here and there with potatoes and small delicacies for side dishes, all of which he washed down with innumerable cups of coffee with milk."

Bryan had since developed diabetes and was meant to avoid

all sugar and starch in his diet. He, once again, ignored his doctor's orders that Sunday—this time with fatal consequences. Bryan suffered a massive stroke and died in his sleep that afternoon, July 26, 1925.

The enormous strain he had been under, especially when shamed by Darrow's barbed questions, led to swift conclusions as to a darker source of death. Fulton Oursler, the respected journalist, argued that Darrow had "cross-examined the helpless William Jennings Bryan into his grave." John Scopes, meanwhile, admitted, "There was a rumor about town that 'the old devil Darrow' had killed Bryan with his inquisition."

Darrow had to confront the charge. Revealing the news of his rival's death, a reporter told Darrow that "people down here believe Bryan died of a broken heart because of your questioning."

"Broken heart nothing," Darrow muttered. "He died of a busted belly."

He soon found a more gracious note in a statement he made to the Associated Press. "I am pained to hear of the death of William Jennings Bryan. I have known Mr. Bryan since 1896, [and] supported him twice for the presidency. He was a man of strong convictions and courage. I differed with him on many questions, but always respected his sincerity and devotion."

Mencken, back in Baltimore, was less conciliatory. "God aimed at Darrow, missed, and hit Bryan instead," he quipped. "We killed the son-of-a-bitch!"

CHAPTER 12

DARKNESS IN DETROIT

O N THE DAY that Bryan died, Mary Field Parton read her morning newspaper through a glistening veil. Yet the tears that glazed her eyes were in celebration of Darrow. It had taken more than thirteen years, a time sparked by the ignominy of bribery and framed by anguish and loss and then reconciliation. His vindication now moved her in a way she had not anticipated.

Mary had followed the Scopes trial avidly during the eleven days it spanned—relying on the *New York Times*'s exhaustive coverage. And now that it was over, at least until it was heard again in the Supreme Court, she could hardly believe it had ended in such personal glory. Holding her section of the *Times* in quivering hands, a movement she tried to conceal as her husband shared breakfast with her at the table where they had entertained a jealous Darrow at a dinner party two and a half months before, Mary felt overwhelming pride. Russell Owen, one of the country's leading feature writers, had produced a full-page profile headlined "Darrow Likes to Fight for Lost Causes." The subheading made her smile: "Self-Educated, a constant student, he enjoys taking unpopular sides—'I Will Defend Any Man.'"

"Clarence Darrow, champion of the minority, has a passion for lost causes. One of the foremost advocates of reason against the might of the majority, an apparently hopeless struggle has an irresistible fascination for him. It was this sympathy with unpopular causes which led him to defend so many labor cases, and it was his instinctive resistance to anything which seems to impair individual freedom that led him to fight against the Tennessee Anti-Evolution Law.

"Darrow is no fire eater. He dominates by sheer intellectual power, and his manner when he is not roused in debate is one of courteous consideration marked by a delightful sense of humor. When he hunches his bony, square shoulders up under his ears and breaks into a smile, any barrier of hostility must melt. . . . A rather awkward figure in baggy trousers upheld by suspenders, without a coat, one shirt sleeve torn half off, and thin hair sticking damply in all directions, he was a person to hold any eye during the Scopes trial.

"The impression of personal carelessness, however, ends with his clothes. His face is gnarled and deeply lined, and sets forward a little on his shoulders. Sunken and piercing eyes shine out from under a bulging brow. His lips protrude a little when he laughs or frowns at an opponent with a look that stabs deep. And his expression can shift in a moment from penetrating scorn to a genial or a quizzical smile."

As the glowing feature rolled on, Mary remembered how different it had been amid their fall in Los Angeles. Then, she had been frightened by the absence of hope in Darrow. "It's over, it's over, it's over . . . ," he had said, as if each repetition shut down another part of his life. His career was over. Their relationship was doomed. His life was finished. How easy it would have been then if she could have displayed this article of faith from the future as something to which he could cling.

But the sweetness of his euphoric return was made all the richer by the preceding desolation. This was how life worked. This was why she sat across a table from another man she loved, Lemuel, reading with a gleaming gaze of Darrow's now famous trials in the *Times*. She almost laughed when the bribery charge was ignored in a single line that skated across the most painful period in Darrow's life: "He later defended the McNamara brothers in the case growing out of the bomb explosion in the *Los Angeles Times* building."

There was no mention of his two subsequent trials. *The People v. Clarence Darrow* was in the process of being airbrushed out of history. "The cases of which Darrow is perhaps proudest are the ones which were little known outside the vicinity where they occurred. He once defended a Negro without fee in a case which lasted a month and was taken to a higher court: and when the man was sentenced to prison, Darrow supported his family because he was certain he was innocent. He defended a little cobbler accused of Communism just after the war, despite the hostility such a defense incurred at that time, and cleared him. 'I guess I have defended more cases for nothing than I was ever paid for,' he said. This interest in other people, which has led him to give of his time and money to their defense when necessary, is one of his most pronounced characteristics. . . . There is a humanness about him which is hard to resist."

Darrow shrugged and mumbled when, the following week, they were reunited in New York and, after she had kissed him, Mary pressed the newspaper into his hand. At least he did not churn out his public line that, having wasted the first part of his career trying to get into the papers, he had spent the rest of his life trying to stay out of their pages. It still mattered deeply to him. Darrow said gruffly he would read it later, but, when she returned soon afterward from the bathroom, she saw he was hunched over the *Times*. He was motionless while he read. She stood some distance away, allowing Darrow the time he needed, and when he looked up she moved toward him.

She laughed softly when he murmured in her ear: "Not bad," he said, "not bad at all . . ."

JUST AS HE did after every major trial over the previous fifteen years, all the way back to his first threat to quit "lawyering" in 1910, Darrow stressed that he had reached the end. Dayton had left him, yet again, drained and exhausted. And, for all their differences, the death of a younger man in Bryan had underlined his own mortality. At the age of sixty-eight it was time for him to depart at the pinnacle. As he whiled away those hazy August days he told Mary he would be content never to work in a courtroom again. His only novel, *Farmington*, published twenty-two years earlier, was about to be reissued on the back of his success in defending Leopold and Loeb and then John Scopes. Darrow yearned to write another, and a better novel at that, and to pen numerous works of nonfiction while also contributing regular newspaper articles. The closing line in the *New York Times*'s tribute had confirmed as much, with Darrow claiming, "I would rather write than do anything else."

He did not know it then, but a brutal series of events had already begun to unfold almost five hundred miles from New York. In Detroit, a black family faced a tragic charge of murder. As the final shattering part of the trilogy that came to define his legacy, Darrow would eventually describe the Sweet case as the most momentous of his entire career. He was about to be plunged, one last time, into the depths of an American nightmare.

ON MONDAY July 13, 1925, the front page of the *Detroit Free Press* had been dominated by two seemingly unrelated stories. The left-column lead was devoted to Darrow's battle in Dayton with Bryan. To the right, in an article given equal space, there was a disturbing insight into the racial hatred spreading across Detroit. The reasons for the Sweet catastrophe, and Darrow's subsequent arrival in Detroit, would be traced back to that very story.

Two nights earlier, in a venomous display, ten thousand members of the revived Ku Klux Klan occupied a vast plot of land. They wore their sinister white hoods and brandished burning crosses in the dark. One of their robed and hooded leaders, the *Detroit Free Press* revealed, came from Tennessee, not far from Dayton. He stood "on a platform illuminated with the red glare of fiery crosses [as he] advocated a law to compel Negroes to live only in certain quarters of the city."

Crisscrossing between their report on the Klan gathering and warnings from the mayor of Detroit that the city's black population should refrain from engaging in further race riots, the story confirmed that people living in the exclusively white neighborhood of Waterworks Parks had been urged to attend a meeting to discuss their "self-defense" at the Howe School, on the corner of Garland Avenue and Charlevoix Street.

A month before, on June 7, Ossian Sweet, a twenty-nine-year-old black doctor, had put down a $3,500 deposit to purchase a two-story house at 2905 Garland Avenue—directly across the street from that same elementary school. The modest house had been valued at around $13,000 but, as Sweet was a Negro, the couple selling him their property were ruthless. Ed and Marie Smith told the smart but subdued young doctor that they would not sell to him for less than $18,500.

As his wife, twenty-three-old Gladys, had set her heart on a move out of the dangerous black ghetto where Sweet had built up a busy practice, he gave in to the demand. The Smiths' mercenary streak did not end there. Knowing that a black man, even a doctor, could not easily obtain such a sizable bank loan, they offered to finance his purchase. After he had paid a 20 percent deposit and moved in that September, he would then make 120 monthly installments of $150 to the Smiths. This arrangement meant that it would take him until September 1935 before he owned the property outright. During those intervening ten years, the Smiths would charge him 18 percent interest on the outstanding amount—a sum far above the existing bank rate.

Sweet knew he was being conned. But his desire to please his wife, and provide a new home in a suburban neighborhood for them and their eighteen-month-old daughter, Iva, made him swallow the financial hardship. He was also swayed by the fact that Ed Smith himself was a Negro—although his complexion was so light that most people believed he was white.

After signing the necessary papers, Sweet asked Marie Smith whether the neighborhood was safe for a "colored" family. It was a strange time to ask the question. But Mrs. Smith, smiling thinly, gave the same answer she would have offered had the black doctor asked her before he put pen to paper. Garland Avenue, she said, was a pleasant street. There were no Klansmen round here.

A month later seemingly ordinary men, without hoods, swarmed all over Garland and Charlevoix as they nailed "self-defense" signs to the wooden telephone poles. The euphemistic Improvement Association name was more reputable than the KKK, so that Tuesday evening seven hundred people wandered through the gates to a school hall meeting. It did not take long for the mood to descend into blunt prejudice. The dominant speaker came from the Tireman Avenue Improvement Association—a group that had succeeded, on three separate occasions, in driving out Negro families who had tried to move into their white neighborhood. Encouraging the people of Waterworks Parks to "put the niggers out," the Tireman man was cheered and applauded.

Ossian Sweet understood the threat. His colleague Dr. Alexander Turner, the chief of surgery at Dunbar Memorial, a black hospital in Detroit, had relived the night in June when he and his wife had been forced out of their new house on Spokane Avenue on the plusher west side. Just five hours after they had moved in, Turner's home had been invaded and trashed by a white mob who then forced him to sign over his property to the local Improvement Society. The surgeon looked as if his heart had been ransacked in the process.

Sweet's desolation was countered by Dr. Edward Carter, who urged him not to give up on his own move. Carter argued that black professionals like themselves needed to withstand such intimidation. They were American citizens, after all, who were better educated and more successful than most of their countrymen. Why should they, in 1925, not be able to choose wherever they wanted to live in a northern city like Detroit?

A troubled Detroit, however, made glib optimism almost impossible. The writer Marcet Haldeman-Julius, following Darrow from the Scopes trial to the Sweet case, argued that "Detroit was more ridden with race hatred toward Negroes than any spot north of the Mason and Dixon line." The city had changed radically. Henry Ford's decision in 1914 to offer a groundbreaking five-dollars-a-day wage to his automobile workers had attracted tens of thousands of migrants, both black and white, from the South. By 1925 over a million people lived in Detroit—as the city doubled its population. The black community had expanded to an unprecedented level with white politicians decrying the fact that an estimated seventy thousand Negroes now lived in Detroit. They were still mostly condemned to the same three city wards that had housed a mere six thousand black people in 1910.

The Ku Klux Klan had also exploded across Detroit. With many uneducated white farm laborers moving from the South, Detroit became a fertile recruitment center for the Klan. By 1924 the KKK claimed that it had more active supporters in Detroit than anywhere else in America—a fact proved by the need to rig the ballot in order to prevent a Klansman winning the mayoral election that year. The next vote was scheduled for November 1925, but the indignant men in white hoods did not look as if they were prepared to wait that long to seize power.

They had already gained a significant hold over a city police force swollen with Klansmen. There was little surprise when between forty and fifty black men were shot dead by the police in the first seven months of 1925. The *Detroit Independent* would eventually claim that "more Negroes had been shot down in the [city] streets without a reasonable excuse than [had] been lynched in the entire South, during the same period."

Against such a bleak backdrop Ossian Sweet turned to his wife. Gladys did not come, like Ossian, from a southern plantation. She was a modern city girl from the North, born in Pittsburgh but raised less than five miles from Garland Avenue. For seven years before her marriage she and her family had been the only Negroes on an entire block of a white street. As a schoolgirl she had often been the only black student in her class. Gladys was smart and cool, but she had only read the damning reports of racism rather than witnessed any violence with her own eyes. She was different from her haunted husband.

Ossian, at the age of five, had seen the burning of Fred Rochelle—a sixteen-year-old Negro who lived close to the Sweet home in Bartow in central Florida. Accused of raping and killing a white woman, Rochelle had been hunted down by a lynching party. They chained him tightly to a tree in front of a few hundred people. Little Ossian, hiding behind a smaller cypress tree on the banks of the Peace River in Bartow, had watched the mob set to work. He recognized some of the white men as they shouted "nigger" and drenched Rochelle in kerosene as if they were pouring water over a withered plant caught in the roots of the tree. The local butcher and jeweler were among those who then summoned the husband of the dead woman. They instructed him to light the match that would burn Fred Rochelle and turn a crying black man into a heap of gray ash. Almost twenty-five years later Ossian could not shake the memory from his head.

Yet his life had changed. When he was just thirteen, his parents had sent him away for an education in the North. He had been lonely then, especially when his father, Henry, instructed him to abandon his yearning to become a violinist. Ossian was told that the family had given up so much for him in order that he might pursue a noble profession. It did not take long for Ossian to settle on medicine as his grand destination.

In between working as a dishwasher, a waiter, and a bellhop, Ossian graduated from Wilberforce University in Ohio. He then completed his medical degree at the prestigious Howard University in Washington, D.C., in 1921 and moved to Detroit. Ossian was understandably smitten when, a year later, he met Gladys Mitchell. She was a striking young woman whose graceful poise and sharp intelligence were as attractive to him as her creamy light-brown skin and almond-shaped eyes. As he fell for Gladys he noticed how she flushed a subtle shade of coppery pink whenever she was moved by a deeply engaging conversation and, also, how irresistible little dimples cupped her mouth in a smile. Gladys epitomized everything Ossian had dreamed of when he imagined finding himself a wife.

Their married happiness was sullied only by the loss of their first child—a son who had been born ten weeks premature and died after three days in July 1923. Yet they conceived again almost immediately—and Gladys was in the earliest weeks of her second pregnancy when they set sail for Europe on October 6, 1923. With characteristic ambition and diligence, Ossian had scraped together the money to allow him to spend eight months in Vienna and Paris at the forefront of modern medicine. In Vienna he attended lectures given by the pioneering surgeon Anton von Eiselsberg, one of the founders of neurosurgery. Dr. Sweet's Parisian experience was even more illustrious for, at the Sorbonne, he studied under Marie Curie, who had already made history by becoming the first person to win two Nobel Prizes—for Physics in 1903 and then, eight years later, for Chemistry. Her scientific innovations in the field of radioactivity were unprecedented.

With Gladys at his side after every lecture, Ossian relished his time abroad. His hopes of creating an even better life for his family blossomed alongside his pregnant wife. Although she was denied permission to enter the American hospital in Paris to give birth to Iva, their baby was still delivered safely. That sole incident of racism, with its American link, also illuminated how dizzyingly free they had felt throughout the rest of their stay in Europe—a large period of which coincided with Mary Field Parton's own European trip in late 1923 and early 1924.

The Sweets arrived back home with baby Iva on June 21, 1924, just as, unknown to them, Mary and Darrow were together again in Chicago before the Leopold and Loeb trial. Ossian Sweet dazzled medical colleagues with stories of his continental adventures. His confidence blossomed and, in less than a year, Ossian's success as a doctor, specializing in gynecology, appeared to be symbolized by his indulgent employment of a chauffeur to drive him around in the Buick he had bought to replace his battered old Model T Ford. He could not really afford to employ his driver, a black man called Joe Mack, but Ossian liked to look good.

He still harbored a secret terror. Even Gladys did not really understand the deepest shadows of his past in the South. Ossian was content to shield her and to give into her sweeter wish to move into a house she could describe as a decent and pretty home. But he also fretted about the danger they might face. If she had wavered, Ossian would have gladly remained in the ghetto.

Gladys remained resolute. While they waited to move into the house of their dreams they could stay with her parents in Detroit. Even here, Gladys said, there was a lesson to learn. Her parents, light skinned in comparison to Ossian's pure black, had shown it was possible for a Negro family to live discreetly among white neighbors. And Ossian, especially as a doctor who wore conservative suits and favored a clipped mustache and horn-rimmed spectacles, exuded a propriety that had often been derided in Black Bottom. He would fit more easily into a suburban neighborhood, for in the ghetto he was regarded as aloof and well-to-do—a self-important doctor who considered himself a distinct cut above his Negro neighbors. It was not a charge that bothered Ossian, since he was dismissively arrogant of even his middle brother.

Otis Sweet, a dentist who had done remarkably well in his own career, was considered a little too crude for Ossian's moderate tastes. Working under the banner of "O.O. Sweet—Painless Dentist," Otis loved baseball, dancing, and chasing women. He left Florida long after his older brother, and Ossian said the marked contrast between their personalities represented "the difference between a Negro who has been educated in the North and one who has been educated in the South."

There was less tension between Ossian and his youngest brother, Henry Jr., a twenty-one-year-old college student. Closing in on his thirtieth birthday, Ossian was an established doctor, a husband, a father, and, at least to Henry, a revered figure. Henry modeled his own mustache and spectacles on those sported by Ossian, but he was far more gregarious and charming. He was simply thrilled at the prospect of spending the last two weeks of summer vacation with Ossian and Gladys in their new home—once they had completed their move on September 8.

As that day approached, Ossian felt increasingly uneasy. He asked both his brothers and some of his friends if they would accompany his move to Garland Avenue. Otis immediately agreed to spend the first few nights in the new house while Henry suggested their cousin and his college classmate, John Latting, would happily join them. Ossian's fellow Howard University graduate Julian Perry, who was now a lawyer, and Edward Carter, the doctor who had persuaded him not to succumb to the racists, also agreed to support him. With his driver, Joe Mack, and William Davis, a Negro former soldier and qualified pharmacist turned federal officer, Ossian was backed by seven able men. He still wondered if it would be enough.

Ossian put in a formal request for police protection, but he remained distinctly uncertain. He persuaded Gladys that, at least for the first few nights on Garland Avenue, they should leave Iva with her parents. Ossian told his in-laws that he did not want the baby distressed by the tumult of moving. Gladys's parents, Rosella and Benjamin Mitchell, nodded their understanding.

The mild-mannered doctor made sure that his .38 Smith and Wesson revolver was in working order and he also arranged for the purchase of an additional batch of nine guns. Ossian Sweet was neither provocative nor violent. But, with the cries of a burning man still echoing in his head, he prepared himself for the worst.

2905 Garland Avenue, Detroit, September 8, 1925

Four miles east of downtown, on a humdrum side street, Garland Avenue was far from the loveliest road in Detroit. The entire area, in fact, was much more downbeat than Gladys and Ossian Sweet liked to pretend. "Let me assure you that the neighborhood—not even a middle-class one—isn't one over which to become enthusiastic," Haldeman-Julius wrote sniffily after she had visited Gladys and Ossian. "The Sweets' house on the corner is the only really attractive one in it. Next to it on the left is a frame cottage owned by a piano-tuner. Across from them is a whole row of two-flat houses . . . and a grocery store. Diagonally across is a public school. Opposite them on Charlevoix is an apartment house. 'An apartment house!' I can fairly hear you exclaim."

There was less snobbery in the Sweets on the day of their move. If Ossian's trepidation was softened temporarily by the quietness of a sunlit street just after ten o'clock on a Tuesday morning, Gladys's delight intensified. The house, which she liked to call a bungalow, looked prettier than she had remembered. And, instead of the angry mob Ossian had feared, it seemed as if their arrival was noted by just a few silently watching housewives.

Ossian still felt relieved that they had arrived in a convoy. He and Gladys, sitting alongside Davis in the back of the Buick, were driven by Joe Mack. They were followed by a delivery truck with four more men—his two brothers, as well as Latting and Norris Murray, the driver, whom Ossian had paid to help them move. The other men would arrive after work to ensure they settled in safely.

There was further good news an hour later when Inspector Bert McPherson rapped on the door. He told Ossian his request for police protection had been received and that numerous patrolmen would keep a quiet watch on the property. Ossian might not have to reach for the guns, after all.

The rest of the day slipped past in a blur of opened boxes and excited chatter as Gladys and the men were joined by two of her friends, Edna Butler and Serena Rochelle, who helped with the unpacking and in advising how best to transform a small brick house into a special home. After such a long stay with her parents, Gladys could at last feel like a young wife with a delicious new life about to burst open in front of her. But Ossian could not relax. Even Serena's surname, Rochelle, matched that of the burning man in his head. Fred Rochelle might not have been related to Gladys's friend, but his fate shadowed Ossian Sweet.

As the afternoon melted away and the light faded, a passing chant that Ossian had heard earlier returned to plague him. The soft cry of "nigger, nigger . . ." had echoed around the house a few hours earlier. He knew it had been made by children on their way home from school and so he had ignored them. But as early evening crept across the street, Ossian's mind was full of worry. If their children could laughingly say the word that so hurt him, what might the men, returning from work, do with their far greater capacity for damage and hate?

Gladys noticed his tension, but she was intent on making their first night in a new home a memorable one. She insisted that everyone join them for dinner. While she and Edna and Serena worked in the kitchen, the men played cards in the sitting room, as purposefully as they could in order to keep the mood upbeat. The tempting aroma of Gladys's cooking helped as she prepared a homely chicken stew.

There was enough for all ten of them, and Gladys produced yet more exclamations of delight when she brought out ice cream for dessert—followed by cake and coffee. Even Ossian seemed less twitchy while they were all huddled together so closely. Yet, as soon as he folded his napkin and pushed back his chair, a yelp of alarm came from the window.

One of the men had pulled back the shades to peer outside. Although darkness had settled over Garland Avenue, he could see a mass of white faces outside. The crowd clogged the street and backed up into the schoolyard opposite the bungalow.

It was just after eight and the mood of the mob appeared strangely festive. People bunched together, talking and laughing, while children chased one another with shrieking glee. But Ossian Sweet shuddered. This was how it always started—the street party before the slaughter. He reached for the phone and called Ed Carter, the doctor who had openly derided Alexander Turner for not standing up to the mob when they had taken his new house. Where was Carter now? He had dropped by earlier that afternoon with a china tea set as a housewarming present. Gladys had been delighted; but Ossian needed him now, at his side, with a gun in hand, to fight off the crowd.

Carter became strangely evasive. He wondered if he might not be more help to them from the outside. Ossian slammed down the phone and raced upstairs, calling the other men to follow him so that they could monitor the movements from the second-floor windows. As he fumed over the cowardice of Carter, the man with the biggest mouth of all, Ossian watched the ghostly faces as they stared at his house. The white people were still joshing and joking but that one word—"Nigger!"—rose up again to remind him of the malevolence lurking beneath the surface.

Gladys and her friends huddled downstairs. She had already told Edna and Serena it would be best if they stayed the night. Gladys remained calm, but she also remembered how one white woman had walked up and down the street all afternoon, pausing in front of their house repeatedly. It had become so noticeable that Gladys counted her passing back and forth sixteen times. Who could tell how many other occasions she might have missed as, in between her counting, Gladys discussed all the colors she could choose for her cushions and curtains? Their animated afternoon, rippling with talk of soft furnishings and subtle decoration, gave way to a downcast hush. But Gladys was determined they would not be frightened away. They just needed to get through that first night.

As the mob milled around, uncertain what they might do after the socializing and name-calling, Ossian settled in for a long wait. An attack could happen at any time—but he thought it was more likely to occur at some point deeper in the night when the children and wives had been packed off home and the men outside became restless and edgy. But Ossian and his boys would be ready. They had each other, and their guns.

It was vital, he stressed, that at least one of them maintain complete concentration—so that they could prepare themselves at the earliest sign of trouble. The men agreed, and an informal rotation was drawn up for them to take turns to keep watch through the night.

Two more hours passed, and, apart from the occasional shout in the street below, the small group in the Sweet house settled down. It became easier to breathe a little more naturally and to even sit in a comfortable chair. Gladys and her friends took turns lying down on the lone bed that had been lifted from the delivery truck less than twelve hours earlier.

Ossian may have even closed his eyes, to briefly replenish himself, when the stones flew through the dark. They clattered against the roof with a rolling metallic sound that, in its abrupt suddenness, shocked them all. Ossian peered down in fright at the street below. The eerie white faces looked shadowed and indistinct. It was just before midnight, and the lights along Garland were muted. But he could see some men running back to the safety of the mob. The voices outside were a little louder in their taunting. Ossian's hands, more used to healing and helping people, tightened around his gun.

The next twenty minutes slipped past, and the crowd gradually fell back into sullen silence. Ossian put down his gun and gave the nod for Henry to resume his watch. Another hour slid by and not a single stone skidded across their roof. No rocks were thrown through the window downstairs, and the crowd did not move any closer.

Ossian walked slowly down the corridor to where the women waited in the dark. Gladys squeezed his hand, as her two friends curled around each other for comfort. He suggested they should all try to get some sleep, and Gladys reminded her husband that he should attempt the same. Ossian nodded and returned to his vigil.

It was just after 4:00 A.M. when Gladys, still unable to sleep, went to find Ossian. He sat near the window and looked up at her. She stared down at the street and saw a solitary policeman on the corner of Garland. The crowd had gone. Everyone had disappeared into the dark, retreating to their own homes.

Dawn was just a few hours away. Perhaps they would, as Gladys always said, be left in peace to live their own ordinary lives. Ossian looked typically serious, but, as she smiled down at him, her dimples showing, his hand covered hers. A man stroking his wife on their first night in a new home, he felt calm at last.

Ossian's spirits were lifted by the passing of that first night, and by spending a few hours with Gladys on another sun-kissed morning. They had wandered around the shops before buying a dining-room suite, a couple of armchairs, and more furniture for their two bedrooms upstairs. Ossian signed a check, made out to Lieberman's Furniture Store, for the startling sum of $1,200. After all they had endured, it felt good to be extravagant. And Gladys was so happy that even Ossian began to feel that it might work out all right in the end on Garland. They drove over to her parents and cuddled little Iva, promising that they would soon be together in the new house.

He went back to work in Black Bottom late that morning, as part of his attempt to return to a normal routine. But, in his office, Ossian struggled to maintain his optimism. Even though he had not discussed his feelings with Gladys, he was certain that at least some of the same crowd would set up camp again outside their house. Who knew what the mob might have in mind this time? That mournful thought took hold of him in the middle of the afternoon. It would not be long before Joe Mack drove him back in the Buick to face another night.

When his phone rang shortly before three at his office, Ossian was in turmoil. But he had recently taken out a $10,000 insurance policy on the Garland house with a black-owned company called Liberty Life. Hewitt Watson, the Negro agent who had sold him the policy, explained on the phone that an administrative error had been made in relation to Ossian's birth date, and Watson needed the corrected form to be signed. Before Ossian really knew what he was doing, he began to relate the ordeal he had just experienced at his new home. Since Watson was sympathetic, the doctor confided further in him. He soon pleaded with the insurance man for help.

Watson must have been moved, for he agreed that he would drive over to Sweet's office with two associates, Leonard Morse and Charles Washington. Then, after attending to the paperwork, they would head across town to Garland and spend the night in his house. Ossian might have ordinarily hated his reaction, but, in such circumstances, he did not mind when he heard himself babbling his gratitude.

2905 Garland Avenue, Detroit, September 9, 1925

Gladys rushed to Ossian as soon as he walked in around four thirty that afternoon. It did not take long for the news to spill from her mouth. Edna Butler, on leaving the house that morning, had taken a streetcar home and she was jolted by a conversation between the conductor and a white woman. After the conductor asked about the disturbance on Garland the woman spoke sharply.

"Some niggers have moved in and we're going to get rid of them," she said. "They stayed there last night but they'll be put out to-night."

Ossian's voice grew husky as he promised Gladys that they would soon be joined by three men from the insurance agency. They were not alone.

Otis had not returned from work, but Henry, the youngest of three brothers, had spoken to a patrolman who warned that a much larger crowd was expected that night. Henry was young and defiant and so he had still walked to the store across the street. Amid the suspicious silence he had bought some cigars. He and John Latting were now smoking and reading on the porch as if they anticipated nothing more than another good meal and a quiet night.

Ossian ordered them inside. There was no need for them to incite anyone by looking so obviously at home. As he had done the previous night, Ossian closed the blinds to the frightening white world outside.

There was a measure of relief in the arrival of some reinforcements. His driver, Mack, had picked up the moving man, Murray. Ossian had promised to pay Murray $5 if he returned for another night. They were followed, just after six, by two of the insurance agents—Watson and Washington. The third man, Morse, had failed to meet them at their pickup point.

Ossian knew they would soon be joined by Otis, but how could eight of them hold back a crowd of hundreds? The first wave of white people, having taken dinner early in order that they might arrive in good time for the night's entertainment, could already be seen outside. At the end of a normal day it would have looked as if a beautiful evening had enticed them out onto the street.

And then, unexpectedly, Morse emerged. Having arrived too late to hitch a lift with Watson, he had bravely taken a streetcar out to Garland. Morse had not been accosted when he headed to the corner house. He had been able to walk up the short drive, press the doorbell, and take Ossian's thankful hand. Surely the whites would have threatened Morse if they planned a lynching?

Gladys was in the kitchen when she welcomed the men. She tried to speak happily, hoping they were all ready to eat a hearty meal of roast pork, baked sweet potatoes, and mustard greens. The men smiled queasily, their appetite soured by the fear running through the house. Ossian took them upstairs and showed them the linen cupboard where, in place of crisply ironed sheets and pillowcases, they saw his stash of weapons.

The ensuing game of whist, which they played while waiting for dinner to be served, was a distracted affair. Ossian insisted on jumping up every five minutes to check the doors and locks. He was also anxious that neither Otis nor his old friend William Davis, who'd been with them the previous night, had returned to the house as planned.

It was just before eight, four hours earlier than the previous night, when the first missile crashed against the roof. Ossian thought they must have thrown a small rock rather than a handful of stones. He walked quickly to the window. The breath caught in the back of his throat when he saw the crowd. Two or three hundred people had massed together in the last twenty minutes. There was an ugly silence around them.

Inside the house the men peeled away in panic. A couple rushed to check the back door before they joined the rest hustling upstairs, swearing softly or breathing fast and hard. Ossian made sure that Gladys had also left the kitchen, switching off the heat under all the pots and pans of a meal that was almost ready to be eaten. He snapped off the downstairs light. The rooms upstairs, now containing all of them, were already cast in darkness. Ossian battled to find his revolver and the bullets he needed. Finally, armed and ready, he entered his bedroom and steeled himself for the assault.

But another five and then ten and possibly even fifteen minutes passed. Ossian Sweet waited in the dark, hearing the hum of voices outside, and flinching at every creak of a floorboard. He kept expecting the next rock to fall, or the first rush of men at his door. Yet, as if they wanted to torture him first, they kept him waiting.

Gladys moved toward him in the dark. They whispered encouragingly to each other, but there was nothing more they could do but pray silently for help.

Suddenly, the next rock was hurled, smashing through his bedroom window. Splinters of glass flew toward him, making him cover his face with his arms. The rock rolled across the floor and came to a stop near his feet. He could have touched it with his shoe had he not been startled again by the sound of a voice from the room next door, yelling at him that Otis was outside. Ossian could hear the rumbling sound of a taxi, so he charged downstairs. The thought of Otis being torn apart by that crowd overcame his fear.

He could hear his brother hammering on the door, desperate to be let in. Ossian flung open the first door. But he struggled to pull back the bolted lock to the second outer door. When he at last managed to open it, and the warm night air hit him in the face, he saw his brother and William Davis, who had returned courageously for another night at their side. And beyond them, ten feet away, he saw the white men, their mouths open with abuse, their faces riven with hate. "Niggers!" one of them shouted. "There they are! Get them!"

A few months later, when he recalled that terrible scene, he looked into the face of a sympathetic white writer, Marcet Haldeman-Julius, and spoke with, in her words, "a quiet, dramatic intensity that even now gives me the shivers." Her chilled reaction was nothing to what Ossian Sweet experienced.

"When I opened that door to let them in," he told her, "I realized that for the first time in my life I stood face to face with the same mob that has haunted my people throughout its entire history. I knew that my back was against the wall, that I was black, and that because I was black and had found the courage to buy a home, they were ready to wreak their vengeance upon me. The whole thing, the whole situation, filled me with an appalling fear—a fear no-one could comprehend but a Negro, a Negro who knew the history behind his people."

Henry Sweet, who had pulled them back toward the temporary safety of the house, spoke more simply to Haldeman-Julius. "It looked like death if we tried to hide, and it looked like death if we tried to get out. We didn't know what to do."

Inside the house, with the door once more slammed and bolted, Ossian echoed his brother's refrain. "What shall I do?" he whimpered as a frightened Gladys joined them. "What shall I do?"

More rocks bounced off the roof. Davis managed to get some words out. He told Ossian they just needed to sit tight and hope that the watching policemen, whom he and Otis had seen, would hold back the crowd.

Ossian nodded dumbly as Henry, carrying a Winchester, ran back upstairs. Only the doctor, his wife, and Davis were left downstairs. All the other men were in rooms on the upper landing. Ossian held Gladys, and waited.

The next rock shattered one of the top-floor windows, but, before he could register the sound, Ossian was flung back by a blazing roar. He had rarely heard such a loud noise before and he fell down, certain that they were being fired at by their attackers. It was only after the next clattering barrage of bullets that he understood. Someone upstairs had used a gun.

Ossian could hear screaming and shouting from outside. Davis yelled at him to make them stop shooting from one of the rooms above them. When Ossian called out to the others, there was a deathly silence. He sat slumped in shock, with Gladys, against a wall. The different noises from the street were hard to identify, but Ossian thought he could hear the sound of men running.

He did not know how long he sat there. It was only when the doorbell rang, long and hard, that Ossian jerked himself upright. He immediately thought of Alexander Turner, the black doctor who, a few months before, had opened his door only to have his house taken from him. But he still shouted out, asking who was at the door.

"The police," a voice boomed.

Ossian instinctively reached toward the locked door. When he pulled it open, his initial relief at seeing the blue shirt of a large police office was blunted by the man's anger.

"Jesus Christ!" Inspector Norton Schuknecht of the Detroit police thundered as he entered the house. "What in hell are you fellows shooting at?"

"They're ruining my property," Dr. Sweet said.

Schuknecht, however, stunned him by insisting that he and his patrolmen had seen nothing disorderly from the white folk outside.

Ossian Sweet did not have the will to argue. He promised that there would be no more shooting from his house.

Schuknecht grunted and wheeled away. It looked to him as if the colored fellers had just had the fright of their lives. He knew there would no more trouble from them.

But once he had stepped back out onto Garland Avenue, the rotund and sweating police inspector knew something awful had happened. You did not need to be a policeman of his vast experience to recognize that the mood of the crowd had changed during the few minutes he had been in the house. There was a manic sense of violence in the way the lines of men rocked back and forth as if it needed just one wild soul to risk another bullet for the whole pack of them to storm the house and tear the people inside limb from black limb.

A patrolman told him there had been two victims of the shooting. One man had taken a bullet in the leg. He was down on the street, howling in pain. The other fellow had been hit far worse. With a trembling hand the patrolmen pointed to a motionless figure. The man looked dead.

Norton Schuknecht swore in the dark. He stood for a moment in shock, and then he got moving. Schuknecht ordered his men to make a human cordon around the Sweet house. No one was to get in or out until the hospital wagon had arrived and he had the backup he needed.

The crowd eyed him jumpily, unsure whether they should test the authority of one of their own people. But their attention was soon dragged away by a cry one block to the right of the Sweet home. "There're more niggers," someone yelped as he saw three black men caught in traffic on the corner of Charlevoix and St. Claire. The crowd surged toward them, and the fastest white men reached the car just as the driver gunned his way through to a less crowded lane of tar. He put his foot down as the attacker clambered onto the running board. The white man fell off as the car gathered speed.

By the time he had been hauled to his feet, another car had been spotted. The black couple inside were helpless while their windows were smashed. As the men punched her husband, the woman begged for mercy. The crowd finally broke away, some darker impulse pulling them back toward the Sweets.

The thin blue line of patrolmen still separated them from the house. In the clammy night, as if the mob had taken on a mind of it own, the throng paused for thought. No one moved, as if they all knew it would be a step too far to attack the scared white policemen.

When his reinforcements arrived, Schuknecht took a dozen of the new men to join his best officers. After he had ordered one of his sergeants to beef up the security ring around the house, he called his men together. "We're going in," he said, "to get these Negroes."

Schuknecht, knowing the depth of fear in the Sweet house, chose not to hammer down the door. He simply pushed the doorbell again. When Gladys opened, he brushed past her without hesitation as he and his men took control. They closed the door and moved swiftly, from one room to the next, rounding up all eleven Negroes. Schuknecht counted them—ten men and one young woman.

Ossian was the first to be handcuffed, a ring of steel snapping shut over his left wrist. He was pulled toward Davis, whose right wrist was then cuffed by the second ring. The two men, looking like two small boys, were now chained together. As the others were cuffed the doctor pleaded with Schuknecht that they might be taken down the alley—away from the waiting mob.

Schuknecht hesitated, and Joe Mack, Ossian's driver, stepped forward. He recognized one of the white policemen, Lieutenant Blondy Hays, and spoke softly to him. Ossian could not make out the words, but Mack did well. Hays nodded and, with unexpected gentleness, he approached the black doctor. He introduced himself and explained that he was one of the officers enlisted by Inspector McPherson to protect the house. Hays might have felt a pang of guilt at their failure to help Sweet because he ordered one of his men to uncuff the doctor. He sent another policeman out the back to ensure that the wagon reversed up toward the alley.

Hays turned next to Gladys. A group of his men would wait with her in the house until the crowd outside had dispersed. They would then take her down to the station. But, now, it was time for Sweet and the others to leave the house.

On only their second night in their new home, Ossian and Gladys looked hopelessly at each other. And then, with a small push, Ossian was steered toward the kitchen door. The cooling roast of pork, still waiting to be carved, rested on top of a work surface. Gladys, who'd had such dreams of meals together with their friends in this house on Garland, followed Ossian with her eyes as he was led away. His brothers, Otis and Henry, and the seven other men who had tried to help them trailed him down the alley.

The driver, having backed his vehicle into the garage where the Sweets' Buick was parked, kept the engine running. Hays knew there could be yet more death if he did not somehow control the furious white gang who had pushed up against the wagon. A low growl went up from the mob when they saw the line of black men. Hays raised his gun and warned that he would shoot anyone who stormed the wagon.

As his words rang out, and some of the crowd backed away, the police bundled the ten men into the wagon and slammed the doors shut. At the sound of clanging metal, the driver began to edge away from the garage. The vehicle prodded its way through the crowd, as people moved back grudgingly and then a little more readily as it gathered momentum. Slowly, the driver pressed his foot down against the accelerator, and the wagon, containing a doctor, a dentist, a pharmacist, three insurance men, two ordinary workers, and a couple of college students, picked up speed.

Inside, swaying on the metal seats, the ten men were silent. They had been spared a lynching, but, behind the iron grill of a police vehicle racing through Detroit, they were headed for yet more trouble. A white man was dead. How many of them might end up paying the ultimate price?

INSPECTOR NORTON Schuknecht's account of the murder fueled the one-sided and sensational news story splashed across the front page of the *Detroit Free Press* early the following morning. "Shots poured without warning and seemingly without provocation last night at 8:30 from the upstairs windows at 2905 Garland Avenue, in which Dr. and Mrs. Ossian H. Sweet, a Negro couple, had moved Tuesday, cost one man his life and put another in the hospital with a bullet in his leg. A police inspector, a lieutenant, and a patrolman had been standing within thirty-five feet of the house which was lighted only in the upper window. They had been talking over the potential trouble that might result from Sweet's moving into the white neighborhood. A rain of bullets ended the conversation. As if the triggers of a dozen guns were being pulled simultaneously, the firing started. It stopped with the same suddenness that it began. While the police ran up the stairs and hammered at the door, the gathering throng counted the casualties."

Leon Breiner, a foreman at Continental Motors, who lived on Garland Avenue, was publicly identified as the dead man. Eric Hough-berg, an even closer neighbor to the Sweets, had been luckier. He would survive being shot in the leg.

Early that same afternoon, on September 10, the *Detroit Times* rehashed the story, only altering the tone so that it became even more accusing of the ten men and lone woman being questioned by the police. "The shooting was not provoked," the newspaper insisted. "A small crowd was in front of the house, on both sides of the street, but no threats were made and no missiles were thrown. Suddenly an attic window was thrown up and a rifle was fired. A score or more shots followed in quick succession."

All four members of the Sweet family, and their seven associates, were duly charged with killing Leon Breiner. They would face a murder trial together.

T HE STORY HIT the New York papers the following day, on September 11, 1925. James Weldon Johnson, the dynamic secretary of the National Association for the Advancement of Colored People, read of the shootings and the Sweets' arrest with acute interest. He instinctively knew that there must be some deeper context to their case, and, as a pioneering civil rights activist, he resolved to uncover the truth.

Johnson, who was also a writer, had yet to confirm his authorship of a book he had written anonymously in 1912—*The Autobiography of an Ex-Colored Man*. But he had become increasingly involved in politics. Apart from passionately supporting the Dyer Anti-Lynching Bill of 1921, Johnson was one of the leaders of the Harlem Renaissance, a loose collective of black writers who worked together to create a better world for the "New Negro."

Johnson persuaded his assistant secretary, Walter White, another emerging black author, to travel to Detroit to investigate the Sweet story and establish whether it was a case that the NAACP could support financially and politically. It did not take long for White to realize the significance. "Dr. Sweet and the others are martyrs to the cause of freedom for the Negro," White wrote in a compelling report. "The eyes of America are on Detroit."

The Sweets were represented by a team of black attorneys—including Ossian's friend Julian Perry—headed by Cecil Rowlette, a brash criminal lawyer. Yet Otis Sweet was uneasy in his prison cell. He was finally galvanized by a visit from his father. When Otis began to apologize for the shame he and his brothers had brought on the family, Henry Sweet Sr. spoke quietly but firmly. "You have nothing to be embarrassed about," he said. "Only rabbits run."

Inspired by his father, Otis Sweet urged White and the NAACP to take control of their case. White knew that he had little time left because, on October 5, Judge Frank Murphy announced that *The People v. Sweet* would be brought to trial in two weeks' time.

The next day the Ku Klux Klan won an overwhelming victory in the Detroit local elections, sweeping one district after another in the wealthy west side suburbs and the downbeat flatlands of the east. N. K. McGill of the black *Chicago Defender* had already written to White at the NAACP. Drawing a direct link between the rise of the Klan and the plight of the Sweets, McGill urged him to accept some help from Chicago. There was one man, a friend of McGill's, who might just be able to save the Sweets. That man was Clarence Darrow.

ON OCTOBER 7, 1925, James Weldon Johnson sent a wire to Darrow's office in Chicago: "Case is dramatic high point of nationwide issue of segregation in which National Association for Advancement of Colored People has case now pending in the United States Supreme Court. This issue constitutes a supreme test of the constitutional guarantees of American Negro citizens. Defense requires ablest attorney of national prestige that we can possibly secure who would be willing to undertake such a case. Please wire us collect if you would consider favorably request that you assume charge of this case."

Darrow was nowhere near his office that day. He was in New York with Mary Field Parton.

They were enjoying an early lunch when Arthur Hays rushed in with the cable that had been forwarded to his office from Chicago. Darrow read the message thoughtfully. He was already a member and sponsor of the NAACP, and, despite his best intentions of staying out of court, the urgent cable tugged at him.

He nodded at Hays's suggestion that they should meet that afternoon with Johnson. And then Darrow looked across at Mary with a faint smile. "I know," he said, "I know . . ."

THREE HOURS LATER, in the middle of the afternoon, four men answered Hays's invitation to meet at his Greenwich Village apartment. Darrow was propped up in bed with a book, wearing all his clothes, as Hays led them in to meet him. Johnson and Walter White had been accompanied by two lawyers who assisted the NAACP—Arthur Spingarn and Charles Studin.

They were taken aback to find a courtroom giant like Darrow in such an unexpected setting. Darrow rolled out of bed with a muffled groan and shuffled over to shake hands with each of them. Once he had slipped on his shoes, he and Hays took them into the study. As the head of the NAACP's National Legal Committee, Spingarn detailed the bare facts of the case, and the trauma of the Sweets, with precision.

Darrow had already made his familiar comic complaint that he was far too decrepit to be the man they needed. But the brightness of his mind, and the intensity of his reactions, were evident as the men noted how closely he listened. The occasional prompting question from Darrow was similarly acute, and he grunted in restrained anger during Spingarn's presentation.

The old attorney had been looking down at his hands, as if wondering whether they might be able to stand another bloody fight in the ring. When he looked up at Spingarn, he spoke softly.

"I know full well the difficulties faced by your race," he said.

Spingarn, who was Jewish, and dark in complexion, smiled quizzically. "I'm sorry, Mr. Darrow," he said, "but I'm not a Negro."

Darrow flushed at his mistake. He turned to Studin, who was just as swarthy. "Well," he said, "you understand what I mean."

"I am not colored either," Studin replied.

Darrow apologized and laughed gently. "At least I wouldn't make the same mistake with you," he said to the light-skinned Walter White.

There was a brief pause as White, not wishing to embarrass an old man still further, hesitated. But he did not want to lie to Darrow. "I am a Negro," he eventually admitted.

Darrow cackled in amusement. "That settles it," he said as he reached out for White's black hand. "I'll take the case."

CHAPTER 13

BITTERSWEET BLUES

MARY FIELD PARTON had found her retreat. Two weeks earlier, on a gorgeous Friday afternoon in New York, September 25, 1925, she had finally discovered a small attic room to rent. It had taken months of slogging around the city—"on the trail of that extinct species: a cheap flat"—and she had frequently felt discouraged when returning to the farmhouse in Sneden's Landing. On the modest amount she had scraped together to rent herself a space to write, and to indulge a more private life, it seemed as if "only the ugly, the commonplace, the noisy, the disagreeable is obtainable." But the attic at the top of a beautiful block in Greenwich Village was perfect.

Comparing herself with her husband, Mary had concluded that "Lem's work is hard, interesting and ends at night. Mine is hard, uninteresting and endless." She hoped that, during occasional escapes to the Village, she might taste the freedom that marked a more stimulating working life.

Mary was at her happiest in the summer months, at the farmhouse, when Margaret's extended holidays meant they could avoid the numbing routine of getting her to and from school on West Twelfth

Street. But a new school year had just begun and Mary felt strangely sentimental on the last weekend they had spent at Sneden's before packing up and moving back to the city and the family apartment on East Eleventh Street.

It seemed a ritualistic departure, for the Sunday of their leaving coincided with the ceremonial grape crushing in the Partons' first attempt to make wine for themselves. As soon as they had completed the purchase of the farmhouse five months earlier, she and Lem had paid immediate attention to rebuilding the grape arbor, which, by mid-September, was draped luxuriously with heavy vines.

Mary had scrubbed her young daughter's feet in the bath and then, surrounded by an excited throng of chanting adults, little Margaret had been carried down to the sweet-smelling arbor. Her bare feet looked very clean and very pink as she was lifted up over the rim of a giant tin tub filled with plump and freshly cut grapes. And then Margaret, concentrating fiercely, had begun to stamp her feet up and down in the sea of grapes until her skin was colored purple and red while men and women around her shouted out exclamations like "How charming!" or "How Italian!"

The wine stains could not be scrubbed away for days, meaning that Margaret had started the school year showing off legs splattered with evidence of their harvest during Prohibition. Mary knew that they would enjoy the illicit taste of their labor the following summer but, for now, back in the city and with the farmhouse to be visited only on weekends in the fall and far less during the winter, she would have felt unbearably melancholic without her new attic refuge.

She went there in her quiet moments, when Margaret was at school, or if Darrow was in town. But their age, and marital status, defined everything. Mary had known for fourteen years that Darrow was simply too self-absorbed to fight for her at the expense of his marriage to Ruby. She had come to that painful realization in Los Angeles in the summer of bribery, after four years of forlorn hope that he might summon the will to leave Ruby for her.

Memories of those lost days had been rekindled when she took an old white shoebox of letters with her on the October morning she moved a few writerly possessions into the attic room. Mary, slowly and hesitantly, reread a few of Darrow's letters to her.

"I can hardly think of you as married, dear Mary," he had written in the summer of 1913, "and yet I presume it is—but I am glad that he is good and manly and big." And while he filled much of that two-page letter with his adulation for Nietzsche—or 'Nietzschie" or "Nieche" as he wrote, for even then he could never quite master the spelling of the philosopher's name—Darrow failed to conceal his emotions. "How often I think of you and our many good times even in the presence of trouble, trouble which was no worse than any other trouble but which was bad enough. . . . I have been very quiet and very good, and very stupid. That was not my reason for writing this letter but I want to see you. But I don't see how I can now."

On July 4 of that year, writing from the Riverside Hotel in Montevideo, Minnesota, he lamented to his lost lover, "I am blue and lonesome tonight and wish I could see you so I would be less blue and not so lonesome. . . . The only feeling in the world that can make you forget for a little time is the sex feeling. . . . I wish you were here with me now. Ever and ever."

By August 8, Darrow thought he had found a solution to his problem of no longer seeing the newly married Mary regularly. He encouraged her to write his biography. Darrow also welcomed, with strained diplomacy on his part, Lemuel's acceptance of their continuing friendship. "I do want you to write the book—to begin it now. I want it partly for me and those who come after me but mostly for you. I am so glad that Lem doesn't want marriage, like death, to end all, for I know you can do wonderful things, things he and all of us will be proud of."

As Mary reflected on these letters, written over a decade before, it became obvious that Darrow yearned for her most when she was beyond his reach. If he could not have her, he suddenly realized how much he needed her. His pining explained why he had sent her so many letters shortly after her marriage. Early in 1914, with Lemuel and Mary having abandoned a wild plan of prospecting for gold in Nevada to return to a steadier life in San Francisco, Darrow pursued her once more with the prospect of his biography. "Don't get tired and contented," he warned her. "You can do anything and when it is worthwhile you tower over others. You must write my biography. I would rather have you do it than Tallentyre. You know I don't think anyone would have heard of Johnson except for Boswell and you might make an unknown [famous] sometime."

On March 4, 1915, he had written mournfully, "For some reason I am feeling in the dumps today, so I write you. I haven't heard from you for a long time and I enjoy your letters and news when they do come. I wish I could see you and hope you are happy. But you are not. No one is happy, not anyone who is built like you and me."

If Mary remembered being flattered that Darrow considered her character akin to his own, the physical distance between them intensified her detachment. While Darrow battled to reinvent himself as a criminal lawyer, Mary had became a mother, giving birth to Margaret in 1915.

Yet they remained friends. In a six-page letter on January 19, 1918, Darrow pointed out, "Mary, I will be sixty-one in April. I don't care a damn—nor do you but well I know that sixty-one is quite a way along the road and I keep thinking of you and that biography, which I would only have you to write. No one else could imagine the good or forget the bad as you, my dear friend, and if you can come this way in the summer I think I can get the money together and the time that I would like to see you. Can Lem spare you?"

The biography was never written—despite Darrow's entreaties. His letters became more subdued in tone, and, once he had eased into his successful new vocation in criminal law, he expressed his philosophical gloom and calm acceptance that their affair would never be rekindled. "I am growing a more confirmed pessimist all the time," he wrote on September 8, 1922. "Nothing is worthwhile except to keep the emotions at work so we can forget life. At least our part of it. How I wish I could talk it all over with you. Sometime I hope I shall. What you say about the cooling of the passions is likewise true. They have a distinct function and when they have finished their work we should rejoice. I do for me. As I grow older I think I grow more reconciled to life and death, and neither expect or hope for anything. . . . When I look in the glass to comb my hair I can see the gray slowly coming on and it gives me pleasure. It is part of the whole premonition. Love always. Clarence."

The fact that he sometimes signed himself as the dreaded "Clarence" was a further sign that their relationship had eased into a calm friendship.

And then, out of the scorching heat of June 1924, he had written with the startling news that he longed to see her again. She had answered his call, and yet that week in Chicago proved that nothing had really changed. "Darrow is Darrow," she wrote wistfully to her old friend Helen Todd.

Darrow is Darrow—to anyone who knew him intimately, that apparently reductive phrase defined an acceptance of everything that made up an extraordinary man. He was selfish yet compassionate; kind to the damned but sometimes cruel to those he loved; sweetly amiable yet relentlessly ambitious; maddening but inspiring; profound yet lecherous; hurtful but human. In the end, bowing to a favorite book by Nietzsche, or even "Nietzschie" or "Nieche," Darrow was human, all too human.

A line from one of those letters—"No one is happy, not anyone who is built like you and me"—had haunted her. She sometimes did wonder if she and Darrow were capable of any sustained happiness in their individual lives, for they were alike in their restless hedonism, in their anger and regret, and yet those darker traits of character also drew them together. But since she had taken the night train to Chicago fifteen months earlier, in the giddy hope that everything was about to change in her life, she had been restored.

Less had changed between her and Darrow than the notable shift within herself. Mary had finally published a book, about Mother Jones, and taken the attic in the hope of writing another. She had also uncovered the extent of a lasting love for her daughter and husband. Mary had once written, with bitter secrecy, of her hapless struggle as a mother. When Margaret had infuriated her one long day when she was very young, Mary had scrawled a harsh and desperate line in her journal: "My darling child, I should love you as one loves a poor little crippled mind, a small heart—misfortunes for which one is not to blame."

Those cutting words echoed a more laconic letter soon after Margaret was born when, after bewailing the ceaseless demands of motherhood, she had concluded to her sister Marion, "Hell, I made the kid so I guess it's up to me to raise it."

The journey she had since taken spoke more of her own recovery than the growth of her daughter. She now understood what was most important in her life, and she could write of her ten-year-old with sweeping tenderness and love. "M is my beautiful girl. Her complexion is like peach petals—both in texture and color. There is a coolness, a calmness about her which gives her, as a child, a distinction, a personality."

And there were days at the old farmhouse when the sultry hours slipped away peacefully, as she and Lem talked and walked down the country road, alongside the meadow they now owned, sometimes stopping to take a seat on the low stone wall that skirted the border of their land. They lost themselves utterly on those afternoons when, as Margaret played in the giant chestnut tree, they nurtured their vineyard or Lem's vegetable garden.

These were small moments, little vignettes of hope and pleasure, and yet they made her feel tranquil in the lengthening stretches between her longing to do something different and strikingly independent. Mary felt that her own fascination with Darrow stemmed from an affinity with his desire to fight new battles and conquer different worlds. His striving for greatness left her almost breathless in its wake.

Mary heard his repeated claims to be exhausted and "just about done," and she could see a real weariness in him amid the shuddering bursts of neuralgia and the steadier ache of his rheumatism. But then he surprised everyone by taking on his third epic trial in the dizzying space of a mere fifteen months. She did not quite understand where he found the energy to follow two "trials of the century" with another of even more sapping cruelty and darkness. He had claimed his redemption from bribery, twice over, but something more meaningful now spurred on his quest to save the persecuted Sweets of Detroit. This case, he promised, would surely be his last; but it would be the one that mattered most to him.

It transformed his mood, making him feel so assured and mischievous that soon after he accepted the NAACP brief he tried to persuade Mary to join him on his latest adventure. When they were still together in Greenwich Village, he asked her to accompany him to Detroit under the guise of a freelance writer covering the Sweet trial for one of the New York papers. She could watch him dazzle the courtroom by day and they would party at night. Mary looked at him as if he had gone completely mad. She pointed out dryly that, on October 18, he would be exactly sixty-eight and a half. They could use that as another reason to celebrate, Darrow said, brushing aside Mary's bruised reminder that she was forty-seven. Her partying nights had ended long before. Darrow kept at her, telling Mary what fun they would have in the city where she had grown up.

Darrow knew her personal history intimately, so she was irritated by his suggestion that they should choose Detroit, of all places, during a bleak trial, for their latest tryst. He could, to her, be ridiculously thoughtless. Her youth in Detroit had hardly unfolded in a festive atmosphere. Darrow understood that her own antipathy for organized

religion had its roots in the grim Detroit house ruled by her puritanical father, a Baptist minister, who bullied her mother and terror-
ized Mary and her three sisters and younger brother. Her mother, a
Quaker, was gentle and kind but her father, George Field, had been
the fiercest of fundamentalist preachers. His stern doctrines and hectoring against drinking and the sins of the flesh had marked Mary.

It had only been in her adolescence that she and her sister Sara had
begun to rebel against him. But their revelry was initially restricted
to a furtive chewing of gum—which their father deplored as "the sin
of all sins." Mary would hide the chewed wedge of gum underneath
the dining room table so that Sara could take her turn and pop it
in her own mouth when his back was turned. It had needed more
courage for Mary to stand up to her father, at the age of seventeen,
to ask him if she could go to college. He refused to answer her for a
few days, and when he finally did, after she had asked him again, he
announced that he would only allow her to attend a Baptist college in
Kalamazoo.

After she had written to one of her mother's friends, who offered
to pay her college fees, Mary ran away from home to begin a new life
in Ann Arbor. She found a place for herself at the University of Michigan, where she studied history, philosophy, German literature, and
sociology. Mary's father wrote to tell her that she should never again
darken the door of their family home. But she was fired by a desire
to change the world and, after she gained her degree, she headed for
Chicago, where she finally discovered radical politics and a burgeon-
ing sexuality for herself. Chicago meant freedom and fulfillment to
Mary; Detroit simply evoked misery and repression.

How could Darrow expect her, a woman closing in on fifty, to exult in a return to the city where she had suffered such torment as a
young girl? And, anyway, even if she joined him in Detroit, what about
Ruby? What would his long-suffering wife say if she saw Mary again?

The old attorney shrugged. It was as if the risk could only sharpen
him further. They would find a way, he said, to be alone together.

Mary shook her head angrily. He had crucial work to do while she had her own family to consider. She could not bear the thought of telling Lemuel and Margaret that she was off to Detroit for a month or more. It was absurd, and cruel, and she would hear no more of it. This was the closest they had come to a bitter argument since they had met again in Chicago. But Mary knew she was right. Darrow would have to find another way to amuse himself in Detroit.

"I might just do that," he muttered, before softening his words with an apologetic hug. "But I'll miss you, Mary Field . . ."

THE SWEET case cut deep into Darrow's past. It rekindled hazy but still powerful memories of his father, Amirus, and the ghostly figures who had drifted through his Ohio childhood. Amirus had spoken often around the dinner table on Sunday evenings of the great fighters against slavery. His stirring words were underpinned by stories of all the famous abolitionists who had visited the Darrow home in Kinsman on their way to various "safe houses" for slaves in the secret network known as the Underground Railroad. As an old man Darrow had allowed the facts to blur with his myth to such an extent that he was no longer sure which of the giants of the anti-slavery movement he had actually met. Frederick Douglass, Wendell Phillips, William Lloyd Garrison, and Charles Sumner were all said to have been among the abolitionists who had gathered under the Darrow roof.

A folktale had also developed in which it was claimed that, at the age of five, Darrow had met John Brown, the most militant of all the white abolitionists. While visiting Amirus, Brown had apparently placed his hand gently under the chin of little Clarence and spoken with somber passion: "The Negro has too few friends. You and I must never desert him."

Brown had since become an American martyr. On October 16, 1859, he attempted to spark a revolution against the slave trade by leading twenty-one men in a bid to seize control of the federal arsenal at Harper's Ferry. Two nights later Colonel Robert E. Lee planned the decisive attack on that ragged army. Brown was captured and tried for treason, with his hanging on December 2 in Charles Town helping to fuel the outbreak of the Civil War.

And so Darrow's admirers lingered over the imagined scene in which a small boy called Clarence looked up into the face of John Brown—and felt the inspiring words burn into his heart. No one apparently noticed the discrepancy in dates from a different century. Darrow was only two at the time of the abolitionist's execution, so if Brown had said those actual words, they must have been less a personal memory than just another story handed down to Darrow by his father. But that seemed an irrelevant detail amid the broad brushstrokes of a past in which, as Darrow would insist in his lectures, "John Brown's love of the slave was a part of that fire that, through the long and dreary night, kindles a divine spark in the minds of earth's noble souls."

Darrow's own concern for "the plight of the Negro" stretched back over the decades. Twenty-four years before he met Ossian Sweet and his brothers, Darrow had stood in a black church in Chicago and said, angrily, "When I see how anxious the white race is to go to war over nothing, and to shoot down men in cold blood . . . I admit I am pessimistic." Yet, in that 1901 address, he rallied the Negro congregation by pointing out that white racists "degrade themselves" with their antipathy and bile. "You may be obliged many times to submit to this, but it must always be with the mental reservation that you know you are their equal, or their superior, and you suffer the indignity because you are compelled to suffer it, as your fathers were once compelled to do. But, after all, your soul is free and you believe in yourself, you believe in your right to live and to be the equal of every human being. I would like to see a time when man loves his fellow man, and forgets his color or creed. We will never be civilized until that time comes."

During the Springfield race riots in 1908, he had written in a similar vein that "all people, high and low alike, should demand for the Negro the privileges and rights of every other citizen; [they] should judge him for what he is, and not on the color of his skin." Such thinking, although groundbreaking in early-twentieth-century America, came naturally to Darrow. He felt the same easy passion for the rights of labor unions and of men to drink alcohol whenever they chose—just as a rejection of the death penalty pulsed fiercely through him.

After capital punishment and evolution, only the festering racial strife of a vast country could galvanize him into a final epic tilt against injustice. His own immortality might just be sealed in the process.

Despite Darrow's soaring ambition, his instinctive gift for empathy was most evident when he and Walter White lit up the gloomy confines of the Wayne County Jail on the Thursday morning of October 15, 1925. His reputation meant the previously desolate Sweet men were immediately transformed. White suggested that "I have seldom seen such joy in the faces of any persons as appeared on those of the defendants when I introduced Mr. Darrow to them." A more lasting hope was also instilled by Darrow's extraordinary capacity to identify with their anguish and to offer a way out of an apparently condemned life. Otis Sweet would explain that "he talked to you and tried to [understand] just how you felt, then he'd fill in to show you there wasn't nothing to be afraid of . . . the whole atmosphere of the jail changed. Even the turnkeys started to speak to you . . . that's when you built up a little hope. Then they started to bet in jail between an acquittal and a conviction. Before it had been 10–1 for a conviction."

The trial was meant to start the following Monday, and Darrow knew it was imperative to secure a delay. Four days with ten frightened men, and the bailed Gladys Sweet, would not be enough time to craft a proper defense. Even White had been reluctant to voice the exact truth when Darrow had first quizzed him. Asked whether the Sweets had shot at the mob, White hesitated. "I am not sure," he stuttered, an unusual affliction for such an articulate man.

What did his instinct tell him? Did he believe that someone in the Sweet house had actually fired into the crowd? "I believe they did fire," White answered uneasily, anxious that he might lose the only lawyer in America who could rescue the Sweets. Darrow, of course, reiterated his certainty that he had been right to take the case. "If they had not had the courage to shoot back in defense of their own lives, I wouldn't think they were worth defending."

Darrow was fortunate that Frank Murphy had been appointed as the presiding judge. The thirty-three-year-old Murphy was the youngest man to have become a judge in Michigan, and he brought an open mind to his court. "I want the defendants to know that true justice does not recognize color," Murphy stressed in his resolve to be seen as a fair judge.

His sense of social justice was rooted in the radical politics of Ireland. Murphy's grandfather, an Irish revolutionary, had been killed for the cause while his father had been imprisoned and narrowly escaped the death penalty. It also helped that Murphy was slightly in awe of Darrow—privately naming the old attorney his courtroom hero—and so he listened with obvious sympathy when the request for more time to prepare the defense was put to him. He told Darrow and White that he would delay the start date for jury selection another fifteen days, to October 30.

White was so delighted he sent Murphy a signed copy of his first novel, *The Fire in the Flint*. "It was a great pleasure to meet you," he wrote, "and a most pleasant surprise to find one with such lofty ideals. It is so seldom that those of us who are trying to secure even-handed justice for Negro citizens encounter one like yourself and you can imagine our joy when the experience comes."

Marcet Haldeman-Julius was smitten by the judge, and her appreciation of his qualities was not limited to his compassion. Frank Murphy, clearly, had floored a writer who had followed Darrow from Dayton to Detroit. "In appearance he is tall, very good looking, inclined toward slenderness," she said of Murphy, "with a long beautifully modeled head, thick curly auburn hair and contemplative blue eyes. Irish eyes they are, full of dreams. For brilliant as his rise in his chosen profession has been, Frank Murphy has the brooding, imaginative temperament of an artist. . . . He can be stern enough, but his most usual mood is one of tranquil thoughtfulness. When he smiles his face has the sparkle of a quiet river in the sunlight."

Yet even a suddenly girlish Haldeman-Julius could control her racing heart long enough to acknowledge that Darrow still rose up as the dominant presence in Detroit. She saw in him "the quality of a great, majestic ship, or some mighty oak, that has weathered many a storm." Closing on the end of his tumultuous career, Darrow could feel the confidence surging through him. When Mary asked him how he would cope with having so little time to master the legal intricacies of a trial that would again test him to the limits, Darrow cuffed her on the arm. Had she not learned to trust him after all this time? He would rely, as he always did, on his instinctive feel for the case and his ability to work himself into the heads of both the defendants and the jurors who would decide if they might be freed or imprisoned for life.

Darrow planned a simple defense that would give him the framework to weave the kind of emotive courtroom magic that had so moved the galleries of Chicago and Dayton. The key legal point for Darrow centered on the definition of a "mob"—which the law clearly stated constituted a grouping of more than fifteen people gathered together with threatening intent. If he could overturn the prosecution's ludicrous claim that people outside the Sweet home had merely been "passing-by" on those two September nights, he would prove as a point of law that the crowd constituted a violent mob. And once the legal semantics had been established, he would find a way to make the jurors feel what it must have been like for the Sweets to have their own home besieged by a lynching mob. He would instill the terror and despair that gripped a brutalized people and, for once, make some normally thoughtless white citizens of Detroit understand exactly what it meant to be a Negro in America in 1925. Darrow did not have to plan such a strategy. After forty years in the courtroom, he just needed to uncork all his emotional intelligence and let it seep into the hearts and minds of a jury he would have selected with utmost care.

He had brought Arthur Hays with him to Detroit, anyway, to handle the more tedious legal complexities and precedents. Hays, twenty-four years younger than Darrow, still had the energy and tenacity to pore over the staggeringly dull law books and state legislation. Darrow was there, instead, for the grandstanding moments of drama that would turn the collective head of the courtroom and send unified gasps and shudders rippling through the jury box. He would rely on a rough old mastery of sentiment and language, which no book could teach.

A MID THEIR pain and separation Ossian and Gladys Sweet felt the seductive tug of unexpected fame. Their trauma could not be forgotten, but at least it assumed a more righteous meaning in the company of Darrow and White. The Sweets were now also heroes, as well as martyrs, and it was typical of coverage among America's black newspapers that New York's *Amsterdam News* should argue, "Perhaps the most important case the Negro has ever figured in all the history of the United States is heard out in Detroit, Michigan. Thank God Dr. Sweet moved in! Thank God that his noble wife moved in with him. And thank God nine of his relatives and friends came in with him. . . . This is the spirit of unity the Negro must more and more evidence if he is to survive. He must face death if he is to live! He must be willing to die fighting when he is right! When police authorities fail to protect him and his family, when courts of law desert him, when his own government fails to take a stand on his behalf, he faces death anyway, and might just as well die fighting!"

Gladys had spent thirty nights in a jail cell with three other young black women who were shut away on charges ranging from murder to jumping bail. She could barely comprehend the hideous nightmare. The prosecution had finally shown some pity toward her when, on October 6, they allowed her release on $10,000 bail. Showing real defiance, Gladys moved into a rented apartment that was only a mile from the fated bungalow and once again on a "white" street. She devoted all her spare time to the campaign to help her husband and his fellow defendants. Relieved to see Ossian again, even if she was restricted to prison visits, Gladys worked closely with Darrow and White. In a letter to the black novelist she confessed coyly, "I consider it quite an honor and of no little importance to be noticed by a person as busy and so much in demand as yourself. If things keep on I shall soon believe myself to be quite a personage."

Ossian found consolation in the tributes that honored his courage and pride. In his worst moments he might still have felt like a terrified boy, but in reading the newspapers and talking to Darrow he began to believe it was possible he might represent his entire race. The thought strengthened him as the trial hurtled ever closer, a trial in which his fate, entwined with his wife, two brothers, and seven other men, rested solely on an old white man called Darrow.

CHAPTER 14

A PECULIAR FEAR

EARLY ON Friday, October 30, 1925, Ossian Sweet's thirtieth birthday, the overcrowded courtroom erupted into spontaneous applause when Darrow moved through the throng toward his seat. His rival, Robert Toms, had already slipped unnoticed into his place at the prosecution table. It seemed very different to the opening day of the Scopes trial, when Bryan, rather than Darrow, had received the rapturous welcome on his arrival in court. But, in Detroit, Darrow's celebrity was at its very peak, and at least half of the room was filled with black spectators.

The Municipal Court Building was a simple but elegant space, sixty feet long by fifty feet wide, with marble floors and mahogany walls. Three large windows, even with the shades pulled down and snow lining the ground outside, allowed plenty of natural light. Toms had decided to use a similarly delicate touch against Darrow. He was thirty years younger than his venerable opponent, and he had monitored Darrow's success in laying waste to the abrasive Robert Crowe during the Leopold and Loeb trial, and in supposedly killing off the vitriolic Bryan in Dayton. And so Toms knew a far more subtle approach was required—especially as the State hoped to prove that all eleven defendants were implicated in the murder of Leon Breiner. But he also told himself that there was no need to be afraid just because he faced, in his words, "a giant" in Darrow.

As the chief prosecutor for Wayne County, the thirty-eight-year-old Toms informed the *Detroit Saturday Night* newspaper that he prided himself on his fairness, claiming that he would be happier to let ninety-nine criminals revel in their freedom than to ever send an innocent person to jail. Tall and striking, with his shock of blond hair and blue eyes leading down to a strong jawline and a long, tapering neck, Toms was highly educated. But he was also tough and ambitious, with a canny courtroom persona that covered his ability to scheme and scrap his way to the occasional unscrupulous victory that boosted his chances of one day reaching political office.

As he began to question the first of the sixty-five prospective jurors, Toms responded with determined politeness whenever Darrow intervened. It set the pattern for an effective prosecution strategy that would increasingly frustrate Darrow. "I treated him throughout the trial with the utmost reverence and deference," Toms revealed years later. "After a week of this he said to me, 'Goddamn it, Toms, I can't get going. I am supposed to be mad at you and I can't even pretend that I am.'"

Darrow, however, had not lost any of his artistry in jury selection. After Toms had settled on his preferred choice of twelve, Darrow spent the rest of Friday, Saturday, and most of Monday dismantling that list of men and selecting his own band of jurors. He had undertaken the task so many times before that he knew exactly what kind of juror he required, or needed to avoid. The questions he posed to each man were so direct in their ease of asking that Darrow had little difficulty in wheedling out all the secret Ku Klux Klan members and other racists who might have sat in judgment of the Sweets.

Even after Toms took much longer the second time around to dismiss half of the defense selection, Darrow did not drag out proceedings. He questioned the new members of Toms's revised group on Wednesday, November 4. Having uncovered another Klansman, and demanded his replacement, Darrow waved his assent. He had such belief in his abilities, and in the truth of the Sweet case, that he declared himself ready for trial.

All twelve jurors were white workingmen who came from neighborhoods not dissimilar to that surrounding the doomed bungalow. They lacked the depth of education and professional success of the physician, the dentist, and the rest of the accused, but Darrow was happy enough with his selection. He would privately tell the judge, hammering home a link with Murphy's heritage, that he was delighted to have six jurors who came from Irish Catholic backgrounds. When Murphy innocently asked if they were less likely to call for the execution of a defendant, Darrow's eyes glittered in the creased folds of his face. "I never met an Irish Catholic yet who didn't think that someday he might be in trouble himself," he said. Murphy laughed, remembering his own politically rebellious father and grandfather, and shook Darrow warmly by the hand.

Once the jury had been agreed upon, Darrow made a great show of muttering his usual mantra: "The case is won or lost now. The rest is window dressing." He still appeared to believe his own sly line the following morning—even if the country's racial divide was underlined in that legal setting. Adjacent to the two sets of lawyers, the black defendants and white jurors faced one another from opposite sides of the courtroom. To Darrow's left, his eleven Negro clients sat mutely together in a line on a stark wooden bench. On his right-hand side, twelve members of the jury were seated in two neat rows, with six men at the front and six seated behind them.

After Toms had made his preliminary remarks, claiming that the Sweet family and their friends had acted with "malice aforethought" and intent "to shoot to kill," Darrow asked the judge if he could postpone his opening comments until the completion of the prosecution case. It was an unusual request but Murphy agreed.

Toms did not object because the impact of his own more evocative words still resonated: "Witnesses will testify that there was no disturbance, that everyone was going about his business, that there was no loud talk, that there were no groups of people together, that no violence was threatened and committed; that, suddenly, without warning, from the front windows of Dr. Sweet's house, there was a volley of shots."

When Toms called up the first of his sixty-nine witnesses, handpicked in close consultation with the police and the Waterworks Park Improvement Association, Darrow feigned indifference. He settled down in his chair and, as he would do repeatedly throughout the opening days of the trial, took a leisurely crack at the crossword puzzle in his morning paper. Between clues and in the midst of the prosecutor's leading questions and the often disingenuous answers from the witness stand, he occasionally closed his eyes. It would have been easy to imagine that the old man was about to nod off had he not, every now and then, thrown out a biting objection that proved just how closely he was listening.

Toms was more obviously busy. He produced a chalkboard and sketched the location of the Sweet bungalow in relation to the two-family flats and apartment buildings in front of which Leon Breiner had been shot. Toms and his police witnesses also described the mostly empty rooms in the bungalow, so devoid of furniture, which he then contrasted with the arsenal of weapons contained in the Sweets' linen cupboard. He made them sound like an armed gang rather than an ordinary couple moving into their new home. Dr. Sweet flinched in the dock, but Darrow kept puzzling over his crossword.

Toms called his first significant witness—the heavyweight figure of Inspector Norton Schuknecht, who had burst into the Sweet home on that fateful night. Schuknecht had also supplied the flagrantly dishonest account of the shooting, which he said was totally unprovoked, that had appeared in the Detroit newspapers. The corpulent policeman again insisted that there had been no mob on the corner of Garland and Charlevoix. Instead, it was "nothing more than the ordinary person walking by the place. I don't believe there were more than two or three at a time."

Looking meaningfully at Schuknecht, Toms asked, "At any time was there as many as ten people gathered in one group in that place?"

"No, sir," Schuknecht lied coolly, "there was not."

"Were there ever as many as five?"

"No, sir."

"Were there any that stopped and stood there for more than half a minute?"

"No."

During cross-examination, and in response to the policeman's blithe concession that "I daresay there were ten or twelve [people], simply standing and talking, as far as I know," Darrow turned the screw. How many people might he actually have seen that night? Could it have been as many as two hundred?

"Where?"

"Around the neighborhood of Dr. Sweet's house," Darrow suggested.

"Oh," Schuknecht said, as if suddenly remembering, "within a block either way of the house, I would figure for everybody sitting on their porches and out in front of their lawns, there may have been a couple of hundred people."

Darrow kept Schuknecht on the stand all afternoon, for almost four hours, and chipped away at the inspector's credibility with unerring if typically folksy accuracy. Schuknecht tried to deny everything, but he ended up sounding surly and defensive. He was eventually trapped when Darrow asked him, again with illusory airiness, if he had seen anything besides a bed and two chairs when he entered Dr. Sweet's front bedroom.

"I found a small stone," Schuknecht conceded, despite himself.

Darrow scratched his head as if he did not quite follow the significance of such an admission. "Did you find any broken glass?" he asked in puzzled innocence.

"Yes," Schuknecht said.

"Why didn't you mention finding broken glass on the floor?"

"Why didn't you ask me?" the policeman said tetchily.

Darrow did not strike back. He merely asked whether the fact that the glass was found inside the bedroom "would lead a policeman to the idea that it [the stone] was thrown from the outside."

"We found the stone on the inside and I believe it was thrown from the outside," Schuknecht said. "I do not doubt that at all."

The *Detroit Times* captured a telling moment by revealing that, as soon as he heard Schuknecht's confession, Toms groaned: "You cannot ask for anything better than that, can you?"

Schuknecht's deputy, Lieutenant Paul Schellenberger, said that on the night of September 9 there had been around six distinct "knots" of people, containing no more than seven in a cluster, who stood separately from each other in orderly fashion. It was still a blatant distortion of the truth, but the talk of fixed groups represented a significant difference from Schuknecht's initial insistence that there were no more than two or three people strolling past when the Sweets opened fire. Schellenberger also inadvertently corrected his boss by admitting there were "more" than two hundred people milling around the surrounding area. But, still, he denied that there had been a mob outside the bungalow.

Deputy Superintendent James Sprott was blunt. He had not seen anyone on Garland, or Charlevoix, or even a single soul either to the north or south. When asked why the decision had been taken to close a supposedly deserted street like Garland to traffic half an hour before the death of Leon Breiner, Sprott bristled: "Precaution."

Darrow turned to the jury and spread his arms wide, in disgust. "When a man comes here as your assistant commissioner and swears that there was no crowd at the Sweet house he is not to be trusted. He lies."

Toms also turned to the Sweets' neighbors. Florence Ware said she only stepped onto the street, "out of curiosity . . . to see the crowd of people—police on our corners."

"Now you first said 'crowd of people,'" Darrow murmured, "and you then changed."

"I meant policemen," Mrs. Ware said stonily.

"You didn't mean people?"

"No, sir."

It was a long struggle to eke out the truth from a line of stubborn and evasive witnesses. There were many good days for the prosecution, as on Monday, November 9, when, according to the *Detroit Free Press*, "a number of police officers testified to arresting the eleven defendants in the house, to finding a quantity of arms and ammunition, and to the fact that there was little disturbance in the district prior to the shooting. . . . The state's case developed considerably in the course of the day's proceedings."

Detective George Fairbairn, among those who arrested the men, provided detailed evidence of all the loaded weapons he had discovered in the Sweet house. As he lingered over descriptions of two .38 Colt revolvers, a black automatic Remington, two more shotguns, a German automatic revolver, and the bags of ammunition, he created the impression that the Sweets were a bloodthirsty gang. William Davis, one of the accused, even carried his own diamond-studded, blue-steel revolver. These were not innocent victims, Fairbairn implied, but heavily armed killers who had moved into a peaceful neighborhood. They had come in search of trouble.

Sergeant Joseph Grohm focused on the victim. As the first policeman to reach Leon Breiner, Grohm spoke of the blood and death he had seen. But the court gasped audibly when Toms elicited a more touching observation.

"Did you see anything in the hands of the deceased?" he asked Grohm.

"Not in his hands, no . . ."

"Did you see anything in his mouth?"

"Yes, sir. He had a pipe."

The difference between that humble pipe and a diamond-studded automatic revolver wounded the defense. It conjured up an unshakable image of an ordinary and neighborly man, puffing on his pipe, as he was cut down in a hail of bullets fired from a darkened house. The impact reduced the courtroom to chaos after Breiner's widow, Leona, listening in the second row of the gallery, slumped to the floor in a dramatic faint.

While two policemen rushed to lift her up, amid cries of alarm and the clatter of seats as people strained to see the fallen woman, Judge Murphy shouted at the jury: "Step right out, gentlemen!"

Murphy had immediately recognized her as Breiner's widow and understood the potential influence her collapse might have on the jurors. Hays was already on his feet, accusing the State of stage-managing her faint. He shouted out his demand for a mistrial, but Murphy instead summoned the rival lawyers to his bench. Darrow wanted to question each jury member to ascertain the impact of the incident in swaying his judgment. Murphy refused, insisting that it would be enough for him to ask how many of them had "observed a lady spectator in the courtroom who fainted, [and] became ill."

When they filed back into court, only four men raised their hand in answer to Murphy's question. They all shook their heads when asked if they could identify her.

Darrow still repeated the call for a mistrial. Murphy might have admired him enormously, but he understood the political ramifications of abandoning a racially loaded trial in a city where the Ku Klux Klan had just narrowly lost the latest mayoral election. "Motion denied," he said.

The trial ground on, with most of Toms's witnesses echoing the outrageous claim of one of their number, Edward Belcher, who insisted that "everything was in peace and quiet until the Negroes started shooting." There was more honesty in the answers supplied by three of the neighborhood's young boys to Darrow's simple but revelatory questions. They were meant to be prosecution witnesses, but Darrow turned them right around.

Seventeen-year-old Dwight Hubbard initially tried to emulate his elders by covering his tracks. "There was a great crowd—no, I won't say a great crowd, a large crowd—well, there were a few people and the officers were keeping them moving."

"Do you know how you happened to change your mind and whittle it down so fast?" Darrow asked.

"No, sir."

"Have you talked with anyone about the case?"

"Lieutenant Johnson."

Darrow paused while the murmuring in court died away. "And when you started to answer the question you forgot to say 'a few people' didn't you?"

"Yes, sir," the boy gulped.

Two thirteen-year-olds, George Suppus and Ulric Arthur, were even more incapable of sustaining deceit. Darrow spoke kindly, as if he were their grandfather asking them about a day at school, and Suppus admitted that the large crowd in the street had gathered purposefully in front of the "colored" family's house. Everyone had been talking about the Negroes.

"Did you see anyone throw any stones?" Darrow asked Arthur.

"Well," the boy replied, "there were three or four kids between the houses. They were throwing stones."

"Where were they throwing them?"

"They were throwing them at the house where the colored people moved in," Arthur admitted.

"Did you hear the stones hit the house?"

"Yes, sir."

"Do you know if any of them broke any glass?"

"I heard glass break," the boy nodded nervously. "I guess it was the stones."

"And the shooting happened right after it," Darrow said, "didn't it?"

"Well . . . ,"

"What?" Darrow demanded more loudly.

"Yes, sir."

Darrow gazed searchingly at the jury, to see if the truth had finally registered.

A T NIGHT, he engaged in a different life. Although there were times when he and Ruby went out for dinner with Arthur Hays and his wife, and occasions when he gave a lecture at the swanky Penguin Club or went down to talk in Black Bottom with Walter White, Darrow loved to gather a group of friends or ardent new admirers around him while he read aloud extracts from Nietzsche or far lighter verse from Newman Levy's *Opera Guyed*. It was then that he could hold court and drink in the flattering attention.

In the absence of Mary he was not averse to keeping his beady old eye cocked for female company. At the Penguin Club he saw a glamorous woman whom he had already observed numerous times in court. There was no need to introduce himself, and he soon learned that her name was Josephine Gomon. She was the secretary of the Rationalist Society in Detroit, a political activist and mother of five. Her good looks and politics appealed to Darrow—as did the revelation that she was also a friend of Judge Murphy's.

As she praised Darrow for his outstanding performances in court, which he accepted with his usual "aw, shucks" shrug, Gomon boosted him further by stressing how Murphy had also been transfixed. The young judge had told her, when she and Murphy had met at the snappily named Miss Lincoln's Dixieland Tea Room, of his certainty that the Sweet trial would end up as one of the most famous in American legal history.

Darrow, forgetting Ruby and Mary in such a giddy moment,

slipped his hand through the young mother's arm and said he would walk out and hail a streetcar for her on such a bitterly cold night.

When Gomon explained that she was driving herself home, Darrow immediately suggested that she might take him back to his hotel. And once they were there, he said smoothly, she might care to come up to his room. He was expecting a few interesting fellows to drop by and they would read some poetry together. It was as clear a pickup line as she had ever heard. But the combination of fame and poetry in a warm room, especially when offered up to a beautiful and earnest young woman, usually did the trick for Darrow.

This time, however, he smiled ruefully when they were suddenly joined by the much older Harriet McGraw, who also claimed to be in need of a lift home. McGraw, who styled herself as the leading progressive in Detroit, had helped Murphy's swift elevation to the judge's chair. Darrow knew her moderately well—and she was not the kind of woman to whom he particularly enjoyed reading poetry. She had the look of an old harridan out to ruin his night of revelry.

"You seem to be well protected," he said, smiling wryly at Josephine Gomon as his hand slipped away from her side. They would no doubt talk again, he said, as he wished them both a pleasant night and turned back to the club.

THE CASE for the defense opened on Saturday, November 14, 1925, with Darrow and Hays presenting a carefully planned motion that Murphy should dismiss the case. Hays argued that, after nine days of evidence, the prosecution had not produced a sliver of conclusive evidence that the defendants had planned to kill anyone. His relentless reasoning was bolstered by an unwittingly emotional charge when the Sweets' little girl, Iva, burst into tears midway through his delivery. "That is the Sweet baby," Hays said on impulse. "We have brought her here as an illustration. Had she been in that house that night she might well have been arrested and tried and the evidence here would condemn her to the same extent it does the defendants."

Darrow underlined the absurdity of eleven people being charged with murder by asking what would happen if someone was shot dead in court that afternoon. Would every single person in the room be guilty of murder? Only one bullet had killed Leon Breiner, and Darrow argued that it was far from certain that Henry Sweet had actually fired the fatal shot. He also pointed out that prosecution witnesses had constantly misstated the facts. The Sweet home had clearly been stoned, and Darrow reminded the judge that he had forced further confessions, however reluctantly, from members of the Waterworks Improvement Association that around six hundred people had pledged to keep their neighborhood exclusively white. All this gave grounds for an immediate dismissal of the case against all eleven defendants.

Murphy said he would consider their request during the weekend recess. Darrow's mood remained serene on that Sunday while the temperature dropped and an ominous gray sky rolled across Detroit. More snow began to fall across the Motor City late that night.

WHEN THEY trudged to court just before nine on Monday morning, three inches of snow covered the ground. It was hard to discern any clear omen in that soft white blanket. Toms was sufficiently concerned to attack Darrow's motion as soon as court reconvened. If Judge Murphy bowed to the defense, "it means that the people of this city, both colored and white, are notified that it is entirely lawful for one or two hundred of them to gather in a house; to provide weapons in a sufficient number to arm them all; to fire volleys from the house and kill one or two citizens on their own doorsteps under the pretext that they believe themselves endangered."

Darrow was unsurprised when Murphy ruled against his plea. Yet a first crack in the prosecution case had emerged earlier that morning, when Toms had approached Darrow and Hays to confirm that he was willing to drop all charges against Gladys Sweet. The proud and defiant twenty-three-year-old refused the offer. As long as her husband and the other men were expected to stand trial she would face the prosecution.

Her resolve strengthened Ossian as he prepared to take the stand as the defense's key witness. As he listened to Hays's long opening address, which Darrow had postponed from the start of the trial, he looked terribly alone. He knew that one moment of weakness, a flash of rage or a mix-up of words, could destroy their hopes. Sweet was not compelled to be a witness in his own defense, but he understood that no one, not even a master like Darrow, could articulate the depth of terror he had felt that night. He would have to speak.

Sweet tried to calm himself as Hays established that two legal precedents in Michigan were pertinent to this trial. The nineteenth-century case of *The People v. Pond*, in particular, had resulted in a ruling that "a man who is assaulted in his own house need not retreat in order to avoid slaying his assailant."

But Hays concentrated on the current trial of "the doctor"—using Sweet's official title on twenty-two occasions in his hourlong address. He wanted to drive a belief into the jury that they were judging a healer rather than a killer. Hays asked Dr. Sweet to stand, and he did the same with his two brothers, reminding the jurors that Otis and Henry were, respectively, a dentist and a college senior. Hays then turned to Gladys and she, too, rose to her feet. She appeared as herself—an obviously refined young woman.

"These are four of the defendants," Hays said, giving the jury time to absorb their smart but scholarly appearance. They were not murderers. Hays argued that the defense would "prove that a Detroit policeman fired the bullet that killed Breiner." Alluding to the "flat trajectory" of the bullet, he insisted, "It could not have come from the upper window in the home of Dr. Ossian H. Sweet, as indicated by the prosecution. The man who fired this shot is from Tennessee."

It was a startlingly specific accusation, as if Hays and Darrow still had a score to settle with the fundamentalists of Dayton and the rest of Tennessee.

Hays also introduced fresh evidence when Philip Adler, a white reporter for the *Detroit News,* was called. Adler explained that on the evening of September 9 he had been driving to a friend's house, where he was expected for dinner, when he saw a large and intimidating crowd of people, "between four and five hundred," who looked as if they "were prepared for something." Sensing a story, he parked his car and pushed his way into "the mob," and he began to take careful note of the threatening words being said around him.

Toms jumped up to object at the use of "mob," but Adler pressed on with his damning account. Hearing the men around him insist that "we're going to get them out," as they gestured angrily toward the house at 2905 Garland Avenue, Adler approached a police officer. He was told to stay out of the way. And then, he said, the pelting of the house with stones began. The reporter tried to emulate the frightening sound by beating his pencil against the arm of his chair. It made only a soft patter in the packed courtroom, but the insistent rhythm he thudded out was enough to convey a sense of the concerted flurry of stones and rocks that had rained down on the Sweet roof.

There were less dramatic contributions from the next three witnesses, who focused instead on the domestic arrangements of the Sweet household. Serena Rochelle and Edna Butler described the chatter they had shared with their friend Gladys about the soft furnishings and color schemes she might choose before darkness and fear descended. And Edna remembered her distressing ride on a streetcar the following morning, when she heard a woman tell the conductor that they would "put the niggers out" that night.

Max Lieberman confirmed that Ossian and Gladys had spent almost $1,200 at his furniture store on that same morning of September 9—thus debunking the prosecution's claims that they owned a house sinisterly empty of much beyond their collection of guns. He had been scheduled to deliver their new furniture the following morning, but the Sweets, by then, had already been jailed.

Alonzo Smith and Charles Schauffner, two black men who had been attacked in their different cars at the nearby intersection, just before the shooting, stressed the racial violence of the crowd. Smith had heard the cry of "Here's some niggers, get a brick!" while Schauffner had been chased and cut after someone yelled, "There goes a Negro, catch him!"

And then, on the afternoon of Wednesday, November 18, it was time for Ossian Sweet to take the stand. He placed his hand on a black bible and swore to tell the truth, the whole truth, and nothing but the truth. His voice wavered only slightly, for the truth, he knew, told everything there was to know about him and that night.

There was a deathly silence as Darrow trudged slowly across to Dr. Sweet. The old white attorney stood in front of the young black doctor for almost twenty seconds. And then, in a gravelly murmur that sounded more like a question than a statement of fact, Darrow said, "You are a Negro?"

"Yes," Ossian Sweet acknowledged. The question of race, as Darrow had underlined with his unusual but subtly powerful opening, was at the very heart of this trial. Darrow allowed the silence to return, as if to give the jury a little longer to notice that, beyond his race, Sweet was a man. He was a human being with a story to tell, a man who had a past, both personal and racial, that needed to be heard. Darrow gestured toward Hays as, in another surprising gesture, he let his colleague step forward and guide Sweet back into that past. There would be plenty of time, once Sweet had spoken, for Darrow to take the floor.

Hays proved a commendable choice for, slowly and sympathetically, he drew the court into the world of Ossian Sweet. He moved the thirty-year-old man from an impoverished childhood in Bartow to his university education in Washington, D.C., from working as a doctor in Detroit to studying under two greats of medicine, Anton von

Eiselsberg and Marie Curie, in Europe. This had been a life of much achievement—but with an undercurrent of almost unbearable horror. Sweet began to relive all that he had seen as a black American.

He recounted the burning of Fred Rochelle in Bartow, just as he described witnessing a young Negro being battered to death during the Washington riots. He spoke of the violence of the Chicago race riots and of the five Negroes shot dead in Rosewood, Florida, when eighteen black homes and a church were burned to the ground. He remembered the four Johnson brothers of Arkansas, one of whom had, like him, been a physician, while another, just like his brother Otis, had been a dentist. He explained how they had been taken from a train and lynched. He hailed Dr. A. C. Jackson, an esteemed Negro surgeon, who had been killed by the police when trying to protect his home from a mob. He paid homage to the three thousand Negroes lynched in America. He spoke of Dr. Alexander Turner, his friend at Dunbar Memorial Hospital, who had thought more positively of white people—so that, when his new home was surrounded, he still opened his front door when asked. He lamented the fact that a brilliant doctor became "dough in the hands of the mob."

Hays took him back, at least in his head, into the Garland bungalow on the night of the shooting. Sweet went calmly through his grotesque ordeal in a way that the *Detroit Free Press* said "held the jury and spectators silent and immovable."

Near the end, Sweet said, "Stones kept hitting the house intermittently. I threw myself on the bed and lay there a short while—perhaps fifteen or twenty minutes—when a stone came through the window. Part of the glass hit me."

"What happened next?" Hays asked.

"Pandemonium—I guess that's the best way of describing it—broke loose. Everyone was running from room to room. There was a general uproar. Somebody yelled, 'There's someone coming.' They said, 'That's your brother.' A car had pulled up to the curb. My brother and Mr. Davis got out. The mob yelled, 'Here's niggers! Get them!

Get them!' As they rushed in, the mob surged forward, fifteen or twenty feet . . ."

Ossian Sweet stopped in midsentence, covering his eyes as if to shield himself from the memory. Then, he looked at his wife and Gladys nodded gently, encouraging him.

"It looked like a human sea," Dr. Sweet said, his composure regained. "Stones kept coming faster. I was downstairs. Another window was smashed. Then one shot—then eight or ten from upstairs. Then it was over."

Hays had one more question to ask: "What was your state of mind at the time of the shooting?"

Ossian Sweet took in a deep breath. And then, looking intently ahead, as if seeing those twisted white faces more clearly now, he said, "When I opened the door and saw the mob I realized I was facing the same mob that had hounded my people through its entire history. In my mind I was pretty confident of what I was up against, with my back against the wall. I was filled with a peculiar fear, the fear of one who knows the history of my race."

Detroit Municipal Court, St. Antoine Street, Detroit, November 23, 1925

They came early that Monday morning, three days before Thanksgiving, to try to find a space in a courtroom that would be packed in readiness for Clarence Darrow's final summary. The bulk of the crowd outside the Municipal Court was black, as the more balanced racial split at the start of the trial had buckled. For the Negroes of America, the fate of ten men and one woman had assumed great significance.

After the dignified bravura of his testimony, Sweet had withstood a fierce grilling from Robert Toms. The precision and power of his words were just as obvious during his cross-examination. Even the conservative *Detroit Free Press*, the city's most popular newspaper, praised Sweet as "a well-educated and an astute student of the race problem." His passion and suffering, leavened by a commendable restraint, provided his testimony with a credibility that Toms had failed to pierce.

Henry Sweet presented Darrow with a more vexing problem. With similar candor to his brother he had admitted that soon after Ossian felt his "peculiar fear," he had raced upstairs with his .38 Winchester. From the front room window, Henry, a more handsome version of his scholarly brother, said, "I looked out and saw the crowd across the street throwing stones. I tried to protect myself. I fired the rifle in the air the first time. The second time I fired at the crowd."

Lester Moll, Toms's assistant, insisted that Henry had indeed shot Leon Breiner. He dismissed defense claims that the trajectory of the bullet indicated that it had been fired from street level—by a frightened policeman. Breiner had, according to Moll, not been standing "exactly perpendicular" to the ground. It was just one more surreal statement from a man who described Ossian Sweet as "a quasi-intelligent Negro." He could barely conceal his antipathy toward the black defendants.

Darrow knew Henry's confession was lodged in the collective memory of the jury box as firmly as a bullet in the head of a dead man. It gave Moll the leeway to say, "The defense has advanced a fear complex theory. It's poppycock. It's bunk. [This is] a premeditated murder." He spoke with controlled fury all morning, in a speech described by the Detroit press as "masterful" and "a gigantic structure of reason."

Hays's response, after lunch, was more garbled, mashing together lines from the Declaration of Independence, the Second Amendment, the Emancipation Proclamation, and America's involvement in the First World War. Its chief attribute was relative brevity—clearing the way for the rise of Darrow at three o'clock that afternoon.

In his habitual way he walked over to the jury so that they could see his face in close-up and hear even his softest words. It would, as always with Darrow, sound as if he was in a private conversation with each of the men. "I wish it was not my turn," he sighed at the start, "that I didn't feel it was my duty to talk to you in this case. It is not an easy matter to talk about a case of this sort, and I am afraid it won't be an easy matter to listen but you can't help it anymore than I can. It has taken much of your time, but the end is in view. You know how important it is."

Darrow gestured to the eleven on trial, each hoping to win back his or her life, and he spoke of "the everlasting problem of color and race. . . . You can change a man's opinion but not his prejudices that have been taught to us and that began coming to us with our mother's milk, and they stick almost as the color of skin sticks." He pointed out, with such chatty good sense that some jurors nodded in agreement, that "if it had been a white man defending his home from a member of a Negro mob, no-one would have been arrested or put on trial."

The Sweets looked suddenly pitiful in their shrunken silence. "My clients are here charged with murder," Darrow noted, "but they are really here because they are black. Here are eleven people, which is about as many as there are on this jury, on trial for killing a white man . . . let's reverse this. Supposing one of you were charged with murder and you had shot and killed somebody, while they were gathered around your home, and the mob had been a black mob and you lived in a black man's land and you had killed a black and you had to be tried by twelve blacks, what would you think about it?"

David Lilienthal, riveted by Darrow in his press seat, would write later that "the old man with the unalterably sad face and the great stooped shoulders seemed no mere lawyer pleading for hire. He seemed, instead, a patriarch out of another age, counseling his children, sorrowing because of their cruelty and hatred, yet too wise in the ways of men to condemn them for it."

Yet there was anger, too, in Darrow's withering description of "these Noble Nordics" lining Garland Avenue. "They were gathered together just the same as the Roman Coliseum [*sic*] used to be filled with a great throng of people with their eyes cast to the door where the lions would come out. They were gathered together just as in the old days a mob would assemble to see an outdoor hanging and waiting for the victim with their eyes set on the gallows."

Darrow shook his heavy head carefully, as if it too might roll from his shoulders in horror at the worst of humanity. "And they were gathered together to awe and to intimidate the poor black family which had brought the corner house and who had a right to buy it under every law of the land."

He gestured at the accomplished group of people he represented—"Half of them are college graduates or attending college." And yet they had chosen to move into a "not especially high-toned neighborhood, [with] nothing swell about it." One of their members, Dr. Ossian Sweet, had studied medicine in Vienna and Paris with some of the world's finest minds, but he had been attacked in his new home by a group of white folk who, despite their European heritage, could not even pronounce the name one of the streets in their suburb. Dr. Sweet, Darrow said quietly, would have been able to tell them Goethe Street was not pronounced "Go-thee." Unlike his neighbors, Sweet knew that Goethe was one of the greatest writers and thinkers of the late eighteenth and early nineteenth centuries.

"We have presented witnesses that are as intelligent, as attractive, as good-looking as any white man or woman and who are as far above the bunch which testified against these men as the heavens are above the earth." Darrow looked out at the hushed court and then, with telling force, he said, "You know it and they know it."

He was flying now, the words flowing from him without a single note in his hands. They came from deep within him, from the very core of a man that, for all his failings and flaws, understood humanity. The words fell from him in quiet murmurs and roaring cries, as he spoke of Ossian Sweet and his people, human beings, who had been "loaded like sardines in a box in the mid-decks of steamers and brought forcibly from their African homes . . . and sold like chattels as slaves. . . . [until] in every state of the Union telegraph poles had been decorated by the bodies of Negroes dangling to ropes on account of race hatred and nothing else."

Ossian Sweet, said Darrow, knew all this, so much so that he felt bowed by the psychology of fear that stalked his race. "He knew they had been tied to stakes in free America and a fire built around living human beings until they roasted to death; he knew they had been driven from their homes in the north and in great cities and here in Detroit, and he was not there only to defend himself and his house and his friends but to stand for the integrity and the independence of the abused race to which he belonged . . . a hero who fought a brave fight against the fearful odds, a fight for the right, for justice, for freedom, and his name will live and be honored when most of us are forgotten."

And so, Darrow continued, "I speak for a race which will go on and on to heights never reached before. I speak for a million blacks who have some hope and faith remaining in the institutions of this land. I speak to you on behalf of those whose ancestors were brought here in chains. I speak to you on behalf of the faces, these black faces, which have haunted this courtroom since this trial began."

Darrow looked from the black row of faces in the dock to the white jury, and he asked a question that seemed shocking in 1925. Which of these two races, black or white, could claim to be superior in its humanity? Stretching his hand out to the Sweets, paying tribute to their reason and forbearance, and then remembering the prejudice and deceit of the people he called the "Noble Nordics," he asked whether a reasonable person would not conclude that the Negro belonged to the superior race.

The courtroom did not emit any dissent or rage. It had been stilled by Darrow, by a man who knew the sheer folly of calling any human, with all his or her foibles and traumas, a superior being. "I know that back of me and back of you is an infinite ancestry stretching back at least five hundred thousand years, and we are made up of everything on the face of the earth, of all kinds and colors and degrees of civilization and out of that come we. Who are we, any of us, to be boastful above our fellows?"

Darrow's voice, wrote Lilienthal, boomed "like a brass gong," his eyes "hard and grim, every muscle of his huge body tense and strained." It had all come down to this, over forty years of defending the lost and the damned, and nothing mattered more than saving the Sweets. He had spoken for seven hours, over the course of two days, and he was now only a few feet away from the jurymen, some of whom were crying as they looked up at Darrow. But he knew that as many of them looked down, away from him, and it was hard to tell what they might be thinking. He lifted himself for one last push over the mighty hill and, then, the passion rolled out of him like a great big boulder gathering momentum as it raced down the other side toward the bottom.

"I ask you gentlemen on behalf of my clients," Darrow cried out. "I ask you more than everything else, I ask you on behalf of justice, often maligned and down-trodden, hard to protect and hard to maintain, I ask you on behalf of yourselves, on behalf of our race, to see that no harm comes to them. I ask you gentlemen in the name of the future, the future which will one day solve these sore problems, and the future which is theirs as well as ours, I ask you in the name of the future to do justice in this case."

Darrow drew back with a muffled sigh. He wiped a hand across his eyes and then returned to his place at the table of the defense. Charles Mahoney, one of the black attorneys working alongside Darrow and Hays, looked up at Frank Murphy and saw an incredible sight. "I've heard about lawyers making a judge cry," Mahoney said later, "but Darrow was the first man I ever saw do it."

JOSEPHINE GOMON, the attractive woman with whom Darrow had been so smitten at the Penguin Club a few weeks before, followed his performance closely in court. "It was wonderful," she wrote of his final plea. "Eloquent, logical. People wept, and jurors were moved."

Darrow himself was moved more by her alluring charms. He asked Josephine to accompany him to lunch immediately after the conclusion of his address. She was flattered and agreed instantly—only to be deflated about an hour later when Ruby Darrow and Arthur Hays unexpectedly joined them. "I was terribly disappointed," Josephine wrote in her diary, "since I know a good many interesting women like Mrs. Darrow—but Mr. Darrow is one of the most intelligent men I have ever talked to."

IT TOOK the judge an hour to sum up proceedings for the jury on that afternoon of Wednesday, November 25, 1925. He made careful note of the rigorous rebuttal Robert Toms had presented for the prosecution in a three-hour address in which the prosecutor urged the jury to remember the wretched fate of the victim. "The trouble with this case," Toms argued, "is that Darrow doesn't want to look at it as a criminal case, but as a cross-section of human emotions. But that's not what we are here for. What an insignificant figure Leon Breiner has been, and yet we started out to find out who killed him." The Sweets, especially Henry, who confessed to shooting directly into the crowd, had taken away the life of a pipe-smoking, neighborly man, a man just like every one of them.

Yet Judge Murphy seemed to sway more, in his instructions to the jurors, toward the defense. "Under the law," he reminded them, "a man's house is his castle. It is his castle whether he is white or black, and no man has the right to assail or invade it." Murphy urged the jury to think calmly. "In your deliberations, try to be reasonable. Be tolerant of the other man's viewpoint, try to understand, for in this way you will have the best chance of reaching a verdict."

At twenty past three that Wednesday afternoon, the day before Thanksgiving, the jury filed out of court. Darrow and Hays remained in their chairs, yearning for a favorable early verdict. There was laughter and hope in everything they said as they sat and waited, their optimism turned to near elation in the faces of the Sweets and their supporters. After all that Darrow had done they were sure that freedom would be theirs in time for a Thanksgiving celebration. Hays and White had already decided that they would hold a small party on the 7:00 P.M. train back to New York.

But seven o'clock came and went and, by then, Darrow and the others had drifted away to a muted dinner. A "not guilty" verdict usually came through quick and early. It was clear that at least some of the jury were still not convinced of the Sweets' innocence.

Darrow came back to court and hung around for another hour, when, just after ten, there was a knock on the locked door of the jury chamber. The expectation was immediately cut down by the foreman, Charles Naas, asking for clarification from the judge on a number of crucial points.

Anxiety soured the mood. After midnight people rested their heads on their arms or stared blankly into the silence. At ten minutes past two, in the dead of night, Murphy had had enough. He ordered the jury up to sleep in the sixth-floor dormitory of the courthouse. Everyone else was sent back to their homes or hotel rooms.

Thanksgiving was ruined, and tempers grew more violent as the following afternoon brought no respite. As they played cards and talked quietly among themselves the lawyers and spectators could hear snatches of loud arguments behind the closed door. "Damn it," one juror yelled, "I am a reasonable man!" Another roared that he would "rot in hell" before he agreed with another. More confusingly, someone else could be heard shouting, "I'll sit here forever before I condemn them niggers." To escape the madness, Darrow, Hays, and White went for a bite of Thanksgiving turkey at the black YMCA in downtown Detroit.

The jurors were ordered out for their own Thanksgiving dinner at six o'clock that night, but their meal was eaten in sullen silence. They were soon locked in again, and the sound of hysterical shouting and furniture being smashed lent an almost comic edge to the fraught tension.

Darrow and his supporters took refuge in a side room where, almost magically, a couple of bottles of whiskey were produced. As the drink flowed, and Darrow lost himself in a dreamy discussion with Josephine Gomon, his spirit revived. When Gomon spoke out against Prohibition, which made her "so damn mad," his rapture was almost complete. He held her hands in his and chuckled deliciously. "We are affinities. To think I should ever hear such an opinion from a woman," he sighed, ignoring the fact that Ruby and Mary had also damned Prohibition for years.

The old devil might have started reading poetry to the lovely Josephine—but then, shortly after 11:00 P.M., Judge Murphy called the jury back into court.

"Are you unable to reach agreement?" Murphy asked, with a touch of exasperation.

"We can't agree," the foreman said sadly.

"Do you think you will be able to reach an agreement at any time?"

"No."

Murphy sighed. "My duty is plain. This trial has been running almost a month at great expense to the city and defendants, and it is absolutely necessary that I keep you here some time more. You may go to bed and resume deliberations at 9:30."

Gladys Sweet could not help herself. She wept bitter tears as court officials went to clear the room where the jurors had raged for so long. They found that most items of furniture had either been cracked or broken.

OUT OF SUCH anger and dissent, Murphy soon realized, justice would never be served. The following morning the deadlock remained. Although they were unanimous in their belief that eight of the defendants should be acquitted, there was no shifting fixed opinions on the fate of Ossian, Henry, and, less obviously, of Leonard Morse, the insurance agent who had lent his support to the Sweets. Five of the jurors believed that all three men should be freed. Seven others, however, were immovable in their belief that the trio should be found guilty of murder.

At one thirty on the afternoon of Friday, November 27, 1925, forty-six hours after they had first begun to consider their verdict, the foreman informed Judge Murphy of the enduring stalemate. He was asked, twice, if there was any chance that some of them might reconsider their view. Each time, Charles Naas replied with a weary one-word answer: "No."

The young judge voiced his regret and concluded that he had no option but to declare a mistrial. They would need to return to court, in the new year, and do it all over again.

CHAPTER 15

SWEET AGAIN

A DARK HATRED persisted in Detroit. As the trial had ground toward its hung jury, an unknown man soaked some bundles of cloth in gasoline, thrown them through a small opening in the garage to Ossian Sweet's house, and tried to set them ablaze with a lit match. At least the fire had been quickly snuffed out by the patrolmen who were guarding the property, but the protection they offered the beleaguered doctor did not last long. When Sweet and his two brothers were released on bail, on December 3, 1925, a week after the trial ended, Inspector Norton Schuknecht withdrew his men. It was, once more, Dr. Sweet's responsibility to look after his house.

The bungalow now stood empty on the corner of Garland and Charlevoix, for Sweet had resisted Arthur Hays's urging that he should move back into the house in a statement of defiance. After eighty-four days in Wayne County Jail, Sweet needed some respite with Gladys and baby Iva. And so they all moved into the small apartment Gladys had rented on her release from prison. Ossian knew that he had to replenish himself before the next trial, which would be a particularly taxing ordeal featuring his youngest brother, Henry, alone in the dock against the full might of the State.

Darrow and Hays had demanded separate trials for each of the eleven defendants, a strategy aimed at securing the freedom of many of the lesser figures in the Sweet drama. But the danger for Henry was amplified. Since he had already admitted firing into the crowd, it was inevitable that Robert Toms should insist on Henry Sweet being put on trial first. Darrow knew that if they could save Henry, then the other cases would almost certainly collapse. But the pressure bearing down on them was even more acute than it had been in the first trial. If Henry was found guilty, then Ossian and many of the others would surely follow him down into a life sentence.

As Ossian adjusted to the world outside prison, and began to work again at his even busier and now celebrated practice in Black Bottom, Darrow returned to New York, where he drummed up yet more support for the Sweets. At a Salem Methodist Episcopal church rally, in Harlem, he was introduced as "The Great Defender" by James Weldon Johnson of the NAACP. The black congregation of two thousand broke out in mass jubilation as the shambling warrior shuffled to the pulpit. Darrow, while flaying the racists of Detroit, made a forceful reminder of his resistance to religious piety. The old devil, booming with mischief, told the congregation that "you're too blooming pious!" His accusation might have unsettled the church leaders, but the crowd greeted it with the same acclaim that accompanied the rest of his well-targeted aims during an hourlong speech.

Harlem was humming. Darrow went out and spoke for another thirty minutes to the fifteen hundred souls who had failed to gain entry to the church. Coins and notes poured into the buckets being carried around in support of the Sweet defense—with Darrow's fiery delivery raising another $1,600 to add to the $3,500 they had already collected in church. With the possibility that eleven different trials would need to be financed, the NAACP was desperate for money.

Darrow might have been let down by Leopold and Loeb's families in their reluctance to pay him, but there were other millionaires in Chicago to whom he could appeal. Julius Rosenwald, chairman of Sears & Roebuck, had helped Darrow before in financing his defense of black clients. He had also been shocked to have had a tangential role in the previous summer's trial of the century—for Rosenwald's immediate deputy was Richard Loeb's father, and his own grandson, Armand Deutsch, had come close to being murdered before Leopold and Loeb settled on Bobby Franks as their victim. And so when Rosenwald heard of Darrow's far more uplifting work to rescue the Sweets, he immediately contributed $2,000 to the NAACP fund.

Walter White, in an effort to produce a money-spinning publicity tour, had charmed Judge Murphy into allowing the Sweets to leave Michigan between trials. They followed Darrow to New York and, on January 3, 1926, Ossian and Gladys were hailed by a large crowd at Mount Olive Baptist Church on Lenox Avenue. Arthur Hays, in a wildly extravagant mood, introduced the couple by suggesting that "no man has ever had a greater opportunity to do more for his race or mankind than Dr. Sweet . . . [in] leading the fight against segregation."

The praise made Ossian giddy, and he became increasingly conceited during a five-city tour that moved on to Philadelphia, Pittsburgh, Cleveland, and Chicago. He made some portentous speeches about his own experiences in Europe and all that he had learned during his travels abroad—which included some brief visits to docks along the coast of North Africa. "Even in the jungles where the so-called barbarians and semi-civilized people live," Ossian said in a strangely reactionary manner, "a man's home is his castle. It was then [in Africa] that the thought came to me that the least we who claim to be civilized and more progressive can do is to uphold this principle at the cost of our very lives."

Such claims led to sniping arguments between Ossian and Gladys. She felt uncomfortable with the way in which he allowed himself to be depicted as a fearless hero standing alone against a baying mob—when the memory of their shared terror and humiliation in the bungalow was still so raw.

Robert Bagnall, a branch director of the NAACP, had been seconded by White to lead the Sweets from city to city. He soon regarded Ossian as almost unbearable, and, having finally seen the Sweets onto the train from Chicago back to Detroit, he fired off a letter to the NAACP head office. "Tell Walter," he wrote bitingly, "never again!" Yet the tour had been such a success that $76,000 had been raised. The defense fund could face the threat of successive trials for the Sweets and their colleagues with renewed conviction.

That injection of funds was soon needed. As the trial kept being pushed back—first by Toms as he fought another case for the state and then by Judge Murphy while he cared for his terminally ill father—Hays informed Darrow that he would be unable to return to Detroit. He was full of regret for letting down the Sweets, but he had already given his word that he would defend Countess Vera Cathcart in March. An English actress who had been denied entry into New York to perform in a play, Cathcart was accused of "moral turpitude" after committing adultery in South Africa. Her reputation appealed to Darrow, and so he accepted Hays's decision with chortling good humor.

He could not go into war in Detroit alone, however, so he instructed White to hire Thomas Chawke. As the toughest and best criminal lawyer in Detroit, and something of a mobsters' favorite, Chawke did not fit the NAACP profile. After meeting him in Detroit, White noted that Chawke had "none of the idealism of Darrow or Hays. So far as I can judge he has no prejudice against Negroes but he is going into this case as a business proposition and because, as he phrased it, it is a case that can be won."

Chawke duly demanded equal billing with Darrow and insisted that his nonnegotiable fee for taking the case was $7,500—$2,500 more than the NAACP planned to pay America's most revered lawyer. Darrow did not care much about earning less money than Chawke or the title of cocounsel. He wanted to save the Sweets, and he knew that with such an uncompromising lawyer at his side his chances would be sharpened considerably. Darrow advised White to accept the terms and close the deal.

WHEN THE TRIAL of Henry Sweet opened on April 19, 1926, the day after Darrow turned sixty-nine, he and his new partner took turns in cross-examining witnesses. Chawke, a large and commanding man, bristled with tough-guy menace. At six foot two, dressed to kill in a flashy suit that would have created envy among Detroit's smartest gangsters, his slicked-back hair gleamed and his blue-gray eyes blazed as he stalked his adversaries with a booming voice. Whereas the thirty-eight-year-old Chawke sounded furiously blunt, the stooped and elderly figure of Darrow gently mocked the increasingly evasive racism.

He confronted Inspector Schuknecht again. The policeman, with his thick neck and bulldog face, was determined to conceal the truth.

"Did you hear any unusual noise or shouting before the shooting?" Darrow asked.

"I did not," Schuknecht said with a small smile.

Darrow made a wordless exclamation—recorded as "Humpf" in some transcripts—and turned to the jury. His appealing expression invited them to share his disbelief.

He looked quizzically at Schuknecht, whose heavily ringed hand was draped casually over the back of his chair. "Notice any automobiles?" Darrow asked.

"Yes," Schuknecht said, suddenly looking more wary. "Traffic was heavy. We had two men to divert it."

But Schuknecht insisted that he had not seen a taxi draw up outside the Sweet home just before the shooting. He had not seen Otis Sweet or William Davis almost lynched by a mob howling "Niggers!

Get them!" Schuknecht had not seen any crowd at all along Garland Avenue. He saw only "a few people." He did not see them throw any stones—although he again admitted, grudgingly, that he had found a stone in the Sweet bedroom.

Chawke battered the next witness, Lieutenant Paul Schellenberger, over the contradictory issues of supposedly heavy traffic and mysteriously missing people.

"Did you see an unusual number of automobiles in that district while you were there that night?"

"I should say not," Schellenberger said dismissively.

Chawke kept at him, stripping away all the lies and hedging until, within another minute, as he trapped the sweating Schellenberger, he asked a similar question.

"Was traffic getting heavy?"

"It appeared to me that people were getting curious," Schellenberger sighed, "and there was an unusual amount of traffic."

"Then there *was* an unusual amount of traffic there, wasn't there?"

Schellenberger flushed a deeper red and looked helplessly at Toms. And then, with a defeated crack in his voice, he said, "There were . . ."

Darrow also confused Patrolman Ray Schaldenbrand. As the attorney probed the inconsistencies in his account, Schaldenbrand hesitated. "I didn't think," he began.

"What do you mean, you didn't think?" Darrow asked incredulously. "Don't policemen think?"

"No," Schaldenbrand said sadly, "a policeman is not supposed to think."

Chawke went after John Getke, a piano tuner who lived in the house next door to the Sweets. After Getke had claimed that "my memory has always been naturally short," he "surmised" that he might have seen a crowd gather on the night the Sweets arrived.

"What do you mean?" Chawke asked.

"I just surmised that I saw [a crowd]," the little piano tuner bristled.

"Why don't you tell the truth?" Chawke thundered. "Was there a crowd in front of Dr. Sweet's?"

"No."

"You 'surmised' didn't you, that Mr. Toms wanted you to say 'no.' You saw between five hundred and a thousand people there didn't you?"

"No."

"Do you know what a crowd is?"

"Well," Getke sneered, "they say that one is company, two is a couple and three is a crowd. I saw three policemen in front of the Sweet home."

"That's all?"

"Yes."

With a gesture of contempt, Chawke dismissed Getke from the stand.

The sullen standoff between reluctant witnesses and indignant attorneys continued, without either Chawke or Darrow nailing the definitive breakthrough they needed to save Henry Sweet. They had to make the jury understand that a large and threatening mob had terrified the Sweets, closing in on the house with intent to cause real harm. The legal significance of these two points had become even more urgent, because, despite often resembling a replica of the first trial, there had been a significant shift in the defense.

Darrow had been forced to respond to the decision of the prosecution, on the opening day of the trial, to display all the weapons found in the Sweet home. The guns were laid out on a table in full view of the jury, as if to plant in their heads the idea that the Sweets had been heavily armed hoodlums who fired into a suburban street. Choosing to make his opening address straight after Toms, rather than delaying it as he had done in the previous trial, Darrow spoke directly to the jurors. "I don't know who killed Breiner," he stressed. "It might have been Henry Sweet. I can't tell and he can't tell."

Darrow waited for the silence to settle after that apparently shocking concession. And then he made the jury recognize the real question of the trial. "It is your task," he said, "to determine if he was in a conspiracy to kill Breiner or anyone else. If Henry Sweet was there to kill someone on slight provocation, Mr. Toms is quite right in saying he would be guilty. But if he killed in defense of that home and that family, he is innocent. . . . We will show that the white people in the neighborhood began to prepare for the reception of Dr. Sweet. They organized the Waterworks Park Improvement Association. It had nothing to do with waterworks or improvement. They organized to keep Dr. Sweet out of his home."

Individual witnesses were made to look deceitful or foolish by Darrow and Chawke, but that was still not enough to establish the grounds for justifiable self-defense. Darrow, nevertheless, enjoyed showcasing the ignorance of the white witnesses in contrast to the erudition of the Sweets.

"What do you do?" he asked Marjorie Stowell, a feisty woman whom he already knew had taught at a high school for the past fifteen years.

"I am a teacher," she confirmed.

"And you live near which street?"

"Go-thee."

"What?" Darrow said, knowing she meant Goethe Street.

"Go-thee."

"You mean to tell me," Darrow said in mock horror, "anyone is fit to teach school in this city who pronounces Goethe that way?"

The *Detroit Times*, in a story headlined "Darrow in Clash with Teacher," lingered over Stowell's embarrassment. But such point scoring did little to advance the case for the defense. Darrow and Chawke, more than anything, needed to expose the racist intimidation of the

Waterworks Improvement Association. They kept hammering away, but the witnesses refused to detail any planned violence. Darrow succeeded only in exposing the extent of their stupidity.

He asked Edward Miller, a foreman and neighbor of the Sweets, to explain the aims of the association. "Oh," Miller said breezily, "we wanted to protect the place."

"Against what?" Darrow wondered.

"Against undesirables."

"Who do you mean by 'undesirables'?"

"Oh, people we don't want."

"Like who?"

"Negroes," Miller conceded.

"Anyone else?"

Miller thought for a moment and then grunted: "Eye-talians." They didn't want "anybody but Americans" living in their neighborhood.

Darrow looked incandescent. Did Miller not know that "Negroes" had lived in the country for over three hundred years? Did he not know that America, at least for white-skinned people, had been "discovered" by an Italian?

But the real breakthrough they needed came from another witness. Alf Andrews, who had attended the Waterworks Improvement Association meeting, was grilled by Darrow. Andrews admitted that they had listened to an address made by a spokesman for the Tireman Improvement Association, a fellow "who called a spade a spade when he talked."

Darrow asked if the Tireman speaker explained how his association had run Dr. Alexander Turner out of his new home.

"Yes, he did," Andrews nodded. "They didn't want colored people in their neighborhood and proposed to keep them out. He was very outspoken."

"You felt the same?" Darrow said quietly.

"If by legal means we could restrict them."

"Did the speaker talk about 'legal means'?"

"No."

"He talked about driving them out, didn't he?"

"Yes, he was radical—I admit that."

"You say you approved of what he said and applauded it, didn't you?"

"Part of his speech."

Darrow paused, as if adjusting to the truth. "In what way was he radical?"

"Well, I don't . . . ," Andrews began and then hesitated. "I myself do not believe in violence."

"I didn't ask what you believe in," Darrow said sharply. "I said in what way was he radical?"

Andrews was silent, but Darrow persisted. "Any more you want to say about what you mean by radical?"

"No," Andrews said as he shifted uncomfortably in his chair. "I don't want to say anymore."

Darrow had his witness dangling on a hook. He was not about to let him go. "You didn't rise at that meeting and say, 'I myself don't believe in violence,' did you?"

Andrews shuddered. "No! I'd have had a fine chance with six hundred people there!"

"What?" Darrow shouted out in apparent shock, his voice echoing around the courtroom. "You wouldn't have dared do it at that meeting?"

Toms, seeing the trap snap shut on Andrews, jumped up and yelped desperately at the witness: "Don't answer it!"

The prosecutor then objected more formally to Judge Murphy—who sustained his claim that it was "a very, very improper question."

Darrow plowed on, dismissing both the judge and the prosecution. He had eyes only for Andrews. "What do you mean by saying you had a fine chance?"

Toms shouted out again in disbelief—"Wait a minute!" He crossed the space separating them to grab hold of Darrow's arm. Toms had completely forgotten his policy, in the first trial, to avoid any conflict with the ultimate courtroom bruiser. "Didn't you get the court's ruling?" he snarled.

Darrow ignored him and stared instead at Andrews. "What did you mean?" he repeated.

Andrews could not stop himself. "You imagine I would have made myself heard with six hundred people there? I wasn't on the platform."

"Did anybody in that audience of six hundred people," Darrow asked, "protest against advocating violence against colored people who moved into the neighborhood."

"I don't know."

"You didn't hear any protest?"

"No."

"You heard only applause?"

"There was, as I stated . . . ," Andrews said, and then changed tack. "This meeting was in the school yard—"

"You heard nobody utter any protest," Darrow interrupted, "and all the manifestation you heard was applause at what he said?"

"Yes."

Toms was in trouble, so he called the witness to answer a key question of his own. There was only one answer he needed, a simple "no," to claw back some of the ground that Darrow had crossed so effectively.

"Did he advocate violence?" Toms asked.

Andrews hesitated again before, as if Darrow had finally uncorked the truth in him, he nodded. And then, to the despair of Toms, the word came: "Yes."

WHEN THE CASE for the defense opened on May 11, Bruce and Mary Spaulding, in admirably restrained accounts, described the mob that had looked ready to lynch them in their car shortly before the shooting. "Both Mr. and Mrs. Spaulding, very light mulattos, well dressed and courteous in the manner," Marcet Haldeman-Julius noted approvingly, "made a striking contrast to the ill-bred, ill-mannered, arrogant and pert white men and women who had preceded them."

Robert Smith, another black man who'd been attacked at the intersection near Garland, had since died. His testimony, as ungrammatical as it was vivid, was read in an eerily dramatic monotone by Chawke—as if he spoke in a voice of truth that reached from beyond the grave. "I goes where I goes, and pays as I goes." Chawke read. "I was goin' to dinner, and when I wants to eat anywhere I goes by any street I pleases." That night, however, Smith and his nephew, James, were slowed to a crawl by a mob who stoned their car. Chawke, still reading Smith's words, repeated the crowd's taunts in court, the impact sounding all the more telling in the voice of a strapping white man more used to defending mobsters: "There's niggers now!" Chawke said. "Get 'em! They're going to the Sweets!" Switching back into the voice of the deceased black man, Chawke read Smith's next simple line: "Who was these Sweets, that's what I wanted to know?"

Darrow decided that it would be too much of a risk to allow the jury to get to know Henry Sweet more intimately. At only twenty-two, looking polite and scholarly and, at five foot eight inches, not especially imposing, he had the appearance of an appealing young man. But a more dangerous certainty loomed. Toms would insist that Henry relive his confession in intricate detail—which meant that the increasingly anxious defendant could implicate himself still further. Darrow, instead, called Ossian to the stand, in the belief that he would again speak the truth with the compelling dignity that had transformed the first trial. Ossian did not let down Darrow, or his young brother, even though he had to withstand a ferocious cross-examination from Toms.

Lester Moll, in a three-hour summary for the prosecution, again derided Dr. Sweet as "a quasi-intelligent Negro . . . who is passing as a second Lincoln. The defense would have you believe that there was a menacing mob outside the house and that an attack was made. I challenge anyone to find, anywhere in the record of this trial, any evidence that would show the home of Dr. Sweet was damaged beyond the smashing of one small window. For the breaking of a window are we to trade the life of Leon Breiner?"

Detroit Municipal Court, St. Antoine Street, Detroit, May 11–12, 1926

When Clarence Darrow began his summary, the court was again crammed. In the last great trial of his life, he approached the jury somberly, as if he understood the gravity of the Sweet situation as well as the slow end to his work in court. All the jurors looked intently at him, even Ralph Fuelling, a former army man turned water board employee who had dozed intermittently throughout the trial—so much so that the whisper of "number 8 is asleep again" had become a familiar lament during the preceding three weeks. Fuelling, however was alert and sharp, eager to listen to Darrow. The other eleven men, ranging in age from twenty-four to eighty-two, leaned forward expectantly.

"If the Court please, gentlemen of the jury," Darrow began in a voice so soft and weary that it seemed hard to believe that he would be able to speak for seven minutes, let alone the seven straight hours that would eventually frame his plea for Henry Sweet. "You have listened so long and patiently that I do not know whether you are able to stand much more."

Some of the older men smiled sympathetically, as if silently promising that they would match him if he could somehow find the energy to continue. "I want to say, however, that I have tried a good many cases in the forty-seven or forty-eight years that I have lived in courthouses. In one way this has been one of the pleasantest trials I have ever been in. The kindness and consideration of the Court is such as to make it easy for everybody, and I have seldom found as courteous, gentlemanly and kindly opponents as I have had in this case. I appreciate their friendship."

Toms and Moll bowed their heads in gratitude, perhaps forgetting Darrow's old trick of showing himself in the most gracious light. He appeared to the jury as a lovable old granddad who liked to think the best of everyone. But Darrow, of course, had already begun to forge ahead, leaving behind the flattering pleasantries to cut to the heart of the matter.

"My friend, Mr. Moll, says that this isn't a race question. This is a murder case. We don't want any prejudice; we don't want the other side to have any. Race and color have nothing to do with this case. This is a case of murder."

Darrow leaned against the railing, holding on to the wooden beam, as he gazed down the length of two rows of jurors. His voice was now forceful, rather than charming. "I insist that there is nothing but prejudice in this case; that if it was reversed and eleven white men had shot and killed a black while protecting their home and their lives against a mob of blacks, nobody would have dreamed of having them indicted. I know what I am talking about, and so do you. They would have been given medals instead."

He nodded sagely, as if they were all wise old friends gathered together against the inequities of the world. "Ten colored men and one woman are, in this indictment, tried by twelve jurors. Gentlemen, every one of you are white, aren't you? At least we all think so. We haven't one colored man on this jury. We couldn't get one. One was called and he was disqualified. You twelve men are trying a colored man on race prejudice. . . . I want you to do all you can to be fair in this case, and I believe you will."

Darrow had stroked them collectively again. But he knew how important it was to make each juror confront his own deep-rooted prejudice. "How many of you, gentlemen, have ever had a colored person visit you in your home?"

The silence stretched across the court, forcing every white person to ponder the question in the privacy of his or her own mind. "How many have you ever visited in their homes? How many of you have invited them to dinner at your house?"

Darrow broke the next hush with a shrug. "Probably not one of you. Now, why, gentlemen? There isn't one of you men who does not know, just from the witnesses you have seen in this case, that there are colored people who are intellectually the equal of all of you. Am I right? Colored people living right here in the city of Detroit are intellectually the equals, and some of them are superior, to most of us. Is that true? Some of them are people of more character and learning than most of us."

He returned to the white victim and remembered that Moll, for the prosecution, "says I wiggled and squirmed every time they mentioned Breiner. Well, now, I don't know. Did I? I have been around courtrooms so long that I fancy I could listen to anything without moving a hair . . . this isn't the first case I was ever in. [But] I don't like to hear the state's attorney talk about the blood of a victim. It has such a mussy sound. I wish they would leave it out. I will be frank with you. I don't think it has any place in a case. I think it tends to create prejudice."

Suggesting that it was not easy to talk about the dead—"unless you slobber over them, and I am not going to slobber over Breiner"—Darrow used an increasingly accusatory tone. It was a tough and difficult stance, but he was determined not to sentimentalize a dead man. "Who was Breiner anyway?" Darrow asked bluntly. "I will tell you who he was. He was a conspirator in as foul a conspiracy as was ever hatched in a community; in a conspiracy to drive from their homes a little family of black people."

The court was stunned, as if no one could believe that Darrow dared speak ill of the dead. But he reminded them that Breiner had been among a crowd of "seven hundred people" who met under the auspices of the sinister Waterworks Park Improvement Association less than two months before the Sweets actually moved into Garland Avenue. "That is why Mr. Breiner came early to the circus on September 9, 1925. He went past that house, back and forth, two or three times that night. What was he doing? 'Smoking his pipe.' What were the rest of them doing? They were part of a mob and they had no rights, and the Court will tell you so, I think. And, if he does, gentlemen, it is your duty to accept it."

Throughout Darrow's long and involving address, juror number 8, the usually sleepy Ralph Fuelling, sat bolt upright, wide awake. After Darrow described the intimidation of the Sweets on September 8, he said, "The mob came back the next night and the colored people waited while they were gathering; they waited while they were coming from every street and every corner, and while the officers were supine and helpless and doing nothing. And they waited until dozens of stones were thrown against the house on the roof. I don't know how many. Nobody knows how many. They waited until the windows were broken before they shot. Why did they wait so long? I think I know. How much chance had these people for their lives after they shot, surrounded by a crowd as they were? They would never take a chance unless they thought it was necessary to take the chance. Eleven black people penned up in the face of a mob. What chance did they have?"

Darrow talked on, hour after hour, stretching into a continuation that covered most of the next day. "Prejudices have burnt men at the stake," he said gravely, "broken them on the rack, torn every joint apart, destroyed people by the million. Men have done this on account of some terrible prejudice which even now is reaching out to undermine this republic of ours and to destroy the freedom that has been the most cherished part of our institutions. These witnesses honestly believe that it is their duty to keep colored people out. They honestly believe that the blacks are an inferior race and yet if they look at themselves, I don't know how they can. If they had one colored family up there, some of the neighbors might learn how to pronounce 'Goethe.'"

He almost laughed, but scorn took hold of him. "It would be too bad to spread a little culture in that vicinity. They might die. They are possessed with that idea and that fanaticism, and when people are possessed with that they are terribly cruel. They don't stand alone. Others have done the same thing as long as this weary world shall last. They may do it again but, gentlemen, they ought not to ask you to do it for them. That is a pretty dirty job to turn over to a jury, and they ought not to expect you to do it."

His passion and lucidity strengthened as he pulled the jury back toward the defendants. "Now let us look at these fellows. Here were eleven colored people penned up in the house. Put yourselves in their place. Make yourself colored for a little while. It won't hurt, you can wash it off. They can't, but you can. Just make yourself black for a little while, long enough, gentlemen, to judge them, and before any of you would want to be judged, you want your juror to put himself in your place. That is all I ask. . . .

"Dr. Sweet has a child. He also has a wife. They must live somewhere. If they could not, it would be better to take them out and kill them, and kill them decently and quickly. Had he any right to be free? They determined to move in and to take nine with them. What would you have done, gentlemen? If you had courage, you would have done as Dr. Sweet did. Wouldn't you? He didn't shoot the first night. He didn't look for trouble. He kept his house dark so that the neighbors wouldn't see him. He didn't dare have a light in his house, gentlemen, for fear of the neighbors. Noble neighbors, who were to have a colored family in their neighborhood."

Darrow swept on, daring to ask the jury some harsh questions in order to depict the state of America in the mid-1920s. "Supposing you had your choice, right here this minute," he said, "would you rather lose your eyesight or become colored? Would you rather lose your hearing or be a Negro? Would you rather go out there on the street and have your leg cut off by a streetcar, or have a black skin?"

There was a muted gasp at Darrow's savagery. He pulled back and spread his hands in apology. "I don't like to speak of it . . . but it is true. Life is a hard game, anyhow. But, when the cards are stacked against you, it is terribly hard. And they are stacked against a race for no other reason but they are black."

He named the doctor and the dentist, the pharmacist and the college students, the respectable insurance men, and thought of them "all gathered in that house. Were they hoodlums? Were they criminals? Were they anything except men who asked for a chance to live; who asked for a chance to breathe the free air and make their own way, earn their own living, and get their bread by the sweat of their brow."

When Darrow paused, it was done in determination, rather than fatigue, as he forced the jury to examine both themselves and the consequences of their judgment. "Your verdict means something in this case," he said near the end as he looked across at Henry Sweet. "It means something more than the fate of this boy. It is not often that a case is submitted to twelve men where the decision may mean a milestone in the progress of the human race. But this case does. And I hope and I trust that you have a feeling for responsibility that will make you take it and do your duty as citizens of a great nation and as members of the human family, which is better still."

Reminding the jurors of how the twenty-one-year-old Henry had gone to the defense of an older brother he revered, Darrow asked the jury one last question. "Do you think he ought to be taken out of his school and sent to the penitentiary?"

Darrow took their quietness as a sign of tacit approval. "All right, gentlemen, if you think so, do it. It is your job, not mine." He turned away, as if he was finally finished, in spirit as much as word, only to stop in midstep. Darrow was already talking as he moved back toward them. "But if you do, gentlemen, if you should ever look into the face of your own boy, or your own brother, or look into your own heart, you will regret it in sackcloth and ashes. You know, if he committed any offense, it was being loyal and true to his brother, whom he loved."

The old man looked more kindly at the jury now. "I know where you will send him," he said softly, "and it will not be to the penitentiary."

After six and three-quarter hours on his feet, and almost fifty years in the courtroom, Darrow hurtled toward his close. His voice rang out, sounding defiant and majestic, and the words rolled from him, as they had done for so many decades, in defense of so many frightened and innocent men and women. He might have been as flawed and fractured as the rest of us but he was also Darrow, the great defender, and so he rose in honor of our battered humanity one last time.

"I do not believe in the law of hate," he said. "I may not be true to my ideals always, but I believe in the law of love, and I believe you can do nothing with hatred. I would like to see a time when a man loves his fellow man, and forgets his color or his creed. We will never be civilized until that time comes. I know the Negro race has a long road to go. I believe the life of the Negro race has been a life of tragedy, of injustice, of oppression. The law has made him equal, but man has not. . . . I know that before him there is suffering, sorrow, tribulation and death among the blacks, and perhaps the whites. I am sorry. I would do what I could to avert it. I would advise patience. I would advise toleration. I would advise understanding. I would advise all of these things which are necessary for men who live together. . . .

"I have watched, day after day, these black, tense faces that have crowded this court. These black faces that are now looking at you twelve whites, feeling that the hopes and fears of a race are in your keeping. This case is about to end, gentlemen. To them, it is life. Not one of their color sits on this jury. Their fate is in the hands of twelve whites. Their eyes are fixed on you, their hearts go out to you, and their hopes hang on your verdict."

Darrow bore down one last time on the jury and spoke his parting words. "This is all." He sighed before, gathering himself, he raised his voice once more and switched his gaze from them to Henry Sweet and back again, his eyes glittering in his creased face. "I ask you, on behalf of this defendant, on behalf of these helpless ones who turn to you, and more than that—on behalf of this great state, and this great city which must face this problem, and face it fairly—I ask you, in the name of progress and of the human race, to return a verdict of 'Not Guilty' in this case."

IN HER DIARY, Josephine Gomon stressed that "I shall never forget that final plea to the jury. One could have heard a pin drop in the crowded courtroom. Everyone listened breathlessly, crowded so closely together that women fainted and could not fall. He went back through the pages of history and the progress of the human race to trace the development of fear and prejudice in human psychology. Sometimes his resonant, melodious voice sank to a whisper. Sometimes it rose in a roar of indignation. The collars of the jurors wilted. They sat tense, in the grip of strained contemplation of historic events and tragic happenings which he made real and present again before their eyes.

"As Judge Murphy left the bench I met him just inside the door of his office. I had never seen him so moved. He took my hand and said, 'This is the greatest experience of my life. That was Clarence Darrow at his best. I will never hear anything like it again. He is the most Christ-like man I have ever known.'"

Robert Toms spoke for almost a whole day himself, in answer to Darrow, and argued that it was not the jury's place to solve the racial problems of America. "It was not such a poor summing up, either," Marcet Haldeman-Julius suggested, "but somehow it reminded one of the clatter of folding chairs after a symphony concert." Toms knew how much he suffered in comparison to Darrow, and so he acknowledged to the ordinary jurors of downbeat Detroit, "I cannot compete with this emotionalist from Chicago, neither verbally nor emotionally, but if you stick to the facts I have no fear of the outcome."

During the morning session of Thursday, May 13, Judge Murphy gave his final instructions. "Gentlemen of the jury," he said, "I consider it my duty to especially caution you and warn you against prejudice or intolerance in your deliberations. . . . Real justice does not draw any line of color, race, creed or class. All charged with crime, rich or poor, humble or great, white or black, are entitled to the same right and the same full measure of justice."

Darrow and Chawke, mindful of the furniture-smashing and intractable deadlock of the first trial, left the courtroom at twelve thirty. There seemed little point in spending a fraught afternoon hanging around for a verdict that might take another day to arrive. They led a dozen people, including both Ruby Darrow and the dazzled Josephine Gomon, out to a boozy lunch in downtown Detroit.

Three hours later Darrow was in his element. Holding court, drinking whiskey and reciting poetry, he would have steamed on into the night had they not been interrupted by an urgent message. A panting officer of the court told Darrow and Chawke that they were needed in the judge's chambers, for the jury had requested a point of clarification.

The two lawyers, breathing fumes of whiskey and port over each other, were back at work soon after four o'clock as they honed the necessary clarification. But, just as they reached agreement with Toms, they were told that the jury had decided there was no need for any further work. Their verdict had been reached.

Chaos filled the courtroom as the lawyers and spectators rushed to their seats. Murphy had to yell to make himself heard: "Don't bring that jury in until we are ready for them."

Henry Sweet sat down in abject silence, his face turned toward the opposite wall. He looked so scared that James Weldon Johnson went to console him. Johnson, clutching Henry's hand, promised him the NAACP would continue to support him, even if the verdict was as bad as they suddenly feared. The comparative swiftness of the decision, after just four hours, convinced Johnson that Henry Sweet was about to be found guilty—and jailed for life.

Darrow did not appear much more confident. He sat at an awkward angle, his gnarled hands gripping the right arm of his chair, as if he needed to stop himself from falling. He clenched his teeth, and his head jutted toward the empty jury box as he waited. He had not felt this sick and desperate since his own bribery verdict—which had ended in a hung jury. Defeat, this time, would be unbearable for Henry and the Sweets, but also for Darrow.

The door to the jury room was unlocked and the men walked into the court. They were led, in single file, by George Small, their young foreman. Gradually, as if extending the pain of a torturous decision, they settled themselves in place. Most of the men looked away from both Darrow and Henry Sweet.

The clerk of the court approached the box of judgment. "Have you gentlemen, in the course of your deliberations, reached a verdict in the case of Henry Sweet?" he asked. "And if so who will answer for you."

Small stood up. "We have, and I will." He paused, as if unsure what to do next, but then, after a nod from the clerk, he said the two words that mattered most:

"*Not guilty.*"

Ossian Sweet reacted first, covering his face with his hands. There was the sound of applause as Thomas Chawke moved next, turning toward Henry to offer his beaming delight.

Darrow went to follow him but as George Murphy, the judge's brother, wrote later, "his great spirit almost seemed to have left his body. He had given his all, body, mind and soul, to the trial."

The old devil, suddenly looking every one of his sixty-nine years, sank back heavily in his chair. Robert Toms, who had already shouted out his disbelief, asking the foreman to repeat his verdict, rushed across at the sight of an apparently falling Darrow. He thought his great rival was about to faint in exhaustion.

Toms reached out to catch him, but Darrow, his eyes glinting darkly, waved him away just as the foreman made the same call: *"Not guilty!"*

"Oh, I'm all right," Darrow murmured to the man he had just beaten. "I've heard that verdict before."

CHAPTER 16

SLIPPING AWAY

E IGHTEEN DAYS LATER, as if destiny might be determined to complete the perfect circle of Darrow's defining trilogy of trials, the
Tennessee Supreme Court met in Nashville to decide the outcome of
John Scopes's appeal against his conviction for teaching evolution the
previous summer. It was less a trial than a hearing set aside for "oral
argument" over a restricted two-day period, on May 31 and June 1.
Although the American Civil Liberties Union still needed his name
to head their appeal against the State of Tennessee, Scopes refused
to travel to Nashville. He insisted that he was not interested in the
outcome and, more fervently, that he wished he could forget the entire surreal episode.

The glare of national attention on his original trial had convinced
Scopes that "I would not live happily in a spotlight. The best thing
to do was to change my life and seek anonymity." Dayton had returned to its own form of obscurity, with the *Nashville Banner* concluding that it "is back where it was before the trial began, a sleepy
little town among the hills." Scopes found life more problematic. He
had been offered a new yearlong contract by the local school board,

with the inevitable condition that he refrain from teaching evolution, but Scopes chose instead to return to college. He initially planned to complete his law degree, and an enthusiastic dean at the University of Kentucky offered the reluctantly famous young man a place that fall. "You could be another Darrow!" the dean gushed.

As much as Scopes revered Darrow as a man, the thought of being compared with a giant unsettled the twenty-five-year-old. He knew there would no escaping his Dayton trial if he worked in the law. And so Scopes decided to reinvent himself by signing up for a degree in geology at the University of Chicago. He would be able to visit Darrow as often as he liked, and of course the old man would appreciate the evolutionary bent to his newfound study of ancient rocks.

If Darrow reveled in his fame, with the Sweet triumph earlier that month confirming his renown as "the greatest one-man draw in America," the ACLU was less impressed. His grandstanding performance against Bryan had turned the organization's stringent legal challenge into a celebrity showdown. Darrow might have been hailed as the moral victor in newspapers across the country, but the reality remained that, in legal terms, the ACLU had lost the Dayton case. There was also serious distaste for Darrow's tendency to hog the headlines. Edwin Mims, an esteemed liberal professor of English at Vanderbilt University in Nashville, sneered that "when Clarence Darrow is put forth as the champion of the forces of enlightenment to fight the battle for scientific knowledge, one feels almost persuaded to become a fundamentalist."

For the previous nine months the ACLU had plotted against Darrow in a bid to remove him from the defense's appeal. In August 1925 the organization's associate director, Forrest Bailey, had written to John Neal, the Knoxville attorney whose official role as chief counsel in Dayton had been totally overshadowed by Darrow. "All of us feel that when the case goes to the Supreme Court," Bailey stressed, "there ought to be more of Neal and less of Darrow."

Neal, creditably, refused to oust Darrow. And when the ACLU in

New York tried to covertly replace the great defender with Charles Evans Hughes, a far more sober attorney, Neal wrote back stingingly. He was not willing for any other lawyer to win the appeal and "take from Darrow and [Dudley Field] Malone the credit to which they are entitled. More than this . . . I am not willing to have conservative lawyers and conservative organizations reap the benefits of work done by liberals and radicals."

When Darrow responded forcefully to the ACLU campaign, Bailey claimed a "misunderstanding" and, in a more blatant lie, that "I never at any time asked that you be invited to withdraw." But Bailey was soon plotting again. "We ourselves have no wish to have Darrow, Malone and [Arthur Garfield] Hays continue," he wrote to a colleague in December 1925. "My only point was that we should not be made to appear as having kicked these men out."

But Darrow was too powerful to be shunted aside. He had a deeply personal stake in the outcome, having fought against the fundamentalists for so long, and he ensured that Hays led the meticulous drafting of a detailed appellate brief to the Tennessee Supreme Court. The prosecution responded with an even thicker four-hundred-page brief supporting the original judgment.

On May 31, 1926, as usual when Darrow appeared in court, "every door and window was blocked by scores who, unable to gain entrance, contented themselves by standing on chairs and tables." The record attendance for a Supreme Court hearing had, the *Nashville Banner* noted, attracted "many women of prominence in the social life of the city." They were bound to be disappointed for, in a two-day hearing rather than a trial by jury, the mood was much more subdued than it had been in the hothouse of Dayton. The *Chattanooga Times* lamented the fact that "there was no rising to the feet to interpose objections, no bickering between members of the counsel and no religious or anti-religious atmosphere." Such dry legal discourse meant the Supreme Court challenge was "a flop as a news story compared with the trial."

K. T. McConnico, prosecuting for the State, came closest to instilling some animosity into proceedings when he warned, "If you permit the teaching that the law of life is the law of the jungle, you have laid the foundation by which man can be brought to accept the doctrine of communism and to the point where he believes it is right to advocate murder." Turning darkly toward the defense, in a move to fuel ACLU fears, McConnico muttered that Darrow was a known protector of communists, murderers, and evolutionists. McConnico had brought in a pile of Darrow's past writings, topped by his pamphlet *Arguments in Defense of Communists*, but he was instructed "to confine himself to the case before the court." He urged for the appeal to be quashed on the basis that such a ruling would "preserve the Bible for all sects." The rival hopes of the defense, he warned, were grounded in "sinister and unclean efforts" to "teach this animal dogma [of evolution]" in the public schools of Tennessee.

Darrow presented a decidedly modern reply the following day. If McConnico argued that the Bible represented universal truth while science was aligned to individual opinion, Darrow believed the opposite. He suggested that religion was "a personal matter," between the individual and God, whereas science was a public activity geared "to the cause of progress . . . and everything that makes civilization today . . . we are once more fighting the old question, which after all is nothing but the question of the intellectual freedom of man." Although he slipped into emotive hyperbole when comparing the conviction of Scopes to the execution of Socrates, Darrow ended on a note of quiet good sense: "As long as man has an inquiring mind, he will seek to know and to find out. And every child ought to be more intelligent than his parents."

Darrow had been by far the most convincing speaker during those two arid days in Nashville, but the absence of drama was accentuated by the Supreme Court's announcement that they would spend the next seven months assessing the conflicting arguments—and announce their verdict only in January 1927. The ACLU, preparing itself for the worst, wrote to Darrow four days after the conclusion of the Nashville hearing. In the event of a second defeat in Tennessee they would have one final resort—fighting the case in the United States Supreme Court. Forrest Bailey suggested to him that whoever led that legal battle "should be utterly beyond the reach of prejudice of certain members of that august body, and we seriously doubt whether you, Mr. Malone or Mr. Hays, for example, would meet this requirement."

Darrow had spoken publicly of his desire to retire after the Sweet victory, but such was his resistance to being undermined that he fired back a withering reply: "Any possible prejudice that might exist toward Mr. Hays or me would be very much stronger against your organization." He might have felt weary, but he would not be denied one last conquest.

ALTHOUGH ROBERT TOMS had yet to formally drop charges against the ten other defendants in the Sweet case, he had made it clear to Darrow that, after the acquittal of Henry, he would not bring anyone else to trial. The joy of Ossian and Gladys, however, was shockingly brief. Their bleakest suspicions had been confirmed. Gladys and baby Iva had both fallen ill with an active strain of tuberculosis. Ossian knew that TB spread virulently though the jails of Detroit, and he was convinced that Gladys had contracted the virus during that terrifying month she had spent in a prison cell with three coughing young women. She would then have passed it on to Iva when they were cooped up together in the small apartment while awaiting the trial.

Iva was desperately sick. Just a few weeks after their freedom had been so spectacularly confirmed, Ossian and Gladys faced the truth. There was little that could be done for Iva, apart from transporting her to a hot dry climate where the quality of air would be more suited to her ravaged lungs. So Ossian and Gladys were separated again as she and Iva moved to rented accommodations in Tucson. Their hopes that the clean warm air surrounding the Arizona desert would help their little girl recover were, at best, tenuous. But Ossian believed Gladys's own chance of survival would, especially in winter, improve in sunny Tucson rather than cold and wet Detroit—where he would remain and work in Black Bottom.

Three months later their tragedy was sealed. Iva Sweet, aged two years and two months, was dead. On an August afternoon her small body was driven to the Roseland Park Cemetery in Berkley on the fringes of Detroit. The Sweet cortege was blocked at the cemetery by two guards. Ossian stepped out of the second car, brought to a standstill behind the hearse. Gladys, still stricken with TB, stayed in the backseat, her head bowed in grief.

The doctor stared at two sullen men as they told him that his cortege needed to go around to the rear of the graveyard. They would find the Negro entrance there. Ossian Sweet stepped toward the men, as if to show the torment etched into his face a little more clearly. They looked back at him, their own expressions unchanging. This entrance, the one man said, is for white folk only.

After everything he had endured, from the two nights on Garland Avenue to the eighty-four days in prison to the loss of his baby, Ossian Sweet reached inside his dark suit. He brought out the revolver. The doctor looked as if he was ready to blow a hole in somebody's head.

He did not, in the end, even have to point the gun at the men. With his voice shaking only slightly, he ordered them to stand aside. They were coming through this gate. They had a tiny girl, his daughter, to bury. The two men stepped away, not daring to look into Ossian Sweet's wretched face. Slowly, the black cortege drove through the white entrance and into the graveyard.

The bungalow at 2905 Garland Avenue, meanwhile, stood silent and empty. Iva Sweet had never been allowed to step inside. She would never see the bedroom her mother had dreamed of painting and furnishing for her in such soft and pretty colors.

THE WORLD was hard and cruel, full of so much suffering and pain that Darrow said he longed for his own passing. So near to seventy, and claiming to relish rather than fear death, he felt he should have been the one closed up in a box beneath the ground. It should not have been a girl so small and delicate. But this was how life mocked humanity with such venom. He read *King Lear* again and imagined the mad old man's blasted misery on the heath as, cradling the dead body of his daughter Cordelia, he cries out in despair that a dog or a horse or a rat might still live in place of her.

Darrow felt ill, and he was so full of gloomy talk that his doctor in Chicago ordered a complete rest. The rigors of two tumultuous years, and his dark triangle of trials, from the overlapping summers of murder and vengeance in Chicago to biblical ranting in Dayton to racial hatred in Detroit, seemed to have finally brought him down. Darrow was stripped of all his energy and spirit.

He claimed to be partly restored in late September and traveled to New York to spend some time with Mary. They had not seen each other for months. But on October 2, 1926, writing again of how "Darrow longs for death," Mary seemed desolate. "Felt all day depressed. A personality can be more impregnating than literal sperm, creating a new mood out of an old. A personality can enter another as the 'Dybbuk'—Oh ya! Oh ya!"

Darrow was now less like a lover than that mysterious dybbuk, the dislocated lost soul of a dead person in Jewish folklore. Mary knew that the dybbuk was meant to have escaped from some form of hell, so that it could attach itself to another living body that it might invade. She also understood how far down into the darkness he had been with Leopold and Loeb, followed by his battle with Bryan, and then on to consecutive trials with the Sweets. He had emerged redeemed and triumphant. But now he seemed utterly hollow. Where was the Darrow she knew and so loved? Where was the old devil?

He was confined again to bed by his doctor in mid-October. Ruby allowed select visitors to their apartment on the Midway, hoping that Darrow would be revived by the good wishes and adoring encouragement. But, on October 26, Fay Lewis of the NAACP wrote to Walter White in New York. She had been to see Darrow but he was barely recognizable as the defiant old lion of the courtroom who, five months before, had saved Henry Sweet. The fierce and lucid fighter who had spoken for seven hours, without hesitation in his masterly closing summary, had slipped away. He had been replaced by a depressed husk of a man who, as Lewis wrote, "seems to be carrying on some sort of propaganda for an early exit from this 'bloody world.'"

D ARROW, OF COURSE, rallied. On December 3, 1926, a snowy midwinter afternoon, he roused himself and went to see someone less fortunate. At Stateville Prison he spent an hour with Nathan Leopold, having brought him tea and chicken and good conversation. Richard Loeb was kept apart from Leopold and remained in the more restrictive prison at Joliet. Darrow did his best to console Leopold but declined to promise that they might consider parole before another twenty years in jail had passed.

It helped Darrow that the looming Scopes decision concentrated his mind again on work. The Dayton teacher turned Chicago college student had been one of his more regular visitors on the Midway, and even if he and Scopes spoke most of geology and university life, Darrow was made indignant by news of the ACLU's tactics against him. It especially angered him that Scopes, the innocent figurehead of their squabbling defense, should be so pressured. Roger Baldwin, the leader of the ACLU, had written to Scopes on August 10: "I want to say confidentially now that if Mr. Darrow and Mr. Hays insist upon staying in the case and arguing it before the U.S. Supreme Court, the Civil Liberties Union will probably have to withdraw." Baldwin drew back temporarily, acknowledging, "You are the defendant. You have the right to employ whom you choose to take up your case." But even this polite regard for Scopes's unique position contained a bullying strategy. "You may be put into the position of making the choice."

If it came down to a straight choice between the ACLU and Darrow, Scopes would always want the old man in his corner. But he was so sick of the whole prolonged escapade that he did not want to stir up yet more trouble by falling out publicly with the ACLU. Darrow advised him to sit back and do nothing. He should focus on his studies and let the Nashville judgment run its course before they decided on anything.

It was wise counsel because, in the end, the Tennessee Supreme Court found a way to fudge the final verdict after the five justices deliberating on the appeal produced an extraordinary deadlock. Two of the men upheld the Dayton ruling while two ruled against it. The latter pair found different reasons to vote with the defense. In a tangled statement one justice argued that the Butler Act did not place a blanket ban over all teaching of evolution; it merely outlawed any lessons that tried to prove that God was wholly absent from the creation of the world. Scopes had not taught anything nearly so extreme and, therefore, he was innocent of the charge. The second supporter of the defense simply decided that the law was too vague to be accepted as constitutional. It was now down to the fifth justice to cast the conclusive vote. Yet, as he had just replaced a man who had died, he claimed that he was too new in the position to pass judgment.

In a slippery piece of legal wriggling, the four remaining justices cobbled together a ruling that did not overturn the law but managed to exonerate Scopes. They decided that the Butler Act was permissible because it applied only to public employees working in an official capacity for the state and did not infringe on the individual liberties of private citizens. Yet they also confirmed formally that Scopes's conviction had been overturned—on the pretext that his $100 fine had been decided upon by the jury rather than, as should have been done, the judge. In effectively declaring Scopes to be now innocent of the charge, they also ensured that there was no conviction against which the ACLU could appeal to the highest court of the land. The circus was over. "We see nothing to be gained by prolonging the life of this bizarre case," the Tennessee Supreme Court pronounced. "On the contrary we think the peace and dignity of the state . . . would be better conserved by the entry of a *nolle prosequi* herein."

That Latin legal term, meaning a voluntary dismissal of criminal charges by the state, brought a deeply ambivalent end to the monkey trial. "Scopes Goes Free, But Law Is Upheld," the *New York Times* confirmed a day later, on January 15, 1927. "The whole matter is left in an unsettled condition," Darrow complained, knowing that the battle between evolutionists and fundamentalists would merely intensify in the years ahead.

He had at least been spared another round of bickering with the ACLU and a sapping Supreme Court test on behalf of the unenthusiastic Scopes, who was far happier returning to the world of ancient rocks. Preparing for retirement, Darrow moved all his favorite books out of the North Dearborn office and back into his study on the Midway. But his office remained open, and he popped in and out a few times a week—in between more rest and recuperative visits to see Mary Parton and Arthur Hays in New York. He also began to reenergize himself with writing.

Darrow produced a series of articles for *Vanity Fair* on "Our Growing Tyranny," striking out against all the familiar evils of capital punishment, fundamentalism, and Prohibition. He also wrote on biology for the *American Mercury* and, more amusingly, on life as a septuagenarian. Darrow would eventually feel serene enough to say, "I have taken life as it came, doing the best I could with its manifold phases, and feel sure that I shall meet final dissolution without fear or serious regret."

But he was continually dragged back to legal work. There was always another pitiful case to consider, and a lost soul to comfort. In the pattern that had marked his life from his first attempt to retire from "lawyering" back in 1912, after the bribery scandal, Darrow was soon surrounded by frantic pleas for help. He eventually decided, with Ruby pushing him hard, that he would spend the summer in Europe. "In the outside office there is always a stream of people, waiting," he wrote to Mary on July 15, 1927, "and I am tired before I can get away. I don't want them to come, but I can't help it. I go somewhere else for relief and there they are—the same crowd forever and forever. The mail is full of letters that I don't want to read and can't answer. . . . Good-bye dear girl. I always think of you and love you."

ON DECEMBER 9, 1927, at the Belmont Hotel in New York, Darrow and Mary took the bridal suite. It might have been a gesture driven more by sentiment than lust, a wistful memory of all they had shared in body and soul over so many years, but there was something poignant about being back in a forbidden place. After his sojourn in Europe, Darrow was attentive and kind, and Mary hoped to make him laugh again. He did, quietly and often, and a tender day passed quickly. They opened themselves to the world again late that afternoon—and the people came. Darrow's old friend Nate Coughlin joined them for dinner, and both before and afterward men and women flocked around the old attorney, hoping for a few words or a touch of his hand. Mary, so used to sharing Darrow with others, watched with affection.

"Darrow looks very old and even more tired," she wrote in melancholic reflection later that night. "Sags like an old bag out of which almost everything is taken. Jokes as usual, simple rustic jokes. Criticizes no one. Asks nothing. People come to strut before him and puff their breast bones and ruffle their feathers. He sits humbly, with head bowed, silent—he, who knows, who feels, so much more than they. Reporters, photographers, tuft hunters all come. They all want something—he waits, asks nothing. A lonely, great man of such imaginative sympathy that the imprisoned bird, the rich 'free' banker with

his load of hidden inhibitions fettering his spirit—all—all he under-
stands. He is the law of suffering made flesh, dwelling among us a
while."

The following day, on a dark and cold December 10, Mary kissed
him outside the Belmont. Their partings had now changed. The
piercing pain she had once felt whenever they said good-bye, which
told her how alive she was with the yearning that they might be together for a little longer as a couple rather than being just two fleeting
lovers returning to different lives, had given way to something jolting.
Mary had begun to frame each farewell with a miserable question in
which she silently asked herself if they would ever meet again. "Saw
Darrow off. Always the sense of great loss, the symbolism of death.
The great station Death with the through sleepers waiting."

There was still hope. He had not quite given up, and Mary's entry
that night trailed away in a mildly optimistic celebration of his enduring compassion. "D. interested in helping some Italians persecuted
by Mussolini—always get his sympathy."

Although Darrow had done little legal work since returning from
Europe, his contact with Arthur Hays during his frequent visits to
New York meant that he was constantly being asked his opinion on
his friend's latest case. Hays had agreed to defend a couple of Italian
immigrant workers, Calogero Greco and Danato Carillo, accused of
murdering two fascists outside a subway station in the Bronx. "You
know, Clarence," Hays said softly to Darrow as family members of
the accused gathered around him in his hotel room, "these men are
innocent." Protesting that he was now over seventy, and tired and
unwell, Darrow resisted. "You can handle it, Arthur," he said firmly.

Greco's brother began to weep in the corner of the room. Darrow
sighed and walked over to him. "All right . . . all right . . . ," he said
as he put a hand on the man's shoulder. "I'll take the case. For God's
sake, stop crying."

On Christmas Eve 1927, after another emotive plea from Darrow,
Greco and Carillo were both found not guilty of murder.

His work in court was not quite yet over. The following month, fulfilling a promise made by his son, Paul, he dragged his aching body to the Vermont Supreme Court, where he appealed against the death sentence that had been handed to John O. Winter after he was found guilty of murdering a young woman in the small town of Windsor. That January of 1928, Darrow saved Winter from the electric chair. He found that not all the evidence had been considered correctly in the original trial, and the death sentence was duly commuted to life imprisonment. "Although handicapped by lack of familiarity with the courts of the state and by his late entrance into the case," the *New York Times* reported, "Mr. Darrow won admiration by the manner in which he handled the exceptions taken by counsel at the time of the trial."

Darrow responded graciously, and then announced his final retirement. His days as an attorney, after exactly fifty years, were apparently over.

H IS DESERVED leisure was made more comfortable by the proceeds he received later that year from his son's sale of a gas plant he owned in Greeley, Colorado. Darrow held only a partial share in the plant, but the money from Paul thrilled him, amounting to more than any fee he had earned in a courtroom. He decided to invest the whole chunk, which was just over $100,000, in rock-solid shares on the stock market. They were expected to generate enough income for Ruby and him to live comfortably for the rest of their lives without Darrow ever needing to work again.

Mary was pleased for him, and content in her marriage to Lemuel, but she still resented Ruby's hold over Darrow. She revealed on January 24, 1928, how she and Helen Todd, who had introduced Mary to Darrow more than twenty years before, lamented his marital position. "We talk of Darrow and the curious fear he has of Ruby, the power she has over him to humiliate, to betray. These last years with their partial recognition of his greatness, these years of financial comfort, he pays for by keeping her placated."

There was something painfully disingenuous in those observations. If there was betrayal between Darrow and Ruby, Mary knew that he was to blame. Darrow had betrayed Ruby with her, and with other women, and he had not often appeared to care much for the hurt he caused his wife. Yet Mary was correct to see that the wielding of power in the relationship was not wholly imbalanced, for Darrow spoke often of wishing to avoid his wife's chagrin over far more minor matters such as a lost umbrella or a torn shirtsleeve.

While Mary celebrated Darrow's love for her, and kindness toward individual women, especially old friends like Helen, she blamed his occasional misogyny on Ruby. "To any woman he is the best, the most helpful, the most understanding friend. But the collective woman! Here he lets emotion color his views. Against the collective woman he rages as he would like to against the little pissant wife whose pettiness and jealousies have galled him for years."

Ruby had a right to be jealous, especially toward his oldest lover, but on February 8, Mary wrote, with some bitterness, "Tonight Clarence Darrow speaks of that which he has no personal knowledge of: personal freedom. Didn't hear him. Cost $1.25."

But late the following week, on February 19, her old love and tenderness were obvious. "Darrow called in [the] morning and I spent most of the day with him *and* the public: colored people, white folks, reporters, interviewers for this and that. He was tired and didn't want to go to the Italian banquet but he let himself be persuaded and went—and received their love if not the fee owed him for saving two lives [Greco and Carillo] from the electric chair. A beautiful life—a brave, fine, noble life and unique in the love it inspires. Glad to hear from his lips the affirmation of friendship and love which the years have not destroyed."

But, as if to admit to herself that their past passion had dimmed,

Mary added one more line that night. Darrow did not belong to her, or even to Ruby, but to the world. "His is a universal love," she wrote, "to all, for all."

That summer, Mary and Lemuel borrowed money from Darrow so that they could complete the purchase of a pretty nineteenth-century brownstone town house on Charles Street in Greenwich Village. As a way of securing their long-term financial future, the Partons decided to convert the town house into a series of apartments. Their own new home would be on the first floor, where the high ceilings, molded cornices, and two white marble fireplaces so appealed to Mary. Darrow himself had no ulterior motive in loaning them money—beyond the desire to be generous and to make Mary happy.

D ARROW HAD always hoped that Mary would write his biography. She knew that it would make a significant and compelling book, and one that she was better qualified than most to produce, and yet the tangled backdrop of their own affair would shadow its writing. "Think how much we would need to leave out," she once told Darrow, when persuading him that it was not a project they could undertake.

There was no one else with whom he wanted to collaborate, so the only alternative remained an autobiography. Darrow loved to write—even if he was a much more colorful stylist with words in his mouth rather than when setting them down on the page—but he complained that the very practice of autobiography reeked of a man standing on a street corner and shouting, "For God's sake, look at me for a minute!"

Nevertheless, the heavy financial losses he had suffered in the stock market with the onset of the Depression made him relent. He accepted a publisher's sizable advance and slowly wrote out *The Story of My Life* in his spidery scrawl. Even more painstakingly, Ruby deciphered his writing and typed out a manuscript that would eventually reach number nine on the nonfiction bestseller list after its publication in February 1932. It was praised for its restraint and humility. "Sitting down to write of himself," the *New Republic* noted, "he quickly forgets himself in its favor."

The book, however, was limited in scope; and it glossed over the most momentous trials of Darrow's life. There was also nothing in the way of personal revelation. Genevieve Forbes, a Chicago journalist who had covered the Leopold and Loeb trial, commented sharply to Ruby that Darrow had no right to call his book an autobiography when he so purposely avoided all mention of his love life. She would have expected at least a chapter on his complex personal relationships.

An angry Ruby told Darrow, and he responded with a dismissive grunt. "A whole chapter? Why, I could do a whole library."

Ruby glared at him and Darrow hesitated. "Well, maybe not a whole library," he muttered, "but at least a couple of best sellers."

II. L. Mencken, his old friend, who knew the ways of Darrow, believed that no one else "was more discussed, more loved and more hated." It was an assessment that Darrow recognized. One day, when he was climbing the narrow staircase in Mencken's Baltimore home, the journalist called out in concern: "Be careful, Clarence. If you fall and kill yourself in my house, the public will crucify me." Darrow responded wryly: "No, they won't. They'll canonize you."

Mencken noticed another harsh truth. "The marks of battle are all over Darrow's face. He has been through more wars than a whole regiment of Pershings. And most of them have been struggles to the death, without codes or quarter. Has he always won? Superficially, yes; actually, no. His cause seems lost among us. Nearly all the imbecilities that he sought to lay live on. But they are not as safe as they used to be. Someday, let us hope, they will be put down. Whoever at last puts them down will owe half his bays to Clarence Darrow."

The old man himself was less certain. "Am I as good as I was five years ago?" he often asked. "Am I doing as well?"

The flattering lies in answer to his anxious questions sounded feeble. Darrow's confidence was dented further by his financial woes. When he failed to return a $1,000 retainer to the NAACP in 1932, he wrote to Walter White: "I am awfully sorry that I could not help out on the thousand that I owe. I haven't been able to do it. The truth is that before these terrible times I had about $300,000 in what seemed perfectly good securities. They are now not worth more than ten and are not paying dividends. . . . This is the first time in fifteen years that I have had trouble about finances, and I am terribly sorry but I will arrange some way to make a substantial payment."

"Here is a most pathetic letter from Clarence," White lamented to Arthur Spingarn of the NAACP. Having seen him in majestic flow while defending the Sweets six years before, White could hardly believe Darrow was so troubled at the age of seventy-five.

Those who were less aware of his monetary plight were even more surprised when, in April 1932, Darrow returned to the court-room, in Hawaii. He defended four wealthy white people accused of murdering a local islander who had supposedly been part of a group that raped the wife of Thomas Massie—one of the defendants. The evidence of rape was dubious, and Darrow would usually have defended the working islanders—two Hawaiians, a Chinese immigrant, and two Japanese men. He was uncomfortable in reconstructing the murder as a justified honor killing for the white defendants when there were so many doubts surrounding their version of events.

Unsurprisingly, he lost the case and the Massie quartet was found guilty of manslaughter—only for the governor of Hawaii to immediately overturn the verdict and set them free. Darrow's own tumultuous career seemed poised to end on that subdued and muddled note.

Yet that October he challenged the death sentence handed down to seventeen-year-old Russell McWilliams after he had killed a street conductor in Rockford, Illinois. Darrow's words did not resonate in the way they had done when he saved Leopold and Loeb eight years before. The Rockford judge upheld the original verdict—and ordered the electrocution of McWilliams.

In despair, Darrow eventually took the train to Springfield, the state capital, to beg the governor to save McWilliams from the chair. On April 11, 1933, the *Chicago American* reported, "Clarence Darrow cried today . . . there was a sob in his throat, tears in his eyes as he pleaded—as a child might ask for something and not be able to understand why it should be denied to him—for the life of 'a boy who never had a chance.'"

A week later, on April 18, Darrow's seventy-sixth birthday, Governor Horner took pity on both the attorney and his client, commuting McWilliams's death sentence to ninety-nine years in prison. "No greater birthday present could have been tendered to me," Darrow said. "Capital punishment is deliberate and cruel. Why can't we have more tolerance in the world?"

Greenwich Village, New York, January 18–19, 1934

Mary Field Parton had not seen the man she had loved for over a year. Darrow's health had deteriorated to such an extent that he had not been well enough to travel. Alongside his ailing body, his mind had also begun to wither. When she arrived at Arthur Hays's apartment for a party to celebrate Darrow's final return to New York, Mary asked Lemuel to accompany her. She was shocked to see that he seemed "broken, bent, shrunken. Besides the younger men he looked small."

At a party studded with literary stars, Mary described the ambitious and then thirty-four-year-old Thomas Wolfe as "a huge creature who could swallow Darrow in one gulp." Other writers ambled over to Darrow to shake his clawlike hand, but he said little beneath the dimmed lights and cackling conversation. Mary was so devastated by his decline that she left early, with "feet aching horribly in party shoes . . . heart aching horribly at the sign of death upon my old friend."

The following day, January 19, she picked up Darrow from his hotel so that she could take him to the Charles Street apartment for an intimate lunch—which she had prepared in advance. Her sadness was palpable when, later, she recalled the pitiful scene. "It wasn't Darrow who went. It was his shadow—physically, mentally. God what a tragedy! He was utterly indifferent to me. I mean nothing to him now. Well, it is better to mean something to a person when his mind and heart function than when they no longer do. Darrow forgets, didn't remember so important a historical fact as the Reichstag fire—walked [him] back and he said, 'Don't come in with me, I'm going to lie down.' His face is shrunken, cheeks hollow, eyes cavernous. . . . I felt so depressed with his 'passing.'"

WITH LOVE, and tenderness, Ruby Darrow cared for her lost husband in the last wretched year of his life. As he slipped in and out of consciousness, dribbling and vacant in a near mindless state, she shut out most of the world. Only three male nurses, working round the clock at his bedside in a strict rotation, were allowed to see him. Ruby wanted the rest to remember the old lion stalking through the courtroom, charming juries and saving lives like no one else had done either before or since his great trilogy of trials. She did not want anyone else to see what Darrow looked like at the end. He died at home, in Chicago, the city he had made his own, on a Sunday afternoon, just as the worst of the winter disappeared for another year.

On the evening of March 13, 1938, Mary Field Parton wrote only a brief entry. "12:30. Darrow died today at this hour. That is, his body died following the earlier death of his brilliant mind. It will not be long before a generation will say, 'Who was Darrow? Never heard of him.' So quickly the waters of oblivion close over the great of a generation. Goodbye, dear friend. We spoke the same language—the inarticulate language of the heart. You, who never knew a moment free from heartache over man's travail, now are free . . . yet death ending consciousness, you do not know you are free. Oh grave, where is thy sting? I know. It is in life itself, for such as Darrow. Farewell. Good night."

CLOSE TO her own death, three decades later, there were dark nights when Mary woke and cried out for Darrow. She was back in the old farmhouse in Sneden's Landing, having returned to live with her daughter, Margaret, and her grandson, Lem, named after her cherished husband.

For many years she had found it impossible to be in the home that she and Lemuel had bought all those summers before. Mary rented it out soon after Lem died, at the age of sixty-three, in January 1943. She had returned to New York then, in the vague hope of writing a little more after her second book, *Your Washington*, had been published, just three months after Darrow's death. Even though it was essentially a guidebook to the capital, she was proud of it and wished that Darrow had been alive, and lucid enough, to revel in the glowing review she had received in the *New York Times*, where she was praised as "wise" and "concise."

Mary never managed to write a third and defining book, and Margaret, who had followed her parents into journalism, was intrigued by her family's personal history. Yet she found her mother to be infuriatingly elusive whenever she asked her about the past. "She was in love with Clarence Darrow all her life," Margaret wrote, "yet she enjoyed a supremely happy marriage to someone else for almost thirty years. This contradictory, delightful and maddening person was my mother. She sang me romantic lullabies and radical chants when I was little, and told me fairy tales and miners' legends when I grew older. She was a Lady, who said 'damn' and 'hell.' She served gin and pineapple juice cocktails during prohibition and puffed at cigarettes which were held gingerly between thumb and forefinger."

In late 1966, while Margaret rocked with fury over the war in Vietnam but delighted in the first few episodes of a new television series, *Star Trek*, which she and her teenage son loved, Mary wandered aimlessly around the farmhouse. She could never concentrate on the screen long enough to watch the emergence of Captain Kirk and Mr. Spock, but Mary was transfixed by the space race as the Soviets and the Americans strove to land the first man on the moon. Darrow would have loved it, of course, but Williams Jennings Bryan and Mary's own fundamentalist father might have despaired of the modern world.

Mary eventually shuttled back and forth between nursing homes and Sneden's Landing as her mind and body crumbled. When she was allowed back to the farmhouse, she would often rise up at night. She would tap her cane in the dark and shout: "Margaret! Margaret, where are you?" When her daughter came to see her, Mary wept that her sins were scarlet and that God would never forgive her. Margaret consoled her and, after a few minutes, Mary would fall asleep again.

It was more frightening on those nights when Mary woke and called for Darrow. She only ever said his surname, just as she had done when he was alive, but she would sing out with startling intensity: *"Darrow! Darrow!"* And in a return to her religious childhood, she would also summon God, beseeching him to forgive the multitude of sins she had committed. Sometimes, in her raving, the two different parts of her life came together and, after pleading for pity on her soul, she would cry: *"God bless Darrow!"* It was enough to turn the old agnostic in his grave.

The only other name she invoked on these tortured nights was that of her sister Sara. Lemuel was never mentioned, and when Margaret tried to question Mary in the calm of morning, her mother would just shake her head mysteriously. "Why did you marry my father?" Margaret once asked. "He was so gentle," Mary answered, "and he loved me so."

But she would not speak easily of Darrow. Margaret eventually turned to Sara, to see if her aunt might understand the reason. Sara said she knew why her sister needed absolution—but it was a private matter. Such talk made Margaret even more fretful, and she began to imagine, wildly, that her mother might have given birth to another child years before. Did she have a half-brother or a half-sister somewhere in the world?

Sara shook her head. The past did not hide anything of that magnitude. She would simply write a letter to Mary, urging her to accept absolution. The letter duly came, and Sara's underlined words were soothing: "Do not reproach yourself for anything in your life. You have fought always for freedom and humanity." Mary found comfort in the message, and she kept Sara's letter at her bedside. Her voice, ragged and hoarse, no longer cried out in the dark.

She was about to be transferred to the Rockland County Nursing Home, and, on the last night they spent together in the farmhouse, Margaret knelt at her bed, just as she had done all those summers ago when she had prayed for Leopold and Loeb, for Darrow, and for her "dear daddy." She looked at Mary and said, "You have been such a wonderful mother."

"My darling," Mary replied, "I always think of *you* as Mother."

In 1969 one death followed another. The first was almost unbearably tragic for Margaret. Her son, Lem, just nineteen, died in the hospital forty-five days after he had been diagnosed with leukemia. Margaret's life lay in pieces, smashed around her, but in her stunned grief she still visited her mother in the nursing home. She resolved not to disclose the death of her grandson, for Mary's mind had already drifted into dementia.

Mary Field Parton fell ill with pneumonia on the first day of July that year. And, two days later, on July 3, 1969, with her daughter at her side, holding her hand, she died in Palisades. Seventeen nights later Neil Armstrong walked across the surface of the moon. Mary and Darrow would have been thrilled at the scientific progress of man in finally reaching the moon—even if the earth below seemed as troubled as ever.

She would also have been proud that leading newspapers acknowledged her life and its merging with the work of Darrow. In the *New York Times*, on Independence Day, in a brief article headlined "Mrs Lemuel F. Parton, 91, Writer and Social Worker," her death prompted a reminder that "Mrs. Parton graduated from the University of Michigan in 1902 and worked for seven years as a social worker in Chicago. She was associated with Clarence Darrow in several investigations."

In her obituary, headlined "Mary Field Parton, A Noted Reformer," it was suggested that "during her reporting days she maintained a warm relationship with Clarence Darrow, and would often investigate prospective jurors for his trials. 'She would go into their homes, see if they had a Bible and whatever else was there and tell Darrow,' her daughter said. She was with Darrow during the famous trial in which he was acquitted of trying to bribe a jury in Los Angeles in 1912."

EPILOGUE: GHOSTS

||||||||||||||||||||||||||||||||||||||||||||||||||||||

O VER THIRTY YEARS before, on January 28, 1936, in Stateville
Prison, twenty miles outside Chicago, Nathan Leopold and
Richard Loeb met as usual for breakfast. In Loeb's cell they munched
on the sweet rolls they had brought from the prison shop. It was an icy
morning, with the temperature plunging below zero. But Leopold,
aged thirty-one, and Loeb, still thirty, had warmed to the idea of a future inside and maybe, one day, even on the outside of those familiar
walls. After being kept in separate jails for a long chunk of their first
six and a half years behind bars, they had been reunited at Stateville,
following a plea from Leopold, in April 1931.

In 1936 they ran a prison correspondence course that they had set
up twelve months earlier. Initially enrolling twenty-two convicts as
students, they had since received sixty-four more applications. With
Leopold in charge of the course's academic direction, and Loeb acting as registrar, they had persuaded three other educated prisoners to
join them as teachers.

Leopold had already studied extensively in jail, learning many
new languages while the years passed slowly. But he learned most,

ironically, through teaching. It had begun in Joliet, within a few months of their imprisonment in September 1924. Leopold had been asked by the prison chaplain to teach some illiterate convicts a couple of evenings a week. Most of his pupils were black, with the odd European immigrant who had ended up in prison before he could read much English. The "genius," as Leopold had been hailed in newspapers and psychiatric reports, found a new humility in front of his jailhouse class.

But his first attempt to teach soon ended. After a newspaper story about his work, a stream of vitriolic letters complained that the "depraved killer" was not morally fit to impart knowledge. His classes were canceled—much to the regret of Leopold and his bewildered students.

There was more lasting success at Stateville. Apart from curbing their boredom, and the sheer pleasure Leopold and Loeb derived from teaching, their educational program had two other appealing features. Such work would look good on their parole applications; and, in the meantime, they could spend time together. On that relaxed January morning, in 1936, they discussed a new course in college-level algebra and then corrected papers in both English literature and history. At around eleven thirty, Loeb picked up his towel and headed for the showers.

Forty minutes later another convict, Abe Belski, rushed in to see Leopold. His hands shook as he revealed that Loeb had been stabbed in the shower and rushed to the prison hospital. The supervisor of prison education accompanied a distraught Leopold to the hospital gate—where he was granted permission to enter by one of the chaplains.

"I hurried into the operating room," Leopold wrote later. "Dick was already on the operating table, with an ether mask over his face. Four doctors were working on him—two on his throat and two on his body. Dick's throat was cut—a series of four deep gashes had almost severed his head from his body. His trunk, his arms, his legs were one mass of knife cuts."

The battle to save Loeb was at such a bloody pitch that the doctors did not even have time to scream at Leopold and order him out of the O.R. He watched mutely as they tried to suture the vessels in Dick's throat while failing to stanch the loss of blood from his abdomen and limbs.

One of Dick's brothers and two other men, both doctors, joined them within the hour. Leopold recognized the more elderly man as the obstetrician whom he knew had delivered both himself and Dick over thirty years before. The small, wizened doctor moved closer to Dick and held his hand. With tears spilling from his eyes he mumbled inaudibly. And then, at seven minutes to three that freezing January afternoon, Richard Loeb slipped away, his hand cradled in those of the man who had brought him into the world.

The small group around the table gradually dispersed, eventually leaving only Leopold and a surgical nurse with the corpse. Although love between two men, in 1936, was still a startling concept, the doctors who had tried to save Loeb showed rare compassion in allowing Leopold to remain in the room. Slowly, tenderly, Leopold and the nurse, a kind man called Leo, began to wash the body. Blood still oozed from the wounds and Leo took time to stitch shut the deepest cuts.

Leopold kept cleaning the cooling flesh that had once thrilled him. The sheer horror of the work sometimes struck him—"I had never, until that moment, really understood what the phrase 'cut to ribbons' meant"—but he gradually lost himself in the ritual. It was a way of saying good-bye to the boy with whom he had descended into darkness a dozen years before. The sponging and drying became easier, and then, an hour later, they were ready.

Leo covered the corpse with a sheet and glanced at Leopold, as if to say it was time to return to life inside the prison. Leopold looked away, and the nurse stepped aside to allow him a little longer in his grief. Peeling back the sheet to expose Dick's face, Nathan Leopold sat down on a stool alongside the table. He gazed at the man he loved and stretched out his hand to touch him, one last time.

Richard Loeb had been attacked in the shower by a fellow prisoner, James Day, who shredded him with a razor. Day alleged that he had been sexually propositioned and that he had "fought Loeb off." After a farcical trial, Day was declared not guilty of murder—although the youngest member of the jury, Joseph G. Schwab, twenty-six at the time, later revealed the depth of his own prejudice. In 1974 Schwab claimed, "Nobody liked a queer, a homo, or a lesbian, which all seems okay today. So it was a good thing to get rid of such people. They were Skum [sic]. Even Clarence Darrow said it was the only verdict we could bring in, which made me feel okay."

Darrow had spent all his life fighting men like Schwab—and it would have angered him to hear his views being misappropriated. Yet, when approached by reporters after Loeb's death, Darrow sighed mournfully. "He's better off than Leopold. He's better off dead." The confused hacks pressed him further—not understanding that Darrow voiced his own longing for an escape from the cruelty of men. He was asked another question: Why had he struggled so hard to save Leopold and Loeb from death row? "They wanted to live," Darrow replied simply. "My duty was to help them do so."

Leopold was placed in solitary confinement, ostensibly for his own protection, but also in a bid to keep him apart from Day. The ache deepened. "I missed him terribly," Leopold wrote of Loeb. "All the more so since the years in prison we had shared and planned everything together. I was very lonely. . . . It felt as if half of me was dead."

When Leopold was eventually released back into ordinary prison life, he received an unexpected visitor. Darrow came to see him one last time. "Physically, he had grown feeble," Leopold wrote. "The mark of death was on his face. But age and illness had not dimmed that piercing inner light. His wisdom, his kindliness, his understanding love of his fellow man shone out from under the wrappings of his flesh as brilliantly on this last day I saw him as it had on the first."

Darrow was too sick to say much, but he conveyed his empathy. He said it didn't matter what the rest of the world thought. There was often great beauty in love that polite society would decry as forbidden or illicit. As they prepared to part, and Leopold stretched out his hand in gratitude, Darrow pulled him close. He knew how Nathan felt, he said, because he too had loved someone not meant for him. He doubted he would ever see her again.

Darrow looked one last time at the young man he had saved. And then, quietly, he nodded his understanding, in recognition of both grief and love. He would soon be dead.

TWENTY-TWO YEARS LATER, on March 13, 1958, Leopold was finally freed from Stateville. That day just happened to be the twentieth anniversary of Darrow's death. As he left jail Leopold was followed by a fleet of vehicles carrying television crews and newspaper reporters. As they trailed him, the fifty-three-year-old Leopold had to persuade his driver to stop the car on four separate occasions so that he could throw up on the sidewalk. And each time the media caravan slowed to a halt as the cameras whirred and reporters babbled about "an unfolding event"—in a manner familiar to all of us who have watched countless "breaking stories" in our different century.

Another group of a few hundred had gathered together to pay their annual homage to Darrow in Chicago that day. Leopold was violently ill near them and the bridge named after Darrow—just as a wreath thudded against the ice in honor of the attorney who had saved him from death row.

PETER ROSE, a professor of sociology, felt acutely distressed when introduced to Leopold in Puerto Rico five years later. Ed Suchman, his boss at San Juan University's Social Services Department, said, "Peter, this is Nate Leopold"—but Rose had already realized he was staring at America's most infamous murderer.

Leopold looked like an ordinary middle-aged man, without any distinctive features beyond his piercing eyes and thick glasses. He stretched out his hand and said, quietly, "Pleased to meet you, Dr. Rose." It seemed to the young academic as if he had just heard a monster speak, and he took Leopold's hand reluctantly. The basis of Rose's work meant that he usually tried not to judge a person on his past. But, that morning, Rose shuddered inwardly, for it felt to him as if the hand he shook were still stained with blood.

Rose had been shocked further when he was told that Leopold would be working directly for him. Even though he was thirty years older than Rose, Leopold was suitably deferential to his highly qualified colleague. As the months and years passed, Rose managed to overcome his antipathy and the two men struck up an understated friendship. They worked together as fellow academics and traveled frequently at night to remote hotels from where, the following morning, they would conduct their rounds of interviews with unemployed manual workers or Jewish exiles—depending on the nature of their sociological study at the time.

Yet, whenever they checked in at a new hotel, Leopold would use his mother's maiden name and sign himself as "N. L. Forman." His real name, almost forty years after the murder of Bobby Franks, was still freighted with infamy. Rose had also watched people he interviewed withdraw in muted horror whenever they realized that his associate, who never asked a question himself, was actually Nathan Leopold. It had reached a point where Rose had been forced to ask Leopold not to enter the interview room because he might be recognized and the subsequent questions ruined by the reaction he received from the people they were studying. Leopold had accepted the gentle instruction gracefully—and restricted his work to background research.

Rose sometimes felt sorry for him—until he remembered how, unknown to Leopold, they were bound together by the past. But he always kept that fact to himself. He could not even begin to guess how an aging intellectual, and former killer, might react to Rose's secret.

One starless night in 1964, as they drove across a long stretch of the island, Leopold turned in the passenger's seat so that he could look at Rose. He asked his colleague which part of America he called home.

"Northampton, Massachusetts," Rose answered.

"I know that," Leopold said. "But where are your family from? New York?"

"Upstate New York," Rose murmured. "Rochester."

The low drone of the engine framed their quiet as Rose decided whether or not it was time. He felt spooked by the fact that, not too many minutes before, he had taken a dark and narrow road that ran along the edge of a cliff. Rose could feel his hands trembling on the steering wheel. And then, as if he could no longer help himself, he added that some members of his family had also lived in Montreal. There was another pause before Rose said two more words: "And Chicago . . ."

"Chicago?" Leopold said in surprise. "Say, maybe I knew them."

"I think you did," Rose replied, with a certainty that belied any ambiguity. There was no going back now. The words fell from his mouth with a husky edge. "The name was Franks."

Rose kept driving through the darkness as emotions churned through him. Long before they had met in Puerto Rico the terrible truth had been passed down to him by his family. Leopold and Loeb had murdered his distant cousin—Bobby Franks.

For the next fifteen minutes, in the strained silence, Rose was stricken with fear that, along the cliff edge, the killer might resort to his old madness. He wondered if Leopold might strike him or force them both off the road and down into the invisible sea below. Who would ever know what had happened to them that black and clammy night?

But then Rose began to calm down. From the corner of a gaze that remained rooted to the road ahead, he saw that Leopold looked more to him like a ghost than a murderer. Rose thought about all they had shared during the past year of work, and he reminded himself that the same old man now sat mutely next to him. He might have killed once, but not again. Rose waited for some reaction from Leopold. And it finally came, over an hour later, as Rose slowed the car and took the gravel road leading to a low-lit gas station. When he cut the engine, Leopold turned to him again and said, "Want some coffee?"

Those words formed a gesture and Rose found it easy to say, "Sure . . . sounds good."

They did not speak again of Bobby Franks, but the following day, after Rose had completed his last interview, he and Leopold stopped for lunch at a beachside café in Arecibo. After eating, they wandered down to the sea so that Rose could take a swim. Leopold sat on the beach and watched the waves. And when Rose finally returned from the sea, with the salty water dripping from him, Leopold began to talk about his years in prison. Rose sometimes asked him a question, but Leopold confined his memory to jail and chose not to mention the actual murder—of which they had both been reminded so graphically the previous night.

Two days later, Leopold presented Rose with a copy of the book he had written while in prison. It was called *Life + 99 Years*—and named after the sentence the judge had handed down after Darrow convinced him that neither Leopold nor Loeb should be hanged from the neck. On the inside cover Leopold had written a message to Rose: "We sure had a swell swim in Arecibo."

Even after Rose returned to mainland America to begin a new stint of work, he still visited Puerto Rico frequently. He always made a point of seeing Nate Leopold and his wife, Trudi, a Jewish florist from Baltimore who had settled years before on the island. One day, while Rose was on business in San Juan, Leopold called to tell him that he had just heard that his parole had been lifted. He was, at last, a free man who no longer needed to report to the police or remain permanently on the island under a form of house arrest.

Rose gripped the phone silently as he thought of poor Bobby Franks while telling himself that Leopold had surely done penance for his crime. And then, at last, he said, "Nate, what shall we do to celebrate?"

It did not take as long for Leopold to answer. "I'd give anything to get off this island."

"You're on," Rose said. "Let's go check out the birds on St. Thomas."

Leopold was still a passionate ornithologist and had spent a chunk of his parole writing a book called *The Birds of Puerto Rico and the Virgin Islands*. He had, in the end, written the bird book that Darrow had always expected from him. Yet, because of his parole restrictions, he had never actually visited the neighboring islands. He had been forced to assume that the birds he had watched on Puerto Rico, after so many gray and silent years behind bars, were the same as those that occupied the Virgin Islands.

Rose made the reservations for two round-trip one-day tickets to St. Thomas—and because this was at a time when passports were not needed for interisland flights, he booked the seats under his own name and that of "N. L. Forman."

The next morning the news of Leopold's parole was splashed across every paper in America, but they spent the day quietly bird-watching on Charlotte Amalie. Although Leopold was the most un-emotional man Rose had ever met, there was something compelling about him as he savored his first taste of real freedom. To Rose it also seemed strangely fitting that Leopold should choose to spend the day with him—a relative of the boy he had murdered as an "intellectual experiment."

A few months later, and on the last afternoon they saw each other, Rose visited Nate and Trudi Leopold in their apartment. Rose was struck most by two photographs in Leopold's small flower-filled home. The first stood on a coffee table in the center of Leopold's living room. It was an unmistakable photograph and, as Rose picked it up, Leopold said: "Clarence Darrow . . . the man who saved my life."

As he headed for the bathroom, Rose passed another photograph. This large blown-up snapshot of a raffish and good-looking young man was placed on a dresser next to Leopold's bed. It was obvious that it had been taken in the 1920s. Rose thought at first that it was a photograph of Leopold just before the murder—because the man in the frame looked around nineteen with the kind of arrogant air Leopold had brought to his crime.

But, as he studied the photo, Rose became aware that Leopold was standing in the doorway of his bedroom—watching him. "That's Dickie Loeb," Leopold said, "the guy who ruined my life."

Rose did not know how to reply but, after a long pause, Leopold himself said, "Still, I gotta tell you something. You know what? He was a really swell guy, the best friend I ever had."

They had killed his cousin, but Rose was moved by those simple words. Yet he also remembered that his mysterious friend carried a different anxiety. Nate Leopold always worried more that people thought he might be gay rather than a murderer.

THERE IS no more tragic relationship in this story than that of the Sweets, engulfed as it was by racism and fear, by tuberculosis and death. Ossian and Gladys Sweet finally moved back into the Detroit bungalow on Garland Avenue in the summer of 1928.

Their decision was shrouded in sadness for, in returning from the warm clean air of Tucson with her health still deteriorating rapidly, Gladys had effectively come home to die. They were not hounded in the house where they had suffered so much on those awful two nights in September 1925. Yet death, rather than a baying mob, closed in on them.

At the age of twenty-seven, in November 1928, Gladys died of tuberculosis. Ossian chose to bury her alongside their daughter, Iva. This time he did not have to reach for his revolver when the hearse drove through the front entrance, usually reserved for white mourners, at the Roseland Park Cemetery. After the loss of Iva and Gladys, Ossian suffered still more. His brother Henry also contracted TB, in the summer of 1939. He lived at the time with Ossian in the Garland Avenue bungalow, but, six months later, early in 1940, Henry Sweet died at the Herman Keifer Hospital in Detroit.

Ossian Sweet married twice more, and was divorced each time, while remaining in the Garland house. It was as if he knew he would never escape its harrowing memories no matter how far he fled. Eventually, unable to go on, the doctor reached for his gun. Ossian Sweet brought it level with his head, pushed the muzzle against his temple, and squeezed the trigger. He shot himself on March 20, 1960, thirty-three years after the end of the second Sweet trial.

In a curious parallel, Nathan Leopold spent just over thirty-three years in jail after his own murder trial. Thirty-three years to repent seemed long enough for Leopold. The same stretch of time, however, became unbearable for Sweet. Leopold was guilty, Sweet was innocent; but life, in its careless brutality, seemed oblivious as to who might find despair or redemption in the end.

AFTERWORD:
SOMETHING PERSONAL

THE OLD LION came to life again with a roaring cry, sweat flying from his face on a steaming Saturday night in a small room. He was a big and shambling man, a giant presence in gray trousers held up by old-fashioned suspenders, as his hair hung limply over his creased brow. Up and down he walked and talked, murmuring and sighing in weary recognition or sometimes shouting in contained fury at the darkness outside. He demanded justice in a spiteful world, and he pleaded for mercy in a country reeking with hate, ruled by racists and fundamentalists. It seemed as if he was speaking directly to us, lost and damned as we were at the bottom of Africa. Clarence Darrow seemed more than a ghost then.

Early in 1984, forty-six years after his death, Darrow came back and took hold of me that night in a cramped theater in Johannesburg, South Africa. He held me in a grip so tight that, for a while, it felt as if I couldn't breathe. I was twenty-two years old, a white South African, dodging military service in a war along the border while working as a schoolteacher in the black township of Soweto, struggling to keep my head straight in a country dominated by apartheid and murder, by theories of biblical retribution and pure white neighborhoods, by the same prejudice and ignorance against which Darrow had raged in his three defining trials sixty years before.

That one-man play, *Clarence Darrow*, had been made famous by Henry Fonda, who had acted out the courtroom life and dramatic monologues of the great attorney on Broadway in the mid-1970s. In a country ravaged by apartheid, and boycotted by the rest of the world, there were no such dramatic stars to light up the South African stage. But this was also a period of fevered vitality and cultural resistance in the theaters of Johannesburg and Cape Town, and Richard Haines, who played Darrow, was perhaps the best actor in South Africa then. He was gay rather than straight, and not an old man like Darrow, but he instilled such vivid life into the part that it seemed as if he gave flesh to the spirit of the twentieth century's most celebrated yet controversial lawyer.

It helped that each of the trials resonated so forcefully in South Africa, for we lived in a country that echoed 1920s America. The Alhambra Theatre, in which Haines became Darrow, was located in Doornfontein, a bleak area in central Johannesburg that mainly housed people known as "poor whites." In 1984 some stray "colored" (as mixed-race people were called) and black South Africans had also begun to live in Doornfontein, contravening the Group Areas Act, which confined different racial groups to separate neighborhoods. My own family lived in the cloistered white suburbs, where the walls were getting higher as armed neighborhood security firms set about their business.

The idea of a black family moving in next door would have elicited in most of our suburban neighbors the furious disdain that fueled the white mob gathered on Garland Avenue to drive out Ossian Sweet and his family. And so the sinister motives of the Waterworks Improvement Association, and the consuming fear suffered by the Sweets, seemed familiar subjects—especially when I spent most of the week shuttling between the white suburbs and the black township. Detroit and Black Bottom in 1925 sounded just like Johannesburg and Soweto sixty years on.

Since I had been taken to church every Sunday throughout my youth, and nailed down in the pew to hear weekly sermons veering between stringent Calvanism as a child and a more folksy variation as a sulky teenager, Darrow's battle against Bryan in Dayton crossed equally recognizable territory. We also lived in a country ruled by a Nationalist Party that often tried to justify apartheid in biblical terms with much talk of "God's will" and the holy vow that had been struck between the Afrikaner people and the Almighty. Our old religious province of Transvaal seemed curiously related to Tennessee and the Bible Belt.

We even had our own celebrity "thrill killers," for I had attended the same school as Charmaine Phillips, who, the previous year, in October 1983, had gone on the run with her lover, Peter Grundlingh, as they murdered four men on the road. Charmaine was nineteen, the same age as Leopold. Her older lover, who had confessed to the murders, was on death row the night I saw *Clarence Darrow*. It was a case that would have intrigued the old lawyer. He might even have rescued Grundlingh—who was hanged in July 1985.

There were tiny glimmers of defiance in white South Africa. Isolated groups of outsiders railed, either publicly or privately, against the imbecilities and injustices of life in our country. It was possible to meet people who read Nietzsche and Dostoevsky, who studied evolutionary theory or worked against apartheid, who were either as bored and spoiled as Leopold and Loeb or innocent bystanders like John Scopes. And if you traveled into Soweto, and spent days and nights in a forbidden area for whites, it was easy to meet black men and women as smart and cultured as Ossian and Gladys Sweet, who faced dangers and humiliations in South Africa similar to those the Sweets had confronted in America.

In my own Soweto classroom, where I learned far more than I ever taught, I knew students just a little younger than me who were steeped in a black consciousness that owed more to Malcolm X and Steve Biko than to Ossian Sweet and Clarence Darrow. Some had been detained, without trial, and one boy, in particular, moved me with his prison story. The most heartbreaking moment of all, he said, had come when he heard the defiant singing of other black political prisoners who were waiting to be hanged, on charges of treason, early the following morning. South Africa, then, would have got to Darrow.

And so that night at the Alhambra transfixed me. I already knew of Darrow's legacy and had read *Compulsion*, Meyer Levin's fictional recreation of Leopold and Loeb's crime. I had seen Alfred Hitchcock's *Rope*, with its stilted allusions to the same case and the premise that a murder could be carried out as an amoral intellectual exercise—as well as the overblown *Inherit the Wind*, a Hollywood mishmash of the Scopes trial, with Spencer Tracy playing the Darrow figure. None of those echoes of his past moved me in the way Haines did as he brought Darrow to life with luminous force. He seemed flawed yet profound, compassionate yet ambitious, a man of enduring light and shade.

I left South Africa a few months later, in August 1984, and carried Darrow in my head. He became a memory of that time and, also, a reminder of how, no matter how many years slip by, the old wounds remain. Darrow, with his humanity and his pessimism, would not have been surprised.

TWENTY-TWO YEARS LATER, in 2006, I felt the same shiver of recognition as this book burst open on the night I interviewed Peter Rose. A venerable academic based in Northampton, Massachusetts, Rose spoke many languages and was comfortable discussing a range of subjects stretching from the "vertical ghetto" in inner-city America to the wines of Italy. But Rose and I spoke mostly about murder and a hauntingly personal account of his curious relationship with Leopold. It seemed as if I had found another way through the immense and often overwhelming material surrounding Darrow's three iconic trials.

Rose, as we now know, was a friend of Nathan Leopold. With his proclaimed IQ of 210, Leopold could speak more than ten languages by the time he left jail. He and Rose seemed to have much in common—which perhaps explained their friendship in the 1960s. As Rose spoke of the unsettling night when he revealed his family links with

Bobby Franks, it almost felt as if I was in the car with them. And, in the same way, I shared his poignant sense when Leopold told him he was more frightened of having his sexuality discussed—rather than his murderous past. I had, by then, read the exhaustive psychiatric reports prepared by numerous consultants during the weeks following Leopold and Loeb's arrest in the spring of 1924. The facts relating to the graphic "sexual compact" they had devised were considered too shocking to be read aloud in court later that year.

Yet the two leading psychiatrists, Harold Hulbert and Karl Bowman, were open-minded in addressing the complexities of Leopold's sexuality—especially in exploring the sadomasochistic undercurrent that shaped his relationship with Loeb. Their interviews offered a startlingly fresh perspective on the boys. There was also, amid the violence and philosophizing, some tenderness between the killers. And it was in Leopold's prison writing, and in the few interviews he gave to friends in Puerto Rico, where he uncovered most about his feelings for Richard Loeb. The image of Leopold cleaning and drying Loeb's body on the operating table always seemed to me one of the most haunting insights into their curious relationship.

In such moments this book seemed to be as much about an interlocked series of moving yet ultimately doomed relationships—as the more obvious trilogy of trials. Nathan Leopold and Richard Loeb. Ossian and Gladys Sweet. Clarence Darrow and Mary Field.

I discovered that Mary Field Parton almost always called him "Darrow," especially when she talked most touchingly of the man she loved. He was "Darrow" to her on the night, early in 1912, when she talked him out of killing himself. He was still "Darrow" to her on the day, twenty-six years later, he actually died—and, again, when she called out his name at night in the ruins of her own old age. She wrote to Darrow in countless letters, and about him in her diary, and she also spoke a little about their past to her sister Sara Field Ehrgott; her daughter, Margaret Parton; and to the labor writer Anne O'Toole. Parton relived the night that her mother steered him away from suicide, and O'Toole transcribed the interviews in which Mary described how she and Darrow rekindled their relationship in 1924.

Much of this material was discussed by Mary in the late 1940s, and there is a time-bound restraint to the language that means she does not take us any closer than the hotel door that she and Darrow approached in Chicago, on the eve of the Leopold and Loeb trial, and again on various occasions in New York in the ensuing years. There is much more detailed evidence of their sexual relationship between 1908 and 1912. But I was always more interested in their compelling bond of heart and mind. If Leopold and Loeb can be regarded as a kind of skewed love story, then Darrow and Mary shared something far deeper and more tangled.

The depth of antipathy between Mary and Ruby Darrow continued after Darrow's death. As I trawled through the letters between Irving Stone and Ruby Darrow it became evident that, in appointing a respected writer as her late husband's authorized biographer, she was determined that a partial but dignified history should be produced. The letters between Stone and Mrs. Darrow are illuminating, especially after he made contact with Mary Field. Ruby cannot bear to write her full name and "M. Field" soon shrinks to the even starker initials of "M.F." as she attacks her husband's lover.

"I dislike feeling forced, for Mr. Darrow's sake, to say more about M. Field," Ruby wrote to Stone on September 4, 1939. "Evidently she is attracting your attention to herself and, like others, is hoping to be featured, immortalized in the life story of C.D. Thank you for not regarding me in M.F.'s category. I couldn't be that! Why don't you ignore her, forget her? Why let her try to fasten herself onto a dead man as a bloodsucker he once shook off, but now is powerless to?"

It is difficult to avoid the feeling that Ruby was just a little too desperate to expunge Mary's name from her husband's history. She warned Stone that she would withdraw her support of his book if he included her rival. "Whatever M.F. may be tempting you with I urge you to not glorify her. I shudder at her indecency and cannot work with you under such circumstances. . . . I am writing all the foregoing to you in strict confidence. . . . Let us draw a veil over M.F. and say no more."

Even a biographer as solid and thorough as Stone submitted to Ruby's will. He made no mention of Mary Field throughout his 1941 biography—*Clarence Darrow for the Defense*—beyond alluding to "a clever and talented woman who caused Ruby many hours of anguish."

THESE DAYS, whenever I read of another person who has been sentenced to death, or executed, I still think of Darrow. I also remember him, and Leopold and Loeb, whenever there is another teenage killer to make 1924 seem a little less far away. And although racism may have become more subtle than that experienced by Ossian Sweet in Detroit in 1925, or under apartheid in South Africa, the scars are still raw. The reality of its existence today remains sharp and pungent across America and the world beyond. "Everything changes," as Darrow once said, "but man stays the same. This is the way it has always been and the way, no doubt, it will always continue."

Darrow would have been more startled by the force of religious fundamentalism across the world today, but he would probably have just snorted, in sardonic recognition, if he had been told of the continuing battle over evolutionary theory or its modern counterpart of intelligent design. After a tub-thumping speech from William Jennings Bryan in that Tennessee courtroom in July 1926, Darrow had asked Arthur Hays: "Can it be possible this trial is taking place in the twentieth century?" And yet, in the twenty-first century, public school board-members in Tennessee campaigned vociferously in the vein of Bryan—insisting that stickers should be placed across all textbooks with a warning that "evolution remains unproven."

Darrow's Shakespearean battles roll on in all their sound and fury, often signifying nothing beyond the limitations of man. But just as he found solace in little mementoes of humanity, so some of the more personal details of lives he touched down the years can be remembered.

Whereas the Sweets were swamped in catastrophe and death, John Scopes became, for him, a pleasingly obscure geologist for an oil company in Texas. He was married, to Mildred, and became a father of two sons. On April 1, 1970, he found himself back in a Tennessee classroom for the first time in forty-five years. *Time* magazine declared, in a claim that now seems hopelessly optimistic, that he was "welcomed like a returning hero and given the long-delayed evidence that his battle for academic freedom had been successful." Scopes died six months later, in October 1970—having maintained that "Darrow was a genius. If he had applied himself to any scientific field he would have been outstanding in it. But he became, instead, the greatest man the law has ever known."

Nathan Leopold married Trudi Feldman in 1960 and they lived together in Puerto Rico for the next eleven years. They became active committee members of the National History Society alongside Irving and Ann Grossman—who relived for me their shared memories of the world's first "thrill killer" as an aging man. Leopold resumed his bird-watching habits in tandem with Grossman—but he appeared curiously unable to express emotion. It was a trait, a numbing of feeling, that seemed to characterize many of his friendships in the years following the death of Richard Loeb. Leopold himself succumbed to heart failure at the age of sixty-six, in 1971.

Mary Field Parton, at ninety-one, had died two years earlier. She lived far longer than anyone in this story. But Mary was often infused with the melancholia familiar to those who believe the defining love story of their lives has never quite been fulfilled. She also feared that, in the end, even the great legacy of Darrow, the love of her life, would be swallowed up by "the waters of oblivion."

Yet, more than seventy years later, beyond the old biblical span of three score and ten, oblivion has still not settled over Clarence Darrow. His name and his memory, the work and his life, live on a little longer. And now, in a dark and reeling world, we miss the old devil more than ever.

ACKNOWLEDGMENTS

‖‖

This book could not have been written without the patient assistance of so many people who endured my seemingly ceaseless requests with both good humor and understanding. There are too many research assistants and librarians to list by name but, without exception, each and every person who dug out or copied another box of ancient correspondence or trial transcripts is owed my gratitude. In particular I would like to thank Joy Austria and Jill Gage at the Newberry Library and Bruce Tabb at the University of Oregon.

In addition, I thank various people for their help during some of the more significant moments of my research and writing. All of the following went out of their way to provide me with vital material, share some of their own memories and research, offer up previously unpublished material, or simply put me in touch with another important contact: Arthur Beer, Christine Bell, Kevin Boyle, Tom Carter, Sarah Dunn, Joan Gartside, Irving and Ann Grossman, Jack Hagner, Katherine Hedin, Elizabeth Jackson, Karen Jania, John Kent, Lillian Kirby, Vincent Martin, Dick Meyer, Linda Montemaggi, Malgosia Myc, Peter Rose, William Sampson, and Jay Savage.

Tim Musgrave and Andrew Gordon, my former editor at Simon & Schuster in London, were both kind enough to read an early draft—and they each managed to make some practical suggestions to improve a very rough first outline while saying all the right things to help me press ahead with the next four versions.

My agent in London, Jonny Geller, was the first person to encourage my interest in this book; and throughout the last few years he has, once more, been a constant source of support. Emma Parry, acting as my agent in New York, was similarly helpful; and I appreciated the way in which Christy Fletcher has taken the book through to publication.

Mike Jones at Simon & Schuster in London patiently accepted my numerous delays and shifts of publication date—and thanks go to him and Rory Scarfe.

Gabe Robinson, at William Morrow in New York, also did a great deal on my behalf as we approached publication. In particular, he was invaluable when it came to choosing and acquiring the illustrations used in the preceding pages. I am grateful for all his generous help.

However, my deepest appreciation is reserved for David Highfill, my editor at William Morrow. David has been the best possible partner in both the planning and the writing of this book—and his combination of unerring clarity and generous encouragement became increasingly important as the months and years slipped past. Without him I would still be lost in the murky depths of a muddled first draft.

If it had not been for my parents, Ian and Jess, I would not have written this book—for I only discovered the raw power of Clarence Darrow through them. They have followed its writing with typical interest and kindness.

My wife, Alison, has yet again lived through each and every word of another book of mine so completely that these pages belong as much to her. From the first vague outline of this idea, through all the proposals and drafts and final-final versions, amid all the teeth gnashing and soul baring, Alison has been with me each step of the way. As always, she was my first reader and best refuge when I needed to have another moan or a laugh about the strange business of researching and writing this book. Thanks go, most of all, to her and, of course, Bella, Jack, and Emma.

NOTES AND SOURCES
||||||||||||||||||||||||||||||||||||||||||||||||||||

The material for this book was obtained from various sources—including archive collections, diaries, correspondence, private memoirs, unpublished biographies, audio and first-person interviews, police records, trial transcripts, psychiatric reports, speeches, newspaper libraries, film footage, television and radio recordings, and many earlier books. A chapter-by-chapter breakdown explains the factual backdrop to each scene, but a basic inventory of sources is listed below.

SOURCE MATERIAL

Archive Collections and Manuscripts

Bancroft Library, University of California–Berkeley
Correspondence between Charles Erskine Wood and Sara Field Ehrgott [sister of Mary Field Parton]

Bentley Historical Library, University of Michigan
Josephine Gomon Papers
Frank Murphy Papers

Chicago History Museum
Nathan Leopold Collection

Huntington Library, California
Charles Erskine Scott Wood Papers—including correspondence with Sara Field Ehrgott

Library of Congress, Washington, DC
Clarence Seward Darrow Collection
NAACP Papers
The Papers of Arthur Barnett Spingarn

Michigan Historical Collection, University of Michigan
Clarence Darrow Collection

Newberry Library, Chicago
Robert Bergstrom Papers
Carter Harrison Papers
Margaret Field Parton–Clarence Darrow Papers; 1909–1975
Eunice Tietjens Papers
Arthur and Lila Weinberg Papers [containing letters from Margaret Parton and Ruby Darrow]

New York Public Library
Letters of H. L. Mencken

Northwestern University, McCormick Library, Illinois
Leopold and Loeb Collection

Princeton University Libraries, Princeton
ACLU Archives

University of Chicago
Nathan Leopold Collection
Julius Rosenwald Papers

University of Oregon, Eugene, Special Collections and University Archives
Margaret Parton Papers [Correspondence, Diaries, Interviews with her mother and father, Mary Field Parton and Lemuel Parton, Articles and Book Outlines and Extracts]

Mary Field Papers [Correspondence, Diaries, Interviews, and Articles]
Lemuel Parton Papers

University of Texas, Austin, Harry Ransom Center
Edgar Lee Masters Literary Collection

Vanderbilt University, Nashville
Edwin Mims Papers

Washington University, St. Louis
Harold C. Ackert collection [Including George Bernard Shaw correspondence]

Oral Histories

Bancroft Library, University of California-Berkeley
Sara Field Ehrgott interviewed by Amerlia R. Fry [with an introduction by
Sara's daughter Katherine Field Caldwell]

Illinois State Historical Society
Clarence Darrow

O'Toole Estate, London
Mary Field Parton's three interviews, from the late 1940s, with Anne O'Toole

University of Oregon, Eugene, Special Collections and University Archives
Margaret Parton's various interviews with her mother, Mary Field Parton

Court and Police Papers, Speeches, and Psychiatric Reports

Darrow, Clarence. *Plea in Defence of Loeb and Leopold.* [Haldeman-Julius, B–20, 1926]
Darrow, Clarence. *Resist Not Evil* [Haldeman-Julius, B18, 1925]

Bryan College, Dayton
Sweet case transcript of statements made to police
Last copy of records held by Kevin Boyle
The World's Most Famous Court Case: Tennessee Evolution Case

Los Angeles County Law Library
People v. Clarence Darrow—Trial Transcript

Michigan Historical Collection/Bentley Historical Library
People v. Henry Sweet—Trial Transcript

Michigan Historical Collection

People v. Ossian Sweet & Others—Trial Transcript

Newberry Library

Hulbert, Harold, and Bowman, Karl—*Psychiatric Examination of Leopold & Loeb*, 1924

Northwestern University

Official Transcript of the 1924 Leopold & Loeb Trial

SELECTED BIBLIOGRAPHY

Books

Addams, Jane. *Twenty Years at Hull-House*. Chautauqua Press, 1911.

Baatz, Simon. *For the Thrill of It: Leopold, Loeb and the Murder That Shocked Chicago*. Harper, 2008.

Blum, Howard. *American Lightning: Terror, Mystery, the Birth of Hollywood and the Crime of the Century*. Random House, 2008.

Boyle, Kevin. *Arc of Justice: A Saga of Race, Civil Rights, and Murder in the Jazz Age*. Henry Holt, 2004.

Bryan, William Jennings. *Shall Christianity Remain Christian—Seven Questions in Dispute*. Revell, 1922.

Bryan, William Jennings, and Mary Baird Bryan. *The Memoirs of William Jennings Bryan*. United, 1925.

Chalmers, David M. *Hooded Americanism: The First Century of the Ku Klux Klan, 1865–1965*. Doubleday. 1965.

Cowan, Geoffrey. *The People v. Clarence Darrow*. Times Books, 1993.

Darrow, Clarence. *Crime: Its Cause and Treatment*. Kessinger Publishing, 1922.

———. *The Story of My Life*. Scribner, 1932.

———, and Wallace Rice. *Infidels and Heretics: An Agnostic's Anthology*. Stratford Company, 1929.

DuBois, W. E. B. *Autobiography*. International Publishers Company, 1968.

Erickson, Gladys. *Warden Rogen of Joliet*. E. P. Dutton, 1957.

Fitzgerald, F. Scott. *This Side of Paradise*. Scribner, 1920.

———. *The Beautiful and the Damned*. Scribner, 1922.

Gertz, Elmer. *A Handful of Clients*. Follett, 1965.

Gompers, Sam. *Seventy Years of Life and Labor*. E. P. Dutton, 1925.

Haldeman-Julius, Marcet. *Clarence Darrow's Two Great Trials*. Haldeman-Julius Co., 1927.

———. *Famous and Interesting Guests at a Kansas Farm*. Haldeman-Julius Co., 1936.

Harrison, Charles Yale. *Clarence Darrow*. Jonathan Cape, 1931.

Hays, Arthur Garfield. *Let Freedom Ring*. Boni and Liveright, 1928.

———. *Trial by Prejudice*. Convici, Friede, 1933.

Higdon, Hal. *Leopold & Loeb: The Crime of the Century*. G.P. Putnam's Sons, 1975.

Hofstadter, Richard. *The Age of Reform: From Bryan to F.D.R.* Vintage, 1955.

Hunter, George. *A Civic Biology*. American, 1914.

Ipsen, D. C. *Eye of the Whirlwind: The Story of John Scopes*. Addisonian Press, 1973.

Jackson, Kenneth T. *The Ku Klux Klan in the City, 1915–1930*. Oxford University Press, 1967.

Janken, Kenneth Robert. *White: The Biography of Walter White, Mr. NAACP*. New Press, 2003.

Johnson, Walter, ed. *Selected Letters of William Allen White*. Henry Holt, 1947.

Kazin, Michael. *A Godly Hero: The Life of William Jennings Bryan*. Alfred Knopf, 2006.

Koenig, Louis. *Bryan: A Political Biography of William Jennings Bryan*. G.P. Putnam's Sons, 1971.

Larson, Edward J. *Summer for the Gods: The Scopes Trial and America's Continuing Debate Over Science and Religion*. Harvard University Press, 1997.

Laurence, Dan H. *Bernard Shaw: Collected Letters, 1898–1910*. Dodd, Mead & Co, 1972.

Leopold, Nathan. *Life + 99 Years*. Doubleday, 1958.

Levin, Meyer. *Compulsion*. Muller, 1957.

Lilienthal, David E. *The Journals of David E. Lilienthal*. Harper & Row, 1964.

Martin, T. T. *Hell and the High Schools: Christ or Evolution: Which?* Western Baptist Publishing Company, 1923.

Masters, Edgar Lee. *Across Spoon River: An Autobiography*. Farrar & Rinehart, 1936.

———. *The Tale of Chicago*. G.P. Putnam's Sons, 1933.

McKernan, Maureen. *The Amazing Crime and Trial of Leopold and Loeb*. Plymouth Court Press, 1924.

Moran, Jeffrey P. *The Scopes Trial: A Brief History with Documents*. Bedford, 2002.
 Oursler, Fulton. *Behold This Dreamer!* Little Brown, 1964.

Parton, Margaret. *Journey Through a Lighted Room*. Viking, 1973.

Parton, Mary Field. *The Autobiography of Mother Jones*. Charles H. Kerr, 1925.

Ravitz, Abe. *Clarence Darrow and the American Literary Tradition*. Western
 Reserve University Press, 1962.

Rodgers, Marion Elizabeth. *Mencken: The American Iconoclast*. Oxford
 University Press, 2005.

Rose, Peter I. *Guest Appearances and Other Travels: In Time & Space*. Swallow
 Press, 2003.

Scopes, John, and James Presley. *Center of the Storm*. Holt, Rinehart &
 Winston, 1967.

Stannard, David E. *Honor Killing: How the Infamous "Massie Affair"
 Transformed Hawaii*. Viking, 2005.

Stone, Irving. *Clarence Darrow for the Defense*. Garden City Publishing, 1941.

Theodore, John. *Evil Summer: Babe Leopold, Dickie Loeb and the Kidnap-
 Murder of Bobby Franks*. Southern Illinois University Press, 2007.

Tierney, Kevin. *Darrow: A Biography*. Thomas Y. Crowell, 1979.

Tompkins, Jerry R., ed. *D-Days at Dayton: Reflections on the Scopes Trial*.
 Louisiana State University Press, 1965.

Vine, Phyllis. *One Man's Castle: Clarence Darrow in Defense of the American
 Dream*. Amistad, 2004.

Weinberg, Arthur, ed. *Attorney for the Damned*. San Antonio: Quadrangle,
 1963.

Weinberg, Arthur, and Lila Weinberg. *Clarence Darrow: A Sentimental Rebel*.
 G.P. Putnam's Sons, 1980.

Weinberg, Arthur, and Lila Weinberg, eds., *Clarence Darrow: Verdicts Out of
 Court*. New York: Simon & Schuster \, 1957.

Wells, H. G. *Experiment in Autobiography: Discoveries and Conclusions of a Very
 Ordinary Brain (Since 1866)*. Macmillan, 1934.

White, Walter. *A Man Called White*. University of Georgia Press [reprint], 1948.

Yarrow, Victor. *My Eleven Years with Clarence Darrow*. Haldeman-Julius Co.,
 1950.

Newspapers and Periodicals

Auk, Quarterly Journal of Ornithology

Chattanooga Times

Chicago Daily News

Chicago Daily Tribune

Chicago Defender

Chicago Evening American

Chicago Herald and Examiner

Columbus Dispatch

Commercial Appeal

Detroit Free Press

Detroit Independent

Detroit News

Harvard Law Review

Jewish Courier

Journal of American History

Knoxville Journal

Los Angeles Examiner

Los Angeles Herald

Los Angeles Times

Los Angeles Tribune

Memphis Press

Nashville Banner

Nashville Tennessean

Nation

New York Times

New York World

New Yorker

San Francisco Bulletin

San Francisco Chronicle

San Francisco Examiner

Saturday Evening Post

St. Louis Post-Dispatch

Vanity Fair

Washington Post

Selected Film/Dvd/Drama Adaptations

Clarence Darrow (Writer: David Rintels, Director: John Houseman, Actor: Henry Fonda, 1974)

Compulsion (Director: Richard Fleischer, 1959)

Inherit the Wind (Director: Stanley Kramer, 1960)

Never the Sinner (Writer: John Logan, Theater Script, Overlook TP, 1999)

Rope (Director: Alfred Hitchcock, 1948)

Swoon (Director: Tom Kalin, 2004)

NOTES

As the bibliography above confirms, various biographies of Clarence Darrow have been written. And yet only one of them, Arthur and Lila Weinberg's *Clarence Darrow: A Sentimental Rebel*, features Mary Field Parton. The Weinbergs briefly discuss Mary's sexual and emotional relationship with

Darrow from 1908 to 1912. That imbalance is partially corrected by Geoffrey Cowan's *The People v. Clarence Darrow*—a remarkably thorough account of the 1911 McNamara case and Darrow's two bribery trials in 1912 and 1913—which features a minor role for Mary Field while stressing her importance to her lover during that specific period. Yet that book, like Howard Blum's more recent and compelling *American Lightning: Terror, Mystery, the Birth of Hollywood and the Crime of the Century*, focuses on the fallout from the McNamara trial. Blum's narrative tracks a detective, William J. Burns, and a filmmaker, D. W. Griffith, alongside Darrow—and, again, Mary Field has only a small part to play in his book.

The difference in the preceding pages is shaped by my discovery that Mary returned to Darrow's life in June 1924. Through new research and some old-fashioned good luck, I stumbled onto a largely unknown corner of the lawyer's tumultuous story. There have been many fine retellings of individual trials, but my narrative, driven as it is in part by Mary, is concerned with the trilogy of epic courtroom cases that enshrined Darrow's legend over two successive summers. In this regard, I hope, I have found a fresh way of looking at Darrow's achievements with Leopold and Loeb, the Scopes monkey trial, and the Sweet case testing him in a way that few other attorneys have endured either before or since those turbulent days.

There were many false starts and dead ends in the long years of research, and, as so often happens in such a process, the work sometimes seemed akin to that undertaken by an amateur detective rather than an ordinary writer. Slowly, strand by strand, section by section, chapter by chapter, the characters and times and dates and facts and feelings unfolded and could be checked one against the other until, at last, it felt as if a mystery had been solved and the once ghostly figures in this book could be seen much more clearly. This process is explained in the following pages by paying particular attention to the key scenes in each chapter.

Mary Field Parton's quote at the front of the book comes from the transcript of her third and final interview in 1946 with the labor activist and writer Anne O'Toole—with permission for its use granted by Mrs. O'Toole's surviving daughter.

Chapter 1: Back in the Loop

The details of Mary's return to Chicago in June 1924 are drawn from a combination of sources—including her private diaries from that period, her correspondence with Darrow and her husband, Lemuel Parton, and her last interview with O'Toole. In a disarmingly frank and intimate conversation with O'Toole, Mary explains how Darrow wrote to her soon after he agreed to represent Leopold and Loeb. Without stating his explicit reasons for making such a blunt request, he urged Mary to meet him in Chicago. She was beguiled by the mysterious invitation from a man she had loved so deeply during their previous affair. At the same time she was uncertain as to Darrow's precise motivation in contacting her.

Although she admits candidly to the powerful feelings she felt for Darrow, she concedes that her motives for traveling to Chicago were also prompted by the hope that writing about the Leopold and Loeb trial would revitalize her journalistic career—which had been on the wane in the eight years since the birth of her daughter, Margaret. In her diary, and in letters between herself and Lemuel, she writes of her fervent wish that Darrow will help her gain an intimate perspective on the trial and its key characters. Her journalistic ambitions also provide her, in relation to Lemuel, with a clear justification for her desire to meet Darrow again in Chicago. Their correspondence reveals that Lemuel plays an active part in intimating to various influential New York editors that Mary has a potential "inside track" to the trial through her links with Darrow.

The crucial context of Darrow and Mary's earlier affair is revealed through her diaries, private memoirs and correspondence, and the account provided by her daughter, Margaret Parton. Mary was interviewed at length by Margaret. Although she was often evasive with her daughter, they both wrote about the night when Mary saved Darrow from attempting suicide in Los Angeles soon after he realized that he was about to be charged with bribery. Margaret, who never liked Darrow, considers her mother's former lover with a mixture of loathing and guilty fascination—but it often appears as if her angry views are colored most by his almost illegible scrawl and his jokey refrains in letters to Mary that they operate on a higher plane of thought than the surrounding masses. It soon emerges that Margaret's feelings are forged by painful childhood memories where the famous old lawyer usually ignored her when visiting the Parton home.

Yet the cruel way in which Darrow sometimes acted toward Mary cannot be dismissed. Even though Sara Field Ehrgott (Mary's sister) and Charles Erskine Wood (Sara's lover and Darrow's former friend) also allow their personal feelings to shape their accounts, they both write explicitly. They address the hurt Mary endured during the latter stages of her first affair—and their correspondence offers an intriguing overview of Darrow's human failings.

The depth of Mary's love, however, is clear to all of them. Her subsequent marriage to Lemuel often seems overshadowed. In her diaries, especially during her early years of marriage, Mary sometimes broods resentfully on her new circumstances as a wife and mother. Even Margaret admits in her memoir, *Journey Through a Lighted Room*, that her mother "felt trapped by marriage, motherhood and domesticity; she loathed housework and resented having to do it. She referred to my father as her 'precious chum' but was, I think, often impatient with his dreaminess and lack of drive. . . . She was unhappy, too, that she was no longer working and blamed this on the dreadful confines of marriage. . . . My father told me that several years after their marriage he had contemplated divorce."

Margaret also highlights the "harmony, affection and love" that characterized the family in later years—and those same feelings are clearly shown in Mary's diaries (now held at the University of Oregon, among the Mary Field Parton Papers, Special Collection 036). Yet there are sufficient undercurrents of yearning, both emotionally and intellectually, to understand why she responded so readily to Darrow's call in 1924.

In her interview with O'Toole, which took place more than a quarter of a century after the Leopold and Loeb trial, Mary is understandably vague about the actual dates of her trip to Chicago. She suggests that it was "a few days" before the trial began—which always seemed unlikely when set against the great strain under which Darrow worked in the last week before he entered court. And so, in a much more practical sense, the diaries were immensely useful. They provided the date of her departure from New York as well as listing various days when she met Darrow over the next week. According to her diary, Mary and Margaret boarded a Chicago-bound train, the 20th Century, on June 16, and some brief extracts from that journey are quoted in this chapter.

It should be pointed out that, in a scrawled letter from Lemuel to Mary, and dated the evening of June 3, 1924, he writes about the possibility of her and Margaret having traveled to Chicago on that actual date—around thirty-six hours after it was announced Darrow would be representing Leopold and Loeb. But this note stands in isolation, and there is no other evidence to support any supposition that they may have arrived in Chicago before June 17—and it clearly does not tally with the dates listed in Mary's diary. The pages from May 21 to June 16, 1924, are blank.

Mary's diary entry of June 17 is so specific that I took my cue for a specific date from these words: "Left for Chicago on 20th Century. Full of hope! Here is my start! Got a story from Darrow on this strange murder in Chicago—Loeb and Leopold, rich boys, precocious, everything to live for. Kill, brutally, a little boy of 14, 'for the thrill' they say. Whole country, foreign countries, avid for news . . . for explanation."

The more poetic and private mix of anticipation and trepidation, with which the book opens, stems largely from Mary's own description of her feelings in discussion with O'Toole. The time of day, the atmosphere of the city, and the sheer bedlam surrounding her as newspaper vendors try to peddle their latest story of the trial are also remembered in that interview.

The newspaper quote—relating to the doctors' search for "a 'skeleton in the closet'"—is from the *Chicago Daily Tribune* of June 17, 1924.

Material relating to the violence of the city has been gleaned from a combination of Mary Field Parton's recollection and my own reading of Chicago newspaper archives covering the first six months of 1924. Hal Higdon also offers a cogent overview of Chicago's violence, and its salacious press coverage, in his admirable 1975 book, *Leopold & Loeb: The Crime of the Century.*

Darrow's famous saying "Hate the sin, never the sinner" supplied the title of John Logan's 1985 stage play about Leopold and Loeb: *Never the Sinner.*

Mary's description of him as "magnificently ugly" comes from her interview with O'Toole.

Margaret Parton notes the comparison with Tolstoy and Christ that Mary repeats in earlier correspondence—as in her August 1910 letter to her sister Sara, where she brackets her lover with Jesus and Tolstoy and "those who have loved and served their fellow men with sincerity and singleness of heart."

Darrow's "philosophy" of life is outlined in his autobiography, *The Story of My Life*—and in numerous of his speeches found in Arthur and Lila Weinberg's *Verdicts Out of Court* collection. Darrow's attitude toward crime and punishment is explained in his book *Crime: Its Cause and Treatment*, first published in 1922.

In F. Scott Fitzgerald's 1920 novel, *This Side of Paradise*, the full quote to which Mary alludes in her June 1924 letter to Darrow is: "Here was a new generation, shouting the old cries, learning the old creeds, through a reverie of long days and nights; destined finally to go out into that dirty gray turmoil to follow love and pride; a new generation dedicated more than the last to the fear of poverty and the worship of success; grown up to find all Gods dead, all wars fought, all faiths in man shaken."

Darrow's urging that Mary should read Nietzsche is repeated on many occasions, but this particular quote—"You must read him. You will love him. With all of his egotism and hardness, he is honest and brave and how we do like that"—comes from his May 15, 1913, letter to her.

His desire that Mary should become his biographer peppers their correspondence. But, as pointed out in the later chapters of this book, it is also telling that some of his most passionate beseeching occurs when, in the first flush of her marriage to Lem, he can no longer see her as often as he had done in the past. On August 8, 1913, he wrote, "I do want you to write the book—to begin it now." On January 14, 1914, Darrow stressed to Mary that "you tower over others. You must write my biography." Four years later, on January 19, 1918, he again said, "I keep thinking of you and that biography, which I would only have you to write. No one else could imagine the good or forget the bad as you." These letters are held in the Margaret Field Parton–Clarence Darrow Papers 1909–1975 at the Newberry Library in Chicago.

Darrow's aches and pains are alluded to in many letters in this same collection—and his neuralgia and rheumatism are examined in Irving Stone's biography.

His mood before meeting Mary is evoked in her interview with O'Toole while the source material for her saving Darrow from shooting himself in Los Angeles are listed in the notes to Chapter 2.

Edgar Lee Masters, the poet, wrote to Carter Harrison Jr., the mayor of Chicago, on July 29, 1944, to insist that "[Darrow] was as crooked as a snake's tail." Such animosity had been even more vehement when Masters's former legal partner had still been alive. In his July 26, 1921, letter to Eunice Tietjens, Masters lamented the fact that Darrow had chosen to represent his wife in their acrimonious divorce. "Darrow is revenging himself for the poems I wrote on him—'On a Bust' in *Songs and Satires*, and 'Excluded Middle' in *Towards the Gulf*," Masters claimed. "But these are just Sunday School persiflage compared to what he will get as I go along. I'll make that son of a bitch the most detestable figure in American history. The material is so abundantly at hand."

An undertow of jealousy and financial squabbling soured the relationship between Masters and Darrow—with the poet being particularly envious of his partner's capacity for fame and his liking for grandstanding cases that took him out for town for months on end. Masters seethed that he had to do the overwhelming bulk of legal work, for poor reward, throughout their uneasy partnership, which lasted from 1903 to 1911 in the form of Darrow, Masters & Wilson. This was not completely true, though, since Masters spent a considerable amount of time in the latter years attempting to write poetry in his company office. He admitted that "my law partnership lasted for eight years but for three years of this time it was a dying thing. . . . I committed the spiritual error of falling into indifference. . . . I was in a half-torpid state of life." Masters's autobiography also drips with such venom toward a long list of fellow writers and lawyers that his account of Darrow's apparent hypocrisy and corruption needs to be read with caution.

A much more detailed account of the tensions between Darrow and Masters can be read in Cowan's *The People v. Clarence Darrow*, Kevin Tierney's *Darrow*, and the Weinbergs' *Clarence Darrow: A Sentimental Rebel*.

Mary spoke to both her daughter and to O'Toole of Darrow's secret quest to be "made immortal" through his legacy of work. In her interview with O'Toole she also explains how Darrow saw Leopold and Loeb as a way of finally redeeming himself from the "stain" of the bribery trials of 1912–1913.

The *Chicago Evening American* accused Darrow of selling out for "a cool million bucks" on June 4, 1924.

Irving Stone touches on the Eugene Prendergast case in his biography of Darrow—but it is considered in more detail by both Tierney and Weinberg. The assassinated mayor Carter Harrison was eventually replaced in his position by his son—who shared his name and struck up a friendship with Edgar Lee Masters just before Darrow's death.

Darrow's letter to Mary—"Chicago is beautiful now . . . but I want to see you"—was written on May 15, 1913.

Mary's reunion with Darrow is relived in her third and final interview with O'Toole.

Chapter 2: Tangled Together

It becomes easier to understand why an apparently happily married woman like Mary Field Parton should answer the unexpected call of her former lover when attention is paid to the illuminating correspondence she and her husband shared during the first few months of 1924. The exchange of letters between Mary, in Europe, and Lemuel, in New York, provides a fascinating insight into their complex relationship. Housed at the University of Oregon (Special Collection 036), these letters provide the backdrop to much of the opening of this chapter.

For example, Lem's letter, written on an unspecified date in January 1924, touches upon her consideration of psychoanalysis: "My own dearest Mary: I hope you have had my cable announcing my job with the North American Newspaper Alliance. I am deep in this work—happy and rejuvenated. . . . I am delighted to hear about your decision to go in for psycho-analysis. The idea is really exciting to me, as I have made such amazing changes in myself, through some inside tinkering, that I feel it is the biggest subject in the world."

Mary and Darrow's first meeting in 1908 is re-created in taped interviews with her daughter, Margaret Parton—held at the University of Oregon. Margaret's unpublished biography of her mother, part of that same Special Collection 036 in Oregon, summarizes this meeting; and its version is corroborated by Mary's interview with Anne O'Toole.

Mary's move to New York, and her smattering of literary success, is documented in letters between her and Darrow in the Newberry Collection.

The sniping response of Ruby to Mary—that she is "the thousandeth and first" to be told she is "the cleverest woman" Darrow knows—features in Margaret Parton's unpublished biography of her mother.

Margaret wrote to Lila Weinberg of the work Mary did on Darrow's behalf when visiting the homes of potential jurors in the company of her six-year-old niece, Kay, in 1911. That letter of April 23, 1975, is part of the Weinberg research papers.

Mary's diary conveys her candid disappointments and frustrations when Darrow, after their evocative reunion on June 17, insists that he is unable to offer her any special favors ahead of the hundreds of other journalists swarming around Chicago. The quoted entries in this chapter are supplemented by Lemuel's letters, which are also held at the University of Oregon.

Mary's daily entries document her movements in Chicago and detail her various meetings with Darrow. But Lem's letters are helpful—as when he writes to her on Saturday June 22, 1924, after she has been "let down" by Darrow. Lem is obviously not aware of the extent of her motivation in rushing to Chicago to see Darrow, and he clearly has no inkling of the intimacy of their reunion. But his entanglement in Mary's planned journalistic "scoop" becomes plain.

Apart from trying to console Mary, Lem reveals the extent of his involvement in her attempt to revive her career through Darrow: "It was clearly understood that this was a short-end gamble and it is understood that Darrow is in a position where it might be ruinous for him to play any favorite. But I am sure you have got something upon which we can build a yarn or two. Be sure and see me before you talk to Pickering. . . . I have not told them that Darrow threw you down—I had to go to the limit in Darrow's indebtedness to you, when I put the trip over, and this would look as if I had been stretching it. Your out is that Darrow is so bowled over that he really hasn't outlined his defense yet, etc, etc.

"I am awaiting your return before doing anything with the Faris and Adams interview. The papers here [in New York] are swamped with interviews with psychologists and we've got to find some novel or exclusive news angle for anything we send out."

Lem's letter combines a hustling newspaperman's eye for "a yarn" with a more compassionate husband's encouragement of his downcast wife. It then takes a more poignant turn. Oblivious to her night at the LaSalle, Lem admits that he is confused by her accommodation arrangements in Chicago. "I am provoked at you to think you are staying at a cheap hotel and eating sketchily. I thought it was clear that you were getting your expenses paid [presumably by Lemuel's newspaper contacts]. That means a decent hotel and decent meals, with allowance for tips and taxicabs, etc. I never dreamed that you didn't understand this."

In an earlier letter Lem admits to having wired Darrow to see if the lawyer could tell him which hotel Mary had chosen for herself. He did not hear back from Darrow—and had no idea that they had been together at the LaSalle. And so it is little wonder that Mary feels remorseful when in that later June 22 letter Lem ends by saying: "I have been awfully disappointed not to have had a personal line or two from you. It seems a million years since I saw you."

In her third interview with O'Toole, Mary relives the morning she left Chicago and lingers over the touching details of her visit to the famous old courtroom with Darrow.

Ruby Darrow, in her September 4, 1939, letter to Irving Stone, warns him that, should he persist with plans to include Mary in Darrow's biography, "[I] cannot work with you under such circumstances." In a later letter to Stone she says: "Should you feel urged to coax M.F. to tell some embroidered account you should then get a long list of rivals and aspirants to the favors of C.D. and let the biography show a catalogue—a sort of directory—of ladies-in-waiting! When C.D. was challenged for not having told anything of his life-love in any chapter of his own biography he turned to me and grunted: 'A whole chapter, eh? I could do a whole library on that subject.' Then, with mock apology to me, 'No, Rube, maybe not that much.'"

Ruby tells Stone, "I am writing all the foregoing to you in strict confidence, for whatever influence it may have in shaping your judgment and attitude about M.F. since she is the one you seem bent on injecting into your story. Please do not—unless you find, and feature, many others equally! But I am certain you will consider the M.F. ilk unworthy of a drop of priceless ink."

Darrow knew how much Ruby loved him—to the point of suffocation— and he often yearned for an escape. When Lincoln Steffens, another of his and Mary's mutual friends, consoled him and asked Darrow how he and Ruby got along, the old attorney spat out his answer: "Fine—because, Ruby and me, we both love Darrow."

He was once asked another leading question: "Is marriage to be considered a lottery?"

"Yes," Darrow snarled. "If only there were prizes."

A young law partner, William Carlin, had seen him open their office at 8:00 A.M. one morning, fully half an hour before the scheduled start. After Carlin had expressed his surprise, Darrow asked whether he was married. "You know I am not, sir," Carlin said. "Well," Darrow murmured, "when you are, you may be down earlier than this."

Chapter 3: Darrow on Trial

At the Newberry Library and, especially, in the University of Oregon's Special Collection 036, details of Darrow's suicidal visit to Mary Field are contained.

Margaret Parton includes some of this material in her unpublished biography of her mother.

Sara Field Ehrgott's description of the apartment she shared with Mary, at 1110 Ingraham Street, as a place where "everything disappears—the bed walks into the wall . . ." comes from a letter she wrote to Charles Erskine Scott Wood on October 25, 1911 (in the Wood collection at the Bancroft Library).

The McNamara and bribery trials have already been covered in exhaustive detail by Cowan and Blum. Cowan is concerned mostly with Darrow's first bribery trial, whereas Blum's attention centers on the McNamara case and its aftermath.

Accounts of Darrow's early years, and his involvement in cases such as the Bill Haywood murder trial, can be found in his autobiography as well as in Cowan, Tierney, and Weinberg.

The wild purple prose supplied by General Harrison Gray Otis—"O you anarchic scum, you cowardly murderers . . ."—was printed in the *Los Angeles Times* on October 2, 1910.

Darrow's resolve held when he was first approached by Samuel Gompers, the president of the American Federation of Labor, with a request to defend the McNamara brothers. "There is no other advocate in the whole of the United States who holds such a commanding post before the people and in whom labor has such confidence," Gompers said in a telegram to Darrow on April 27, 1911. "You owe it to yourself and to the cause of labor to appear as the advocate of these men so unjustly accused."

Ed Nockles, a Chicago unionist and close friend of Darrow's, said, "The whole world is expecting you to defend the boys; if you refuse you convict them before they come to trial."

In regard to Darrow's craving for the national attention such a trial would bring, Francis Wilson, his law partner at the time in Chicago, told Irving Stone, "It would be the international limelight again, a lure he couldn't resist." But Darrow had "a heavy heart" when he finally accepted the brief and his hopes of persuading Wood to join him as cocounsel were soon ended. In a series of letters to his mistress, Sara, Wood had outlined his certainty that corruption would result. He predicted that Darrow "will use bribery where safe, perjury where safe."

Darrow's lament to James McNamara—"My God, you left a trail behind you a mile wide"—comes from Stone's biography.

Sara recounts Darrow's two attempts to seduce her in letters to Wood—held at the Huntington Library. On October 25, 1911, she wrote that, returning to the apartment early one evening, "C.D. was hugging me like an old grizzly to my unspeakable discomfort. I told him I was 'not eligible.' . . . He then made the rather strange remark that if I were out of funds I should come and see him." A few days later, at his office, he locked the door, "and proceeded to *make love* to me in high style." Sara confirmed she had rejected his attempt to seduce her.

Wood warned Sara on October 26, 1911, that "Darrow is above all things Selfish—with a not necessary adjunct to selfishness, vanity, and a side development—avarice." Ironically, it had been Darrow who had introduced the unhappily married Sara to Wood when she and her puritanical husband, a Baptist minister, had moved to Oregon.

Darrow's letter to Mary in New York—"I miss you all the time . . ."—was written on March 5, 1910.

Sara's jaundiced views on Mary's "third-rate" mind, under the spell of Darrow, come from both her letters to Wood and Margaret Parton's account.

Mary's "saving" of Darrow—and his quoted agreement that he will "go on"—is recorded by Parton.

Eugene Debs's response to Darrow—suggesting that the lawyer "loved money too well to be trusted"—was written on February 19, 1912, and is from the Debs Papers quoted by Cowan.

Sara writes of Mary being "widowed" in her letter to Wood on April 17, 1912.

The *L.A. Herald* report on Ruby and Mary arriving together at court was printed on May 15, 1912.

Courtroom scenes are drawn from newspaper accounts—in particular the *Examiner*, the *Herald*, and the *Times* in Los Angeles—while quoted text comes from the trial transcripts.

Mary's polemical report on that first day in court was published in *Organized Labor*.

The district attorney, John Fredericks, confirmed that "the charge against this defendant is that he gave a bribe to George N. Lockwood on or about November 28, 1911. . . . We will show you that in mid-November, about two weeks before the bribe, Bert Franklin went to Mr Lockwood, whom he knew, and offered him a certain sum of money. We will show you that Bert

Franklin was acting at the request and under the direction of this defendant.
. . . We will show you that the money was the money of Clarence Darrow, that
Clarence Darrow gave it to Franklin that morning."

Mary's glowing appraisal of Darrow as "the calmest person in the room"
ran in the *San Francisco Bulletin* on August 3, 1912.

She also wrote about her uplifting day in the countryside with Darrow
to Wood—and her determination to avoid Ruby at any celebratory party—on
August 6, 1912.

Darrow's extraordinary performance in court is drawn from newspaper
accounts, Mary's writing, and the trial transcripts. The same applies to
Fredricks's response and, on August 17, the announcement of the verdict.

The victory party was held at the Café Martan—and the photograph of
Mary between Darrow and Ruby is held at the Newberry Library.

The correspondence between Darrow and Mary, in Indianapolis, is found
in the Newberry Archive. Cowan covers Darrow's second trial in some detail,
while Mary, in her interview with O'Toole, depicts her lost lover as a broken
and bankrupt man.

Chapter 4: The Bridge of Sighs

Irving Stone touches on Darrow's desolation in his biography and recounts
his desperate need to sell his first-edition books for cash. His belief that his
"lawyering days" were over comes from a conversation with Ruby that is
quoted in Stone. The Weinberg biography also offers a sympathetic account of
him being lured back to the law by Peter Sissman.

Few knew how Darrow really felt then—but Mary did. Even if they were
no longer lovers, she understood that he still harbored the hurt and shame of
Los Angeles. Mary also reveals his liking for comparisons to King Lear in her
O'Toole interviews.

Leopold's and Loeb's quotes on meeting Darrow for the first time are
from Stone's biography. Similar accounts of Darrow's reaction to the tension
between the boys are produced by both Stone and Nathan Leopold in his
prison autobiography—*Life + 99 Years*. In that book, Leopold also reveals his
reaction to Darrow's unkempt appearance.

Leopold's brutal justification of the murder as being akin to "an
entomologist killing a beetle on a pin" was one that he made repeatedly to

newspaper reporters—as when he was quoted in the *Chicago Daily Tribune* on June 2, 1924: "It was just an experiment. It is easy for us to justify as an entomologist in impaling a beetle on a pin."

Darrow's reactions are charted in his autobiography and in the book by Stone.

Leopold's discussion with Darrow about Byron and "The Bridge of Sighs" was conveyed to me by Vincent Martin—who had interviewed Leopold in Puerto Rico. There are numerous references to local coinage of the Bridge of Sighs in Chicago newspapers of the day—with one typical example coming from the *Chicago Daily Tribune* of July 24, 1924: "Two young college graduates, still wearing a smile on their lips, despite the ringing in the ears of the cries of their prosecutors that they are 'America's coldest-blooded murderers' hurried across the 'bridge of sighs' connecting the Criminal Court building and the county jail."

The quotes from Leopold to Robert Crowe about his planned translation of Pietro Aretino's *I Ragionamenti* come from the trial transcripts and psychiatric reports.

Hal Higdon quotes the two men, from Nebraska and Michigan, who offer to execute Leopold and Loeb.

Mary Field Parton, in conversation with O'Toole, highlights Darrow's fascination with Nietzsche—and the role the German philosopher played in the Leopold and Loeb case.

The long section on Leopold's and Loeb's family, academic, and sexual backgrounds is all based on the first-person accounts they gave to Hulbert and Bowman at Cook County Jail in June 1924—supplemented by the additional details they passed on to the other defense psychiatrists (White, Healey, and Glueck).

Darrow's decision to switch Leopold and Loeb's plea to guilty is explained in his autobiography.

The opening scenes in court are drawn from newspaper accounts and the trial transcripts.

Ironically, in later years Ruby Darrow told Irving Stone that Mary Field Parton had turned up unannounced on the opening morning of the trial. "At the end of the first day C.D. came home weary and, of course, anxious about the situation," Ruby wrote. "He told me that as court was about to open, in walked M.F. requesting him to arrange a seat for her among the reporters as she had just arrived from New York for one of the papers. He had refused, escorted her out, and at the door gave instructions to attendants that she was not to be admitted at any time—and, somehow, he managed to have someone take her and her suitcases to the next train for N.Y. and to make sure that she was on when the train pulled out. I was astonished that he would and said so to him. But he answered that in LA, he couldn't help himself—but now he wasn't going to put up with the annoyance of M.F. anymore. That was the end of it between them."

As we now know Mary and Darrow were not yet at an "end"—and this is just another example of Ruby Darrow trying to persuade Stone to drop "M.F." from his authorized biography. Mary's detailed diaries place her firmly in Sneden's Landing, at the farmhouse, when the trial opens. Far from being in court that morning, she follows the trial from a distance. Her feelings of longing and loneliness are recounted in her diaries from that period.

She writes of Lemuel leaving for Greenland and Labrador on July 19, 1924—and of writing "all day on Franks stories" on July 21 and 24. The allusions to her "terrible loneliness" and the "tinkling piano" and fireflies are made on July 21. Mary notes that Margaret "takes much patience" on July 22 and, a day later, of how lonely the country is compared with the city—an observation she makes repeatedly as when, on August 9, she complains that "the country is inexpressibly lonely." She completes reading Darrow's book on crime on July 28, when she notes his many contradictions and him being "bigger than his pessimism."

Margaret's prayer for Leopold and Loeb is found in Mary's entry of July 29. "M, with a flower or a bit of parsley leaf in her hair, kneels on the floor before her candle, clasps her little hands in prayer and says, 'Dear Jesus and God. Watch over dear daddy. Please put Leopold and Loeb in the right place. Take care of my cat and doll and bring daddy home safely. Amen.'

On August 24, Mary notes that Margaret "wishes Darrow was 'not so lonely' so that 'when he looks sad at the judge he would look beautiful sad— not lonely sad'—like Douglas Fairbanks who is her masculine ideal of beauty."

These moments are supplemented by my reading of Lemuel Parton's letters to both his wife and daughter in Sneden's Landing during various stages of his trip in search of the Arctic explorer Donald MacMillan.

Yet Lem's trip, undertaken with a band of other New York reporters, is not a success. On August 31, 1924, writing from Indian Harbor, Labrador, he tells Mary, "I can think of nothing except that I never was so homesick in my life and I want to be with my dear ones more than you can know. Somehow, in this loneliness and isolation, I have kept going back, over and over again, over the last years—starting with our first little nest at Lake Street [in San Francisco]. . . .

"This story is a flop, so far as the chance of writing anything is concerned. Communications have been so hopeless that we have had to join on joint dispatches." But he is intensely proud to have been selected to write those syndicated accounts—"It is quite an ad for me to be picked by twenty star New York newspapermen to write the joint yarn."

But he ends by lamenting the fact that, "Mary dearest, I just can't say what's in my heart. I never want to be away from you dear ones [Mary and Margaret] again. I keep wondering if either of you have changed, and a chilly fear strikes me that maybe you won't love me as much as you did."

Chapter 5: Beyond Good and Evil

A gauge into shifting American attitudes toward psychiatry in the 1920s is encapsulated by the Leopold and Loeb trial—and, in particular, the *Chicago Tribune*'s attempts in June 1924 to entice Sigmund Freud from Vienna ahead of the *Herald and Examiner* and *Chicago's American*. Stone highlights the links Darrow made with the trauma felt by soldiers after World War I.

The courtroom scenes are drawn from newspaper reports of the time and the trial transcripts.

Leopold's quips about the hot weather, and playing tennis, were reported in the *Chicago Daily Tribune* on July 26, 1924.

The long section on the murder of Bobby Franks, both its planning and execution, is drawn from the psychiatric reports and trial transcripts. Leopold and Loeb's reaction in the aftermath of the murder, and their bizarre engagement with ensuing press coverage, are from the same sources.

In later years there were also some useful journalistic re-creations of the day of the crime—such as the *Chicago Daily Tribune*'s article "Eleven Years After the Crime of Loeb and Leopold," published on May 19, 1935. That article also contains a detailed map of the area and a time line of events. But I relied on the confessions—only turning to alternate sources on the few occasions when there was a discrepancy in the two accounts. The few lines of dialogue are in either the confessions or the trial transcripts.

Armand Deutsch's fortunate escape is relived in his article, "My Murder: Loeb & Leopold's Intended Victim Recalls the First 'Crime of the Century,'" published in the Sunday magazine section of the *Chicago Tribune* on June 23, 1996. Deutsch had worked as a film producer with stars like James Stewart and Elizabeth Taylor. He later became an executive for both MGM and Warner Brothers, and he was close friends with Humphrey Bogart. He died, at the age of ninety-two, in August 2005.

Leopold and Loeb's ransom note and intricate plans to obtain $10,000 from the father of the boy they abducted are detailed in the same sources listed above—as is Loeb's interaction with the three young reporters. I also studied the reports written throughout the case by Alvin Goldstein and James Mulroy—the *Chicago Daily News* reporters who eventually shared the Pulitzer Prize for their work on the case. It is unlikely that they would have reached such exalted heights but for the help they received from Richard Loeb.

Julius Rosenwald, Salmon Levinson, and Susan Lurie's quotes were sourced from various newspapers—most notably, the *New York Times* and the *Chicago Daily Tribune* of June 1 and June 2, 1924. The first Susan Lurie quote is from the June 2 edition of the *Tribune*, and the longer insight she gives into Leopold's character comes from a June 1 article, entitled "Joked with Girl About the Murder," in the *New York Times*.

After he had been questioned for the first time by the Chicago police about the murder, Leopold still went on a date with Susan on a sunny Sunday afternoon. After lunch at a small country inn, they hired a canoe and drifted down the Des Plaines River. "Unbelievably callous at it now seems," Leopold wrote later, "that afternoon was the happiest of my life. Perhaps the fact that it was my last wonderful day adds to its enchantment now. . . . [Susan] had brought the blue vellum-covered volume of French poetry I had given her for her birthday the week before. And as she read the liquid verses to me, I laid my head in her lap."

The emotion of the afternoon carried through into Sunday night. After he returned Susan to her North Side home, he picked up Dickie Loeb in the red Willys-Knight. They went to a party in Kenwood. Leopold yearned to dance with Loeb, but, of course, such a sight would have been even more shocking in 1924 than the murder no one then suspected them of committing. And so they jived, as usual, with a dazzling series of bob-haired flappers. Dick spent most of the night with his choice girl, the supposedly wayward "Patches"

Reinhardt. Babe cast envious glances their way, but found some comfort in the laughing features of Alice Churchill.

Dick Meyer told me how his grandmother, Alice Churchill, danced with Leopold that Sunday evening. In her old age Alice had often told Meyer this story—and of how she and her cousin Adolf "Bal" Ballenberg had spent many leisurely days through their childhood at both the Leopold and Loeb family homes. Even Babe Leopold, who supposedly had eyes only for Susan or, with more furtive accuracy, Dickie, considered Alice a drop-dead gorgeous and seriously bright party girl. Early the previous year, she had married Daniel Guttman, and she had given birth to a daughter on December 23, 1923.

Five months had passed since she had become a mother and so that night, her first night of freedom since the birth, meant much to Alice. She danced delightedly with Babe, telling him breathlessly between songs how happy she was with her life. Alice pulled him close and shouted out that, approaching her twenty-first birthday, she felt "on top of the world." Babe looked at her dazedly and then, unable to help himself, allowed her to draw him in for a hug.

Alice held the boy she did not yet know was a murderer. She was shaken when he buried his face in her shoulder and began to sob uncontrollably. The music played on, and couples kept dancing, oblivious to his crying. When she had calmed him sufficiently, Alice asked Babe what was wrong. He looked into her concerned face, his own features streaked with tears as his body heaved. He shook his head. The words would not come as he cried on the dance floor.

Alice led him discreetly away so that he might compose herself. She guessed he was just drunk, and she offered him her damp shoulder again. "You're a queer bird, Babe," she said softly.

Chapter 6: Courtroom Whispers

Courtroom scenes are re-created from daily accounts in the various Chicago newspapers, the *New York Times*, and especially the trial transcripts—with the latter supplying the quoted exchanges.

In keeping with the reluctance of William Healy and Darrow to publicly reveal the graphic nature of Leopold and Loeb's sexual relationship, the first published book about the case also tried to ignore the scenes reported in the psychiatrist's courtroom whispers. Maureen McKernan's *The Amazing Crime and Trial of Leopold and Loeb* contained a foreword written by Darrow and his cocounsel Walter Bachrach—but it remains a curiously chaste and sanitized piece of reportage.

Irving Stone writes about the "stolen" copy of the Hulbert and Bowman report—and the way in which Darrow used this supposed theft to allow him to distribute the psychiatric findings to every newspaper in Chicago. He also recounts Darrow's observation of the summery hats on the iron hook and evokes the suffocating atmosphere in court.

Genevieve Forbes, remembering the stuffy heat of the court, wrote years later in the *Tribune*, on July 29, 1957, that "the air was sometimes so foul from sweating humanity that when the dirty windows were raised, the aroma from neighboring fish markets was refreshing."

Leopold, in his prison biography, remembers his fearful question to his brother: "My God, Mike, do you think we'll swing after that?"

Chapter 7: The Book of Love

Mary Field Parton's diaries supply the background to the opening section—most notably on July 31 and August 14, 16, 19, 20, 22, and 24.

On July 31, 1924, Mary notes that Margaret "grows with praise and stiffens and becomes ugly with blame." As Margaret points out as an adult, when rereading her mother's diaries, and quoting from them in her memoir, these warmer reflections contrast with pricklier observations from preceding years—when Mary talks of her daughter as "the maggot" and of "Margaret's cold, icy attitude toward people."

In her interview with O'Toole, Mary conjures up Darrow's mood before he made his final plea on behalf of Leopold and Loeb—while recounting the significance of both Omar Khayyám's *The Rubaiyat* and A. E. Housman's poetry to him. She also remembers receiving the books as gifts from Darrow.

Housman, meanwhile, would quip in later years that Darrow had saved most of his defendants from the gallows by often misquoting his poetry.

The scenes before and during Darrow's address are recounted in the Chicago newspapers and the *New York Times*—while the quoted passages of text can be found in *Attorney for the Damned*, Arthur Weinberg's collection of Darrow's greatest courtroom speeches. In the official trial transcript this section is missing—as Darrow had borrowed it while writing his autobiography. He failed to return it. However, numerous newspapers in 1924, like the *Chicago Herald and Examiner*, reprinted the entire original speech.

Mary's response to Darrow's speech, which she reads in the library in Palisades, is from her diary entry of August 24. She also describes it as a "beautiful, tender, noble plea"—while Margaret reads L. Frank Baum's *The Wonderful Wizard of Oz*. In a curious link with the later chapters, covering William Jennings Bryan, some believe *The Wonderful Wizard of Oz* is a parable about money reform and the 1890s midwestern political movement—led by Bryan with his Cross of Gold speech at the Democratic Party convention in 1896.

Mary's wistful musings in early September are drawn from her diary. On September 3, 1924, she describes Margaret as "a dear little girl" and, a day later, she writes of their lunch at the Waldorf. She mentions her "longing for my mate" on September 6 and of "an early Fall coming" two days later. Margaret's quote—"This has been such a happy summer, Mother"—is from Mary's diary entry of September 13, 1924.

Judge Caverly's verdict is recorded in the transcripts while the atmosphere in court is drawn from various newspaper accounts printed on September 11, 1924.

Chapter 8: The Pinch Hitter And the Politician

The long opening section detailing Mary's preparations for Darrow's visit is based on her diaries and Margaret Parton's unpublished biography of her mother. Margaret's poignant memoir of her own life, *Journey Through a Lighted Room*, is particularly good in outlining her warm relationship with her parents' bohemian circle of friends—which stands in such contrast to more prickly memories of Darrow's interaction with her.

Stone describes Darrow's frustration at the Leopold and Loeb families' refusal to pay him adequately.

Darrow's December 23, 1924, speech against Prohibition was quoted in the following morning's *New York World*—while the anecdote about Mary outwitting the agents of Prohibition with the strange pulley and bottles of brandy under her dress is remembered in Margaret's memoir.

Mary Field's diary notes detailing her dinner with Dreiser and Darrow, and then meeting Ed Nockles, are held by the Newberry Library.

For background I read Louis Koenig's *Bryan: A Political Biography of*

William Jennings Bryan and Michael Kazin's much more recent *A Godly Hero: The Life of William Jennings Bryan.*

In his autobiography Darrow remembers his early years with William Jennings Bryan—as do Stone, Tierney, and the Weinbergs. Edgar Lee Masters, who befriended Bryan, recounted the details of Darrow's scathing denunciation of Bryan's lack of a literary education in McAlister Coleman's 1930 book, *Eugene V. Debs: A Man Unafraid.*

Darrow's personal background, and his father's attitude toward religion and learning, are depicted in his *Story of My Life*—as well as Stone.

In his 1922 tract, *Shall Christianity Remain Christian—Seven Questions in Dispute*, Bryan drew the quoted links between Nietzsche and Darwin. Edward J. Larson, whose excellent *Summer for the Gods* is the definitive book on the Scopes trial, pinpoints Bryan's reliance on the writing of Benjamin Kidd in linking Nietzsche back to Darwin.

The quote from the *Nashville Tennessean* was printed in that newspaper on June 21, 1925.

Darrow's indignant views on the "purity of the Anglo-Saxon race" are quoted by Stone.

William Jennings Bryan's endorsement of the Bible's "divine authority" was published in his 1923 essay "The Fundamentals."

Chapter 9: Hellfire Preachers And Biology Teachers

Darrow's questions to Bryan were printed in the *Chicago Daily Tribune* of July 4, 1923. Bryan's response comes from the July 5 edition of that same newspaper.

T. T. Martin's quote comes from his *Hell and the High Schools: Christ or Evolution: Which?*—published by the Western Baptist Publishing Company in 1923.

After interviewing John W. Butler, Marcet Haldeman-Julius, a writer from Kansas, argued that "it is impossible to conceive of a more glaring, startlingly plain example of the danger of sheer ignorance. Here is a man, innately good, doing all but irreparable harm. A man who stands for progress in farming and in his community generally, attempting without malice to stop thought."

This comes from Haldeman-Julius's *Clarence Darrow's Two Great Trials* booklet—which she and husband, Emanuel, originally published in separate installments in their *Halderman-Julius* monthly magazine during the summers of 1925 and 1926. They then complied her various essays on the Scopes and Sweet trials into a single publication, which they released as part of their innovative Big Blue Book series.

The *Chattanooga Times* editorial in response to Butler was published on February 1, 1925—while the *Rockwood Times* article, called "Monkey Business," was reprinted in the *Knoxville Journal* on February 11, 1925.

The Tennessee Senate's reaction to John Shelton's motion is recorded in Kenneth M. Bailey's "The Enactment of Tennessee's Anti-Evolution Law" found in *The Journal of Southern History* (Volume 41)—and, more recently, by Larson.

Billy Sunday's quotes are from Larson's book, which, in turn, relied on *Commercial Appeal* reports stretching from February 6 to February 23, 1925.

Bryan's cautious advice to remove any penalty was made in his letter of February 9, 1925, to Senator John Shelton.

Darrow's reaction to the Butler law comes from Mary Field's interview with O'Toole.

Governor Peay was so convinced that the Butler bill would remain dormant that he said, soon after signing it, "Nobody believes it is going to be an active statute."

A comprehensive overview of the ACLU, and George Rappleyea's innovative response to their advertisement, is found in Larson's book.

John T. Scopes's account of the famous scene in which, at Robinson's Drugstore, he is drawn into the ACLU-sponsored test case against the law banning the teaching of evolution in Tennessee is from his memoir—*Center of the Storm*. Scopes also remembers his earlier exposure to William Jennings Bryan at university.

On May 5, 1925, the *Nashville Banner* ran its front-page story.

The *New York Times* report on Bryan's speech in Pennsylvania was printed in its May 13, 1925, edition—which also contains his quotes about assisting the prosecution in Dayton.

Darrow wrote about his reaction to Bryan's involvement in the Scopes trial in his autobiography—from where we find his "At once I wanted to go . . . ," assertion.

H. L. Mencken's praise of Bryan as a speaker comes from Marion Elizabeth Rodgers's biography.

Irving Stone writes of H. G. Wells's enthusiastic memory of first meeting Darrow—while the quote from Wells about Darrow's qualities is from his own 1934 *Experiment in Autobiography: Discoveries and Conclusions of a Very Ordinary Brain (Since 1866)*.

The wording of Darrow's cable to the Dayton defense team, offering his and Malone's help for free, is from Scopes's memoir.

Larson explores the ACLU's antipathy to Darrow's "showmanship."

Mary's quoted extracts are from her diaries: "Death in a cruel form . . ." was written on May 16, 1925, while "M barefoot and stripped . . ." and "creatures eating creatures" are from June 4, 1925.

She writes about Carl Sandburg's visit to the farmhouse—"good stories, good drinks, good friends!"—on June 1, 1925.

Lemuel's repeated absences are mentioned in Mary's diary on many occasions—but, for example, during this period on May 22 and May 25, 1925. Again, on June 9, she writes: "Lem leaves for Canada. 'Two days,' says my dashing optimist. 'A week,' at least, says his pessimist wife."

The loneliness of the country at night, and her yearning for the company and excitement of the city, are evoked in her diary entry of June 10.

Three days earlier Mary, quoting Darrow, writes about his admission that he cannot "save" himself from his marriage. She also suggests then that "the mysterious forces of attraction baffle me."

The *New York Times* report on the dinner held in honor of Scopes was printed on June 11, 1925—while Mary Field discusses this evening with O'Toole and writes of it in her diary entry of that same day. It is here, on June 11, that she also writes about the "interminable Russian meeting" and the "Grand Opera ghosts"—as well as her humiliation when Darrow "gave me a little silver—not enough—for taxi fare" and the refuge of Freddie O'Brien's apartment.

Scopes's later arrival at the dinner is mentioned by the *Times*—and he admits to having missed Darrow's car and walking to the Civic Club in his memoir.

The contrasting example of Darrow's generosity toward Mother Jones is passed on from Mary to Margaret Parton—who includes it in her unpublished biography.

Mary's breakfast meeting with Darrow, and their walk through Central Park, where he talks of "good people" and "the danger to liberty," are captured in her diary entry of June 12, 1925.

Chapter 10: Monkey Business

Larson's *Summer for the Gods* is the landmark text on Dayton's monkey trial—but eyewitness accounts supplied by Marcet Halderman-Julius and H. L. Mencken are invaluable. Scopes, Stone, Tierney, and the Weinbergs all cover the trial, and I also read Jeffrey P. Moran's useful *The Scopes Trial: A Brief History with Documents* and D. C. Ipsen's *Eye of the Whirlwind: The Story of John Scopes.*

The *Chicago Daily Tribune* article on Bryan's arrival in Dayton was printed on July 8, 1925.

Descriptions of Dayton's carnival-like atmosphere are compiled mainly from newspaper reports of the time—with particular emphasis on Mencken in the *Baltimore Sun* and reports printed in the *Chicago Daily Tribune*, the *Chattanooga Times*, and the *New York Times.*

Haldeman-Julius captured some of the cold fury within Bryan: "As he sat there in the court room, day after day, silent, fanning, fanning, his face set, I was appalled by the hardness, the malice in it. No one who has watched the fanatical light in those hard, glittering black eyes of Bryan, can doubt that he believes in both a heaven and in a hell . . . it is a face from which one could expect neither understanding nor pity."

Marion Elizabeth Rodgers writes about H. L. Mencken and T. T. Martin in her intricately researched book *Mencken: The American Iconoclast.*

The *New Republic* quote is from July 1925. Mencken's charmed description of Dayton appeared in the *Baltimore Sun* on July 9, 1925.

Darrow's speech to the Progressive Club, soon after his arrival in Dayton, was reported by various newspapers—including the *Commercial Appeal* and the *Knoxville Journal.*

A copy of George Bernard Shaw's "Where Darwin Is Taboo: The Bible in America," originally published in the *New Leader*, is held at Washington University in St. Louis, among the Harold C. Ackert collection. Shaw's instinctive liking of Darrow is quoted from his letter of August 19, 1903, to Henry Salt—included in Dan H. Laurence's *Bernard Shaw: Collected Letters, 1898–1910.*

The description of the court comes from Darrow's biography and is supplemented by Naomi Fein's visit to the historically preserved location in the *New York Times*, on October 13, 1996.

Bryan's and Darrow's arrivals in court are drawn from newspaper reports and photographs.

The courtroom quotes are from the transcript printed in *The World's Most Famous Court Case: Tennessee Evolution Case*, which has been reprinted and made available again by the Bryan College in Dayton. This same transcript also confirms the vocation and religion of each juror.

The courtroom scenes, and the atmosphere on the lawns outside, are evoked in newspaper reports and by Halderman-Julius and Mencken—and additional depictions can be found in Ipsen, Larson, Stone, Tierney, and Weinberg.

Background on Arthur Garfield Hays, Dudley Field Malone, Darrow, and Bryan is also found in two books by Hays—*Let Freedom Ring* and *Trial by Prejudice*.

Mencken's suggestion that the jury was "hot for Genesis," as well as his description of the blast-furnace atmosphere inside the courtroom, is found in his "The Monkey Trial: A Reporter's Account," which is included in *D-Days at Dayton: Reflections on the Scopes Trial*, edited by Jerry Tompkins.

Rodgers writes about Mencken's evening with Nellie Kenyon—while his original article on the Holy Rollers was printed in the *Baltimore Sun* on July 13, 1925. Kenyon's own memories of her time with Mencken appeared in the same paper on September 18, 1977.

Scopes remembers how he was set up in *Center of the Storm*. The *Chicago Tribune* article on the most hectic Sabbath in the history of Dayton was printed on July 12, 1925.

The quotes from the second day in court, dominated by Darrow's great speech in defense of religious liberty, come from the trial transcript—with this section also reprinted in Jeffrey P. Moran's *The Scopes Trial: A Brief History with Documents*.

Rodgers quotes the onlooker's view of Mencken as he stands on top of the table at the side of the courtroom. His reaction to Darrow's speech was printed in the *Baltimore Sun* on July 14, 1925. The *New York Times*, on that same date, assessed Darrow's impact on the Tennesseans: "Clarence Darrow, as champion of evolution and chief counsel for John T. Scopes, bearded the lions of Fundamentalism today. . . . While he was talking there was absolute silence in the room except for the clicking of telegraph keys. The

Tennesseans were undoubtedly amazed at what Mr. Darrow said, and just as amazed at his temerity in saying it . . . [but] his words fell with crushing force, his satire dropped with sledgehammer effect upon those who heard him."

Halderman-Julius's comparison of Darrow as "a giant among mental pygmies" is from *Two Great Trials.*

Ipsen writes of the monkey dressed up like a man, and carrying a cane, as well as the loss of electricity while the water mains were repaired.

Mary Field Parton's "I am a Robot" diary entry is from June 27, 1925. Her anger toward the young man who knows nothing of Darrow and the Scopes trial is from July 12. And two days later she celebrates Darrow's "wonderful plea for truth."

The *New York Times* quote—with Bryan "silent and grim . . . like a warhorse ready for battle"—is from July 15.

All courtroom exchanges are from the trial transcripts.

Scopes's closing quote in this chapter comes from *Center of the Storm.*

Chapter 11: A Duel to the Death

Mencken's analysis of Bryan's bitter hatred and his last days in Dayton comes from his obituary of the Great Commoner—published in the first edition of the *Baltimore Evening Sun* on July 27, 1925. That obituary was less an epitaph than one last stark attack on Bryan. The critic Alfred Kazin suggested, "It was significant that one of the cruellest things [Mencken] ever wrote, his essay on Bryan, was probably the most brilliant." Yet its contents were so searing that, in later editions, the obituary had been drastically edited—with large chunks of Mencken's vitriol edited out.

Bryan's quotes are taken from the court transcripts while Darrow's question—"Can it be possible that this trial is taking place in the twentieth century?"—is reported in Stone's biography.

Malone's address is also from the court transcript. Scopes mentions the poignant exchange between Bryan and Malone in his book on the trial—while Mencken's report was printed in the *Baltimore Evening Sun* on July 17, 1925.

The reaction of the table-thumping policeman is recorded by Scopes in *Eye of the Whirlwind.*

Darrow's wry reaction to the electrical storm that night is recounted by Ipsen.

The *New York Times, Nashville Banner,* and *Chicago Daily Tribune* reports on a momentous day in court were all printed on July 17, 1925.

Darrow's angry exchange with Judge Raulston is recorded in the trial transcript.

John W. Butler's comments are noted by Halderman-Julius.

The absurdity of a circus trial left a sour taste. Mencken, who had already been threatened, decided to get out of town. "I hope nobody lays hands on him," A. P. Haggard, Dayton's chief commissioner, warned the *New York Times* in a July 17 article headlined "Mencken Epithets Raise Dayton's Ire." "I stopped them once but I may not be there to dissuade them if it occurs again."

Bryan's attack on Darrow, in the midst of the prosecution's apparent triumph, appeared in the *Commercial Appeal* and the *Nashville Banner* on July 19.

Mencken's final report from Dayton was published in the *Baltimore Evening Sun* on July 18. It should be noted that not all reporters in Dayton were as certain as Mencken that the trial was over that weekend. Bill Perry suggested cautiously in the *Nashville Banner* on Sunday, July 19, in an article headlined "Scopes Defense Facing Defeat," that "rumors got about that the defense is preparing to spring a coup d'etat."

Darrow's apology for his alleged "contempt," and Raulston's acceptance of it, are from the transcript.

Scopes's quote about the "man-killing heat" is from his book on the trial.

The *New York Times* description of the crowd was printed on July 21, 1925.

The legal battle over the READ YOUR BIBLE sign is recorded in the transcript of the trial.

Similarly, Darrow's detailed cross-examination of Bryan is contained in that same transcript. The occasional descriptions of their actions between questions and answers are drawn from various newspaper reports and photographs published in the aftermath of the trial.

Judge Raulston's ruling the following morning, the responses of Darrow and Bryan, and the jury foreman's verdict are all drawn from the trial transcripts. The closing words from Bryan, Darrow, Hays, and Raulston are also in the official record.

Darrow's letter to Mencken—"I made up my mind to show the country what an ignoramus he was, and I succeeded"—was written on August 15, 1925, and is held in the Mencken Collection at the New York Public Library.

Mencken's private fear of the rise of fundamentalism is recorded in Rodgers's biography.

His assessment of Bryan's fall from hero to "tragic ass" was printed in the *Baltimore Evening Sun* on July 22, 1925. That same paper suggested, in an editorial on July 21, that Darrow's cross-examination "brought about a striking revelation of the fundamentalist mind in all its shallow depth and narrow arrogance." For the *New York Times*, spectators at that open-air humiliation for Bryan showed "no pity for his ignorance of things boys and girls learn in high school."

John Scopes's description of the rumor that "the old devil Darrow" had killed Bryan is included in his book of the trial. Darrow writes of his own reaction to Bryan's death—and his comment that a busted gut was a more likely cause than a broken heart—in his autobiography.

Mencken's reaction to the death of Bryan is recorded in Rodgers's biography. Yet Mencken also admitted years later that it was "not really just" for him to have derided Bryan "as a quack pure and unadulterated." He still joked that Bryan had probably ended up in hell—"Trust the devil," Mencken cackled, "he takes care of his own."

The closing quote from Mencken—"We killed the son-of-a-bitch!"—is from Larson's *Summer of the Gods*. Larson attributes its source as Robert D. Linder's "Fifty Years After Scopes: Lessons to Learn, a Heritage to Reclaim"—published in *Christianity Today* on July 18, 1975. "Summer for the gods," meanwhile, is a quote from Darrow's *The Story of My Life*.

Chapter 12: Darkness in Detroit

Mary Field Parton, in the last of her interviews with O'Toole, describes reading Russell Owen's feature on Darrow in the *New York Times*. She also talks of meeting Darrow in New York and of pressing the story on him.

The *Detroit Free Press* front page, linking Darrow to Dayton and eventually Detroit, is from July 13, 1925.

Kevin Boyle's superb *Arc of Justice: A Saga of Race, Civil Rights and Murder in the Jazz Age* is not only the quintessential account of the Sweet case, it is also, arguably, the best book yet to include Clarence Darrow. The focus of Boyle's masterpiece, however, is the Sweet family.

Boyle partly relies, as do I, on Marcet Haldeman-Julius's account

published in the form of two articles in the *Haldeman-Julius* monthly
journal—and also on Alex Baskin's first-person interviews with most of the
principal characters among the Sweet defendants and their friends (held at the
Bentley Historical Library in the Baskin Collection) and David Lilienthal's
reportage in the *Nation*.

The first Haldeman-Julius essay was printed in June 1926 in volume 4,
number 1, and the second appeared in July 1926 (vol. 4, no. 2). She provides
sympathetic eyewitness accounts in the heat of courtroom strife, as well as
interviewing Ossian, Gladys, and Henry Sweet in their doomed bungalow
before the second trial. I also appreciated Kevin Boyle's generous help
in sending me the only surviving copy of the files containing the police
interrogation of the Sweet family.

Phyllis Vine writes thoughtfully about the same two Sweet cases in *One
Man's Castle*—and she offers a fresh analysis of the Sweet material. She also
illustrates the significance of Walter White and James Weldon Johnson.

White, who did so much to rouse the commitment of the NAACP to the
defense of the Sweets, provides an overview in both his autobiography and
correspondence. Many of his key letters are held at the NAACP Papers at the
Library of Congress. On September 16, 1925, for example, he outlined in a
letter to James Weldon Johnson the exploitative financial conditions of the
"loan" given to Ossian Sweet by Ed and Marie Smith.

Darrow and Thomas Chawke, by means of their forensic cross-
examination, coerced a variety of witnesses who had attended the seven-
hundred-strong gathering on the corner of Garland and Charlevoix to detail
the mood surrounding that meeting—and the address by the speaker from the
Tireman Avenue Improvement Association.

Boyle writes movingly about the trauma visited upon Dr. Alexander
Turner on the night he moved into his new home in a white neighborhood—
while quoting Cecil Rowlette's oral history interview carried out on August 1,
1926. Journalistic accounts of the Turner case were printed on June 24, 1925,
in both the *Detroit Free Press* and the *Detroit News*.

Haldeman-Julius's quote about racism in Detroit being worse than "any
spot north of the Mason and Dixon line" is from her *Clarence Darrow's Two
Great Trials*.

The *Detroit Independent* claim was printed on September 30, 1927, and
covered the preceding two years.

Background into Ossian Sweet's life before he and Gladys moved is contained in the trial transcripts—but, for a full account, *Arc of Justice* should be read. Vine also explores, in some detail, Sweet's youth in Bartow, Florida.

Haldeman-Julius writes flamboyantly about the characters and personal histories of all the Sweets—while making her rather snobbish dismissal of the neighborhood around Garland.

The two torturous nights spent in the bungalow by the Sweet family and their friends is re-created from the trial transcripts and Haldeman-Julius's interviews—and certain sections of the police interrogation transcripts.

On September 10, 1925, the skewed reports in the *Detroit Free Press* and the *Detroit Times* were both printed.

Mary Field Parton remembered Darrow's introduction to the Sweet case in her interview with O'Toole—while Darrow's confusion over the racial background of the three men who ask him to defend the Sweets is recounted by Arthur Garfield Hays in *Let Freedom Ring*, as well as by Stone and in White's autobiography.

Chapter 13: Bittersweet Blues

Details of Mary's hunt for a cheap room to rent are found in her diaries—where she describes the maternal grind in getting her daughter to and from school and the refuge that the farmhouse and the Greenwich Village attic offered. She also writes about the ceremonial grape crushing.

Margaret Parton's memoirs re-create the scene of her barefooted stamping of the grapes in the bathroom and the quotes of "How charming! How Italian!"

Copies of Darrow's letters to Mary are held at the Newberry Library.

Mary's resigned "Darrow is Darrow" comment to Helen Todd was made on October 26, 1926.

Darrow's suggestion to Mary—"No one is happy, not anyone who is built like you and me"—was made on March 4, 1915.

Margaret Parton relives Mary's unflattering diary entries about her in *Journey Through a Lighted Room*.

Mary describes Darrow's assured and mischievous mood before the trial—and her own irritation at his invitation for her to join him in Detroit—ıring her conversation with O'Toole. The background to Mary's puritanical :th comes from her interviews with Margaret—held at the University of

Oregon—and is supplemented by the oral history provided by her sister Sara when she was interviewed by Amerlia R. Fry. Margaret also writes about the family history in her memoir.

Darrow's half-serious quip that he would find another way to amuse himself in Detroit, even if he would miss her, is reported in the O'Toole transcript. His flirtation with Josephine Gomon features in the following chapters. It should also be noted that Kevin Tierney, in his diligent and often skeptical view of Darrow's near mythic status, mentions another passing liaison with a married woman named Sally Russell around the time of the Sweet trials. Tierney quotes a letter from Darrow to Russell, who lived in Chicago: "I have not forgotten you. . . . It was mighty nice to see you but the time was all too short. A girl like you will always have sympathy and understanding. Really, I hated to leave myself. . . . It has been lonely all afternoon. I don't care how you fix your hair. You are all right and dangerously loveable."

Tierney also quotes Darrow's letter to Russell after the trial of Henry Sweet in 1926: "Was in Detroit four weeks and won my case. It was really exciting and very gratifying. . . . I have thought of you many times. . . . I really want to see you again."

Lemuel Parton is aware of Mary's disenchantment during this period for, in a long and sometimes agonized letter to Sara, on August 3, 1926, he writes that "nothing ages and embitters one as the abandonment of creative purposes—if they have ever had them. That is where Mary and I made our great mistake. We have followed the opportunism of a job, instead of doing what we wanted to do. Write Mary when you get strength and time. She needs your love—desperately."

Irving Stone recounts Darrow's apparent meeting as a five-year-old with John Brown—a claim that is, as pointed out, undermined by the fact that the abolitionist was executed in 1859. Darrow's quote about Brown's "love of the slave" is also from Stone.

Darrow's 1901 address is quoted in Weinberg's *Attorney for the Damned* collection, while he wrote about the Springfield riots in the *Chicago Evening American* on August 19, 1908.

Walter White, in his book *A Man Called White*, describes Darrow's impact on the Sweets. Boyle quotes Otis Sweet from the latter's oral history interview with Professor Alex Baskin on August 1, 1960.

White reveals his conversation with Darrow in his autobiography—as well as the impression Frank Murphy made on him and the words he wrote to the judge.

Haldeman-Julius's entranced description of Murphy is from her *Clarence Darrow's Two Great Trials*—which contains her comparison of the old attorney to "a great, majestic ship."

Darrow's tactical confidence is outlined in the O'Toole transcripts.

The *Amsterdam News* claims for the Sweet case were printed on November 18, 1925.

Gladys Sweet described her thirty nights in prison as "a hideous dream" to Haldeman-Julius. Her almost flirtatious letters to White are mentioned by Boyle and are part of the NAACP Papers.

Chapter 14: A Peculiar Fear

Haldeman-Julius describes the courtroom in detail in the section of her book called "Clarence Darrow's Defense of a Negro."

Robert Toms outlines his deferential treatment of Darrow in Stone's biography. Background on Toms can be found in the writings of Haldeman-Julius and Lilienthal, as well as the more recent *Arc of Justice* by Kevin Boyle and *One Man's Castle* by Phyllis Vine.

Darrow's satisfied assessment of the Irish-Catholic base to his jury, as related to Judge Murphy, is recounted in Lilienthal's reportage of the trial—"Has the Negro the Right to Self-Defense?," published in the *Nation* on December 23, 1925, and also in the 1964 edition of his seven-volume collection called *Journals*.

The quoted courtroom extracts are taken from the Sweet trial transcripts—with descriptions of the varied reactions from the lawyers, witnesses, and spectators drawn from newspaper coverage.

On November 6, 1925, the *Detroit Times* reported Toms's despairing aside once Schuknecht admitted that a stone, thrown from the street below, had shattered the Sweets' bedroom window.

Josephine Gomon writes extensively of her friendship with Frank Murphy in her diaries held at the Bentley Historical Museum, at the University of Michigan.

In her November 1925 diaries, on an undated page, Gomon recounts

Darrow's apparent attempt to seduce her—foiled by the intervention of Harriet McGraw.

Hays remembers the cries of the Sweet baby in court in *Let Freedom Ring*.

The next two sections of courtroom battle—including Darrow's final plea—rely on the trial transcripts. Lilienthal's description of Darrow as "the old man with the unalterably sad face," with a voice like "a brass gong," is from the *Nation* on December 23, 1925.

Charles Mahoney's memory of the crying Judge Murphy was made in his oral history interview with Alex Baskin—held at the Bentley Historical Library.

In her diary, on Wednesday, November 25, 1925, Gomon writes of her invitation to lunch from Darrow in the immediate wake of his "wonderful" and "eloquent" plea.

White and Hays describe the jury deadlock in their respective autobiographies—while the *Detroit News* and the *Detroit Times* report on the deadlock in their November 25, 26, and 27 editions.

The Gomon diaries chart her boozy diversion with Darrow—where he proclaims dreamily that "we are affinities."

The exchanges between the judge and the jury foreman are from the trial transcripts.

Chapter 15: Sweet Again

Hays and White write of the arson attack on the Sweet bungalow.

Darrow's talk in Harlem was reported by the *New York Times* on December 11, 1925.

White's successful bid to bring the Sweets to New York for publicity is detailed in the NAACP Papers.

Boyle charts the way in which Ossian Sweet became seduced by his own fame—while Robert Bagnall's lament ("Never again!") is from his January 11, 1926, letter to James Weldon Johnson, now held in the NAACP collection.

Vera Cathcart's trial and Hays's involvement in it are relived in his autobiography. It is also documented in an article called "Crimes Involving Moral Turpitude" in the *Harvard Law Review*, Vol. 43, No. 1 (November 1929).

Darrow writes to White on January 2, 1926, to insist that they find an adequate replacement for Hays.

Vine provides a detailed assessment of the mood of the NAACP, and its financial situation after the first trial.

In White's NAACP correspondence he details Chawke's demands—and includes these in his report of March 21–24, 1926. Chawke's letter to White on April 6, followed by Johnson's reply to Chawke the following day, complete the arrangements.

Marcet Haldeman-Julius's contrast of Darrow and Chawke is from *Clarence Darrow's Two Great Trials.*

The courtroom exchanges are contained in the transcripts to the trial of Henry Sweet.

On May 4, 1926, the *Detroit Times* recorded Darrow's clash with Marjorie Stowell.

Extracts from Darrow's summary are taken from the transcripts—supplemented by eyewitness accounts of his delivery as recorded in various newspapers in the days following the trial.

Irving Stone also includes Gomon's words about Darrow's final speech. Haldeman-Julius's comparison of the two speeches made by Toms and Darrow is in the same collection cited above.

In his Darrow biography, Stone recounts the murmured response when Toms helps the victorious old lawyer to his feet: "Oh, I'm all right. I've heard that verdict before." But this observation was made first by Josephine Gomon in her private papers.

Chapter 16: Slipping Away

The *Nashville Banner*'s description of Dayton's return to obscurity was printed on July 22, 1925.

John Scopes's reflections on his desire for anonymity are found in his biography.

Edwin Mims's criticism of Darrow is made in his essay "Modern Education and Religion"—held in the Mims Papers at the Vanderbilt University Library in Nashville.

Forrest Bailey wrote to John Neal on August 12, 1925—to suggest that "there ought to be more of Neal and less of Darrow." Bailey repeated this same view to Darrow in his letter of September 2, 1925. On December 23, Bailey wrote to Franklin Reynolds of the ACLU to stress that "we ourselves have no wish to have Darrow, Malone and Hays continue. My only point was that we should not be made to appear as having kicked these men out." Neal's condemnation of Bailey's plan is contained in his February 15, 1926, letter. All these letters are from the ACLU Archives.

The *Nashville Banner* report, noting the number of women in attendance to hear Darrow, was printed on May 31, 1926. That same day the *Chattanooga Times* underlined the dreary tone of the hearing. The same newspaper also quoted K. T. McConnico in full as he attacked Darrow's "communistic" background.

Darrow's reply is recorded in the *Nashville Banner* on June 1, 1926—under a headline of "Darrow Makes Fervid Plea."

The final stinging exchange of letters—from Bailey to Darrow on June 3, 1926, and Darrow's reply six days later—is also found in the ACLU Archives.

Boyle recounts the death of Iva Sweet—and her father reaching for a gun on the day of her burial. His sources include a *Detroit Free Press* article on November 3, 1975, and an oral history interview with Claude Woodruff.

In her interview with Anne O'Toole, Mary Field Parton evokes Darrow's increasing fascination with *King Lear* during this troubled period for him. And, of course, Darrow often evoked comparisons with the play—most memorably, when he said that only the actions of the fool in *Lear* could stand compare with the lunacy of Leopold and Loeb.

Mary's diary entry of October 2, 1926, contains her gloomy musing that "Darrow longs for death" and prompts her analogy with the dybbuk of Jewish folklore.

Fay Lewis, a close friend of Clarence and Ruby's, wrote to Walter White on October 14, 1926, to suggest that the old lawyer was "carrying on some sort of propaganda for an early exit from this 'bloody world.'" This letter is part of the NAACP collection.

Nathan Leopold, in his prison biography, *Life + 99 Years*, writes about Darrow arriving to visit him with chicken, tea, and "good conversation."

Roger Baldwin's August 10, 1926, letter to John Scopes is from the ACLU Archives.

Scopes writes about Darrow's advice in his biography while the *New York Times* and the *Nashville Banner* of January 16, 1926, were among many newspapers reporting on the fudged final verdict. Larson, Stone, Tierney, and Weinberg summarize the backdrop to the legal ruling.

Darrow's assertion, that "I have taken life as it came . . . ," is from his biography. A copy of his July 15, 1927, letter to Mary—"In the outside office . . ."—is held at the Newberry Library.

Mary writes in her diary of her meetings on December 9–10, 1927, with Darrow in New York.

Just over six months before, on May 30, 1927, two fascists dressed in their full regalia of black uniform and accompanying insignia had been stabbed outside a subway station in the Bronx. The case was covered widely in Italy after Mussolini expressed his outrage and paid tribute to the slain fascists. Their funeral in New York was attended by the Italian ambassador and pressure on the police to catch the killers escalated.

Eventually, two Italian immigrant workers, Calogero Greco and Danato Carillo, were arrested with a dozen other men. Greco and Carillo, known for their opposition to fascism, were charged with murder. Hays, having agreed to represent them, persuaded Darrow to at least meet their families. It was then that the sight of Greco's weeping brother coerced Darrow into defending them.

He was embraced by the tear-streaked man and, in mid-December, they went to trial. It was a jerky affair with much of the testimony, from both prosecution and defense witnesses, either having to be translated from Italian or delivered in halting English. Darrow and Hays still worked their old routine and a key witness declared that, far from being certain that he had identified the murderers correctly, he was "not sure why I should send an innocent man to jail."

Darrow reached into himself to find one more speech of salvation. "I ask you, gentlemen, to take this case as you find it," he told the jury. "These children of Italy who came here to better their condition, who have lived honest lives, who have worked for their families, worked at plain, manual toil, who bore the bullets of the enemy in defense of America and in defense of Italy, who loved freedom and hated despotism, and, therefore, hate Mussolini, because the name Mussolini is only another name for despotism—I ask you for their sake, gentlemen, for a verdict of 'not guilty.'"

He won the verdict for which he pleaded on December 24, 1927. But Mary would note in her diary the following year that he still awaited payment for his services.

Darrow also saved John O. Winter because of a family promise made ...nty-four years earlier. When Darrow's son, Paul, studied at Dartmouth

College in New Hampshire, there had been a fatal accident involving a horse-drawn carriage. Paul Darrow's horse had bolted in fright and run over and killed a four-year-old boy. In his attempt to console the mother, Paul had written her a note in which he promised that his family would always be willing to help hers in a time of need.

The woman produced the old note in the summer of 1927 after Darrow had delivered a lecture against capital punishment at that same Dartmouth College. Darrow knew all about the family vow and so he immediately agreed to help her nephew—John Winter.

In her *Journey Through a Lighted Room* Margaret Parton writes of Darrow's decision to put up security for Mary and Lem's purchase of their brownstone town house in Greenwich Village—and she also details the interior of their new home.

Mary tells O'Toole that she often reminded Darrow of "how much we would need to leave out" when he pursued her about writing his biography.

His attitude toward autobiographies is recounted in *Story of My Life*— while the Weinbergs quote the *New Republic* review and Darrow's relative success on the bestseller list.

Ruby Darrow's exchange with Clarence about his love life, after Genevieve Forbes's tart remarks about the absence of anything personal in the book, is recounted in her correspondence with Irving Stone.

Mencken's reflections on the lawyer's ambivalent public standing near the end of his life—and that "the marks of battle are all over Darrow's face"—are from his *Vanity Fair* profile in March 1927.

When Mencken married in 1930, Darrow wrote jokingly to him to say: "I knew it would happen. It always does. Still, I want to compliment you on the long brave fight you made. That is all a man can do, for the foul race must be preserved, and this is the way that has been ordained for its preservation. I hope to see you and Mrs. Mencken before very long. Give her my kindest regards and best wishes; also my sincere sympathy."

Mencken delighted in such comic asides, and, on April 23, 1937, a year before Darrow's death, he wrote to his old friend: "I gather from public prints that you are claiming to be eighty. I refuse to believe it. Your actual age is what it has always been—about thirty—which is prime for homosapiens."

Darrow's self-doubt—"Am I as good as I was five years ago?"—is charted in Mary's diaries and interviews.

His letter to White was written on May 3, 1932—and White, in turn, lamented its "pathetic" tone to Arthur Spingarn on June 9 of that year. Both form part of the Spingarn Papers.

The Massie case is documented by Stone, Tierney, and the Weinbergs. There have also been stage plays and television films based on the trial—but by far the most comprehensive account is found in David E. Stannard's 2005 book, *Honor Killing: How the Infamous "Massie Affair" Transformed Hawaii*.

The Weinbergs write about the McWilliams case—while the *Chicago American* quote is from April 11, 1933.

Mary Field Parton writes about Darrow's mournful and faded last visit to New York on January 18–19, 1934, in her diary.

Darrrow's last days are traced by Stone and the Weinbergs. Mary Field Parton's diary entry on the day of his death, wishing him "farewell" and fearing that the "waters of oblivion" will soon close over him, is held at both the University of Oregon and the Newberry Library.

The last section, tracing Mary's own final years, is drawn from Margaret Parton's memoir and her private writings held at the University of Oregon. In particular, the haunting scenes in which Mary cries out for Darrow in the dead of night are from Margaret's private journals, while the more darkly comic moments in which Mary struggles to grasp *Star Trek* are recorded in *Journey Through a Lighted Room*.

On July 4, 1969, the *New York Times* reported Mary's death. The Margaret Parton Collection in Oregon also includes her quoted obituary.

Epilogue: Ghosts

Nathan Leopold describes the prison correspondence course he and Richard Loeb established in his surprisingly moving book *Life + 99 Years*. He also explains how he first started to teach in Joliet in late 1924—and that, to him, "the most interesting pupil was old Uncle Jim Terry, a leathery-skinned Negro of indefinite age. He had a *First Reader* and he demonstrated his ability for me. On the first page there was a picture of a bird sitting on a twig singing. Under the picture was the text. Pointing with his gnarled ebony finger at each letter in turn, Uncle Jim read: 'S-e-e, see, t-h-e, the, b-i-r-d, bird—See the ·rd. T-h-e, the, b-i-r-d, bird, s-i-n-g-s, sings—The bird sings.'"

Once he had read the entire book Leopold complimented him and reached for the *Second Reader*. Uncle Jim shook his head. No, he liked the *First Reader*. He would read the *First Reader* again, and nothing else. The multilingual ornithologist could not persuade him to move to the next level and, reluctantly, he allowed a beaming Uncle Jim to return to the words about a singing bird. "I salved my conscience with the reflection that Uncle Jim was doing a life sentence," Leopold remembered. "Perhaps fluency in reading wasn't too important in his case."

In that same prison memoir, Leopold remembers the day when Loeb was killed. All the details of that eerie day—from their munching contentedly on sweet rolls at breakfast to Leopold losing himself in the ritual of tenderly wiping away the blood from the dead body of the man he loved—are found in a book that has been out of print for decades.

Joseph G. Schwab's quote is contained in a November 1974 letter he wrote to Hal Higdon—and is reprinted in Higdon's *Leopold & Loeb: The Crime of the Century*.

Darrow's bleak response—"He's better off dead"—was reported in the *Chicago Daily News* on January 29, 1936.

Leopold, in his prison writings, recalls Darrow's final visit to him. He also told Vincent Martin, in their Puerto Rican series of interviews in the 1960s, that Darrow had alluded to his own forbidden love of a married woman.

Arthur Weinberg, who was part of the society that gathered every year on March 13 to honor Darrow, relived Leopold's release on that same day in *Clarence Darrow: A Sentimental Rebel*, which he wrote with his wife, Lila. At that exact time, in those same Chicago streets, another group of a few hundred had gathered together in an attempt to pay homage to Darrow on this landmark twentieth anniversary of his death. As the *Chicago Tribune* confirmed, they also waited to see if a "magician" from Detroit could make contact with the great attorney's spirit. Claude D. Noble, that amateur magician and a small-time businessman, had struck an agreement twenty-five years earlier with both Darrow and Harry Houdini that he would try to make contact with them "beyond the grave" on every anniversary of their death. Darrow, whose work on the Scopes trial had established him as America's definitive agnostic, would have obtained a wry pleasure from the showman's predictable failure to communicate with him.

Weinberg remembers how, as the media convoy passed that huddle of people, Leopold vomited again with nerves and shock as he was tailed by a posse of reporters and cameras.

Peter Rose, in his vivid interview with me, recalled his relationship with Leopold in Puerto Rico—including meeting him, telling the former killer about his family links to Bobby Franks, as well as sharing his celebrations on his release from parole and seeing the photographs of Darrow and Loeb in Leopold's apartment. Professor Rose then subsequently sent me a beautifully written account in his collection of essays—*Guest Appearances and Other Travels: In Time & Space.*

Afterword: Something Personal

The one-man play *Clarence Darrow*, made famous by Henry Fonda, and re-created in South Africa in 1984 by Richard Haines, was written by David W. Rintels—who helped establish the Clarence Darrow Foundation in California with Geoffrey Cowan, the author of *The People v. Clarence Darrow.*

Charmaine Phillips and Peter Grundlingh carried out a spree of apparently senseless killings in 1983. Although they were already known as "South Africa's legendary 'Bonnie and Clyde' killers," there seemed a curious parallel to Leopold and Loeb in the way that, for all their desolate violence, they were invested with a kind of deadly glamor. Charmaine was nineteen, the same age as Leopold, but she had always been known as an intriguing "bad girl" to us in Germiston rather than a more refined ornithologist or linguist. She was finally released from prison in the largely Afrikaans town of Kroonstad in August 2004—and emerged as a born-again Christian.

As stated earlier, Ruby Darrow's correspondence with Irving Stone about Mary Field Parton is part of the Weinberg Collection.

Darrow would have been appalled in 1925 if he had been told that, in the twenty-first century, a president of the United State of America, George W. Bush, could drawl, "Well, the jury is still out on evolution, you know."

In D. C. Ipsen's *Eye of the Whirlwind*, he quotes the *Time* article about John Scopes's return to Dayton in April 1970.

Nathan Leopold's last years in Puerto Rico were related to me, in a series of interviews, by Irving Grossman. They were both active members of the Natural History Society in Puerto Rico in the late 1960s.

INDEX

||||||||||||||||||||||||||||||||||||||||||||||||||||||||||